The Fungi of India

The Fungi of India

E. J. BUTLER

BIOTECH

2024
BIOTECH BOOKS®

ISBN 978-81-7622-007-1

Published by: **BIOTECH BOOKS®**
4762-63/23, Ansari Road, Darya Ganj,
New Delhi - 110 002
Phone: +91-011-43003222
E-mail: biotechbooks@yahoo.co.in

PRINTED IN INDIA

THE FUNGI OF INDIA.

I.—INTRODUCTION.

The flora of India has attracted botanists since plants first received scientific attention. The study of the Phanerogams made such progress that they were largely known a half century ago. The fungi, also, were not infrequently collected by early explorers and botanists who visited the country, and gradually increasing interest in them has resulted in the scattered enumeration of many hundreds of species; nevertheless they are still most inadequately known to science. This publication is made to gather together the records of Indian fungi isolated in more than five hundred separate papers, in the hope that it will facilitate the work of those who are carrying on the study of mycology in India.

The general physiographic features of India are well known. The Himalayas exist as an apparent barrier between India and the rest of Asia; but they are not a real barrier, for even the rivers cross them, and not only Chinese but also European plants have penetrated the range to its southern slopes and sometimes into the plains beyond. Sir Joseph Hooker,* in his excellent "Sketch of the Flora of British India", emphasizes the great variation in climate, rainfall, humidity, altitude, soil, and other factors, and the opportunities for immigration of plants into India, so that it is not surprising that "the Flora of British India is more varied than that of any other country of equal area in the eastern hemisphere, if not in the globe". Hooker divides India into three primary botanical areas: the Himalayan, with many European and Siberian genera; Eastern India, with Chinese and Malayan genera; and Western India, with European, Middle Eastern, and African genera. These areas are further subdivided, but though it is obvious that the character of the fungus flora must be markedly influenced by the nature of the supporting Phanerogams of any region, it is premature to attempt to apply these floristic subdivisions to the fungi. Some considerations of the nature and distribution of the Indian fungus flora are presented in the pages which follow.

II.—HISTORICAL.

Linnaeus † appears to have been the first to publish a definite record of an Indian fungus when he described his *Lycoperdon pistillare*, 160 years ago. This species was transferred by Persoon (Syn. Meth. Fung., 1801, p. 150) to *Scleroderma* and by Fries to *Podaxon*. It was collected by Koenig, who was a Danish missionary in Tranquebar, Tanjore, from 1767 to 1785, and some of whose collections of Indian fungi

---

* Imperial Gazetteer, 3rd edition, 1906.

† Mantissa plantarum, altera Generum editionis VI et Specierum editionis II, p. 313, 1771.

were later mentioned by Fries (237, 239). Fries subsequently (240) examined collections made in the Nicobar Islands by Didrick Ferdinand Didrichsen, who spent two months there when the corvette " Galathea " visited these Islands on her trip around the world in 1845-1847.

Early in the Nineteenth Century Wight collected a number of fungi in India, which were later examined by Klotzsch (273, 274) and by Berkeley (51). Bélanger was in India in 1825-29, and Montagne (350) records a few fungi which he collected. Jacquemont then visited India in 1829-32, and Léveillé described two or three of his fungi which came to France. Perrottet followed about 1840 and his collections, which were chiefly in the Nilgiri Hills, were worked over by Montagne. A few other specimens of fungi had found their way from India to Europe before 1850 (Corda (185), for example, cites a few specimens from Dr. J. Helfer), but the total number of species of Indian fungi known to science before 1850 would scarcely have exceeded one hundred.

Dr. (afterwards Sir, recorded in later pages as " Hooker f.") Joseph D. Hooker, and sometimes his co-worker Dr. Thomson, found time, during their survey of the Indian Flora, to collect hundreds of specimens of fungi. These collections were made principally in 1849 and 1850, although some bore an earlier date, and were more especially from Sikkim or elsewhere in the Himalaya and from the Khasi Hills. They were often accompanied, in the case of fleshy fungi, by sketches made from fresh specimens by Dr. Hooker. The specimens and drawings were sent to Berkeley, who began their publication in 1850 (Decade 25 ; 57, No. 241). Berkeley felt constrained to begin his account by remarking that " every species has been most severely scrutinised, and it is only after long study that I have felt myself compelled to propose so many new species. In the Agaricini, though several species have close allies in Europe, a vast portion of the forms are altogether new and peculiar, and often on a scale of the greatest magnificence......The Agarics seem to form far the greater portion of the Fungi of the country ".

It is easy to understand that Berkeley must have had a very difficult task in attempting to match the drawings and dried specimens communicated by Hooker with previously described species, and when, as usual, that proved impossible, in drawing up a new description. Berkeley would have been surprised, doubtless, could he have foreseen that eighty years after his labours, most of the Agaricaceae of India would be known only from the specimens he examined. Lack of opportunity to follow up the work he started, together with the inherent difficulty of the task, are largely responsible for the fact that we would still have had practically no idea of the gill fungi which occur in India were it not for his efforts. Whether the new species he described can be recognised again, remains to be seen ; many of them doubtless can be, with the aid of the specimens and drawings filed at Kew.

The first extensive collection of Burmese fungi was made by Sulpiz Kurz, Curator of the Royal Botanic Garden, Calcutta, whose specimens (which also included many from Bengal) were reported on by Currey in 1874 (196). About this time, and for many years after, Cooke was receiving at Kew and describing considerable numbers of Indian fungi, and specimens were sent also to Rabenhorst, Brefeld, Dietel, Patouillard, and various other mycologists abroad. At the close of the century collections were made by Gollan in the United Provinces and studied by P. Hennings (260, 263), and Massee began the series "Fungi Exotici" in the Kew Bulletin, which he and Miss Wakefield carried on for a number of years and which included descriptions of various Indian species. Theissen (467) reported on a collection of Bombay fungi received from his colleague Father E. Blatter, while Bresadola and Saccardo examined certain Indian fungi to which they had access.

During the past quarter of a century H. and P. Sydow, particularly the son, H. Sydow, have examined critically hundreds of species of Indian fungi communicated to them mainly from Pusa and Coimbatore. The late C. G. Lloyd had about a score of correspondents in different parts of India, and his racy notes on the larger fungi contain many references to Indian species. Specimens of fungi from India have, indeed, found their way to most of the larger herbaria, as is indicated by the list of publications cited in the Bibliography.

The historical summary just presented deals with the studies based on specimens of Indian fungi sent to the herbaria in England, the Continent, or America. In the meantime, work was begun in India itself, where alone the often necessary observations on these plants in the fresh condition and studies of their life-histories could be made. Two Indian army medical officers were the pioneers in this field. D. D. Cunningham published between 1871 and 1897 the results of his studies, carried out in intervals of medical and sanitary work, on certain fungi, especially Mucoraceae and Uredinales, observed in Calcutta and elsewhere in eastern India. About 1885 A. Barclay began his critical work on the rusts near Simla, and during the following seven years worked out the life-histories of many of the Himalayan Uredinales. Mr. (later Sir) George Watt made collections of and observations on a few of the economically important fungi found on tea and some other crops after Barclay's death. The senior author arrived in Calcutta in 1901, was transferred to Dehra Dun in 1902 and to Pusa in 1905, and was officially concerned until 1920 with the mycology of India, especially the parasitic fungi, assisted by W. McRae, F. J. F. Shaw, J. F. Dastur, and others. Later on the Provinces of India extended their agricultural departments to include the investigation of plant diseases, Government mycologists or plant pathologists being attached to the headquarters of several of the provincial departments, while A. C. Tunstall and Leslie Coleman were doing valuable work at other stations. Meanwhile the Indians themselves took up the problems, particularly those of applied mycology (although S. R. Bose has devoted much attention to the Polyporaceae), and are now carrying on most useful work in many places in India and Burma.

Mr. T. Petch, Botanist and Mycologist to the Department of Agriculture, Ceylon, accomplished a tremendous amount of valuable mycological work while in that neighbouring Colony, and has made possible the determination of many Indian fungi, such as the entomogenous species and those cultivated by termites. Many of the fungi occurring in Ceylon, and still unknown in India, will probably be found when more thorough collections are made in the south of the Peninsula.

This brief sketch covers but imperfectly the development of the knowledge of Indian fungi. The Bibliography, however, tells the story more fully.

III.—THE DISTRIBUTION OF THE FUNGI.

From an examination of the List of Species it may be seen that the following numbers and percentages of the fungi recorded for India were first described from India :—

	Per cent.
Phycomycetes 18 species	24
Ascomycetes 233 "	49
Ustilaginales 51 "	51
Uredinales 219 "	57

We have not checked the remainder of the Basidiomycetes, not the Fungi Imperfecti, but similar figures would be obtained in these groups. Of the four groups mentioned, 521 species out of a total of 1,035, or approximately 50 per cent., were first described from India. This does not mean, however, that India has a high percentage of endemic fungi : many of these species, though first described from India, are now known from other countries. As a matter of fact, since India, as Hooker mentions, has a low percentage of endemic Phanerogams, so it has also doubtless a low percentage of endemic fungi. The figures presented only indicate that, poorly though the fungi of India may be known, they are nevertheless better known there than in most other similar tropical or subtropical areas.

The distribution of fungi over the world is still so imperfectly known as to render uncertain an attempt to make a general analysis of the distribution elsewhere of the fungi recorded for India. In any attempt of this sort, one is hampered not only by a lack of knowledge of distribution of the vast majority of fungi, but also by a feeling that misdeterminations entered in the records vitiate any conclusions. These errors of determination are, however, of two kinds which tend to neutralize each other : either that of a previously known fungus having been given a new name, or that of a new fungus having been referred erroneously to an old species. In view of the importance and interest of assembling such data as are available, we have made the following comparisons, with full realization of the uncertainties involved.

Comparison of the fungi of India, of the Dutch East Indies, and of Manitoba. Among the few areas outside of Europe which have brought together lists of fungi comparable in extent with the present list of Indian fungi, we have chosen the Dutch East Indies and Manitoba. The former is a tropical area geographically not far distant from India, but with a phanerogamic flora quite different, especially from that of the parts of India in which mycological collections have chiefly been made. Van Overeem * has brought together a list of the fungi recorded there up to about·

* Van Overeem-de Haas, C. et D. Verzeichnis der in Niederländisch Ost-Indien bis dem Jahre 1920 gefundenen Myxomycetes, Fungi und Lichenes.—Bull. Jard. Bot. Buitenzorg, Sér. 3, IV, fasc. 1, 146 pp., 1922.

ten years ago. Manitoba is a north temperate area in Canada, far removed geographically and with a flora of vascular plants differing almost completely from that of India, except that a very few similar or identical plants are found in Himalayan regions and in Manitoba, and a few cultivated plants and weeds are common. A recent book* treats of the fungi of this area.

TABLE I.

Comparison of the fungi of India, the Dutch East Indies, and Manitoba.

Group	Number in Dutch East Indies	Percentage also in India	Number common to India and Dutch East Indies	Number in India	Percentage also in Dutch East Indies	Percentage also in Manitoba	Number common to India and Manitoba	Number in Manitoba	Percentage also in India	Number common to all three areas
Archimycetes	2	..	..	12	..	8·3	1	4	25·0	..
Chytridiales	3	..	..	4	..	..	..	..	..	..
Oomycetes	17	58·8	10	45	22·2	33·3	15	26	57·7	5
Zygomycetes	10	20·0	2	14	14·3	21·4	3	18	16·7	1
Total Phycomycetes	32	37·5	12	75	16·0	25·3	19	48	39·6	6
Hemiascomycetes	17	..	..	13	..	15·4	2	6	33·3	..
Discomycetes	181	6·1	11	69	15·9	23·2	16	146	11·0	4
Pyrenomycetes	650	8·9	58	394	14·7	7·1	28	265	10·6	5
Total Ascomycetes	848	8·1	69	476	14·5	9·7	46	417	11·0	9
Ustilaginales	23	34·8	8	100	8·0	11·0	11	36	30·6	2
Uredinales	106	14·2		384	3·9	7·3	28	134	20·9	1
Polyporaceae†	352	21·9	77	318	24·2	8·5	27	112	24·1	14
Other Hymenomycetes	478	10·7	51	462	11·0	15·4	71	650	10·9	22
Gasteromycetes	72	22·2	16	75	21·3	13·3	10	37	27·0	5
Total Basidiomycetes	1,031	16·2	167	1,339	12·5	11·0	147	969	15·2	44
Hyphomycetes and Mycelia Sterilia	236	5·9	14	196	7·1	11·2	22	191	11·5	3
Melanconiales and Sphæropsidales	127	6·3	8	265	3·0	5·7	15	262	5·7	..
Total Imperfecti	363	6·1	22	461	4·8	8·0	37	453	8·2	3
Total Fungi	2,274	11·9	270	2,351	11·5	10·6	249	1,887	13·2	62

† Including Boletaceae.

* Bisby, G. R., Buller, A. H. R., and Dearness, J. The Fungi of Manitoba, 194 pp., Longmans, Green & Co., 1929.

Table I presents the comparison. The number of supposedly identical fungi in the three areas is obtained with some difficulty, owing to variations in mycological nomenclature, but the tabulation is approximately correct. The arrangement of the figures follows the groupings used in the present list.

From Table I it may be seen that 11·5 per cent. of the fungi recorded from India have also been recorded in Java, and 10·6 per cent. are listed also in Manitoba. It may be noted further that 62, or 2·7 per cent., of the fungi of India have a distribution which includes all the three areas under consideration.*

The Gasteromycetes, Polyporaceae, Discomycetes, and Phycomycetes show the widest distribution; the Fungi Imperfecti and the Uredinales a more restricted one. This is not surprising, in view of the fact that the rusts are necessarily limited by their hosts, and the Fungi Imperfecti that have been identified in the three areas are principally the parasitic forms, likewise dependent upon specific hosts. A comparison of the fungi recorded from India with those existing in Europe will make the conclusions more definite.

The Fungi of India which occur in Europe. Table II gives the figures and percentages:—

TABLE II.

Fungi common to India and Europe.

Group	Number in India	Number also in Europe	Percentage in Europe
Archimycetes	12	5	41·7
Chytridiales	4	1	25·0
Oomycetes	45	27	60·0
Zygomycetes	14	5	35·7

* The following are the 62 fungi recorded for the three widely different areas: *Pythium de Baryanum, Cystopus bliti, Phytophthora infestans, Sclerospora graminicola, Plasmopara viticola, Rhizopus nigricans, Leotia lubrica, Peziza aurantia, Helotium citrinum, Rhytisma acerinum, Nectria cinnabarina, Phyllachora graminis, Daldinia concentrica, D. vernicosa, Xylaria hypoxylon, Ustilago nuda, U. tritici, Puccinia pimpinellae, Hirneola auricula-judae, Exidia glandulosa, Tremellodon gelatinosum, Guepinia spathularia, Hymenochaete rubiginosa, Stereum hirsutum, S. purpureum, Clavaria stricta, Hydnum coralloides, Merulius corium, Fomes fomentarius, F. igniarius, Ganoderma applanatum, Polyporus adustus, P. arcularius, P. cinnabarinus, P. gilvus, P. sulphureus, P. varius, Polystictus hirsutus, P. pergamenus, P. versicolor, P. zonatus, Russula alutacea, R. emetica, Schizophyllum commune, Panaeolus campanulatus, P. papilionaceus, Hypholoma appendiculatum, H. fasciculare, Agaricus campester, Mycena galericulata, M. pura, Clitocybe laccata, Armillaria mellea, Amanitopsis vaginata, Lycoperdon gemmatum, L. giganteum, L. piriforme, Geaster fimbriatus, Crucibulum vulgare, Aspergillus niger, Botrytis vulgaris, Helminthosporium gramineum.* Most of these species appear to be indigenous to the three areas: only *Phytophthora infestans, Plasmopara viticola,* and the two smuts can be considered definitely to have been introduced into the areas concerned. Mycologists will probably agree that most of the remaining 58 species are widely distributed fungi, and that there is doubtless a considerable number of other fungi with a distribution including India, the Dutch East Indies, and Manitoba. But very few or no vascular plants have a range which would make them indigenous to these three regions.

TABLE II—*contd.*

Fungi common to India and Europe—contd.

Group	Number in India	Number also in Europe	Percentage in Europe
Hemiascomycetes	13	4	30·8
Discomycetes	69	29	42·0
Pyrenomycetes	394	49	12·4
Ustilaginales	100	25	25·0
Uredinales	384	74	19·3
Polyporaceae	318	53	16·7
Other Hymenomycetes	462	143	31·0
Gasteromycetes	75	23	30·7
Hyphomycetes and Mycelia Sterilia	196	49	25·0
Melanconiales and Sphæropsidales	265	58	21·9
TOTAL	2,351	545	23·2

It is now quite evident that the fungus flora of India can be considered, despite our very incomplete knowledge of it and the erroneous and questionable identifications involved in many instances, to consist of a fairly large percentage of species occurring more or less widely in the North Temperate Zone, and a larger percentage of fungi limited to tropical regions or to tropical host plants. Nearly one quarter of the known Indian fungi are recorded also in Europe. This figure is not especially high when compared with the percentage common to Europe and Manitoba (about 60 per cent.), in spite of the fact that Manitoba is more distant from Europe than India is. But the general character of the phanerogamic flora of Manitoba is much more similar than is that of India to the flora of Europe; and more than 20 per cent. of the indigenous species of Phanerogams of Manitoba, but only 6 per cent. of those of India, occur also in Europe.

The distribution of the Phanerogams compared with that of the Fungi. Hooker * observed that of some 17,000 species of Phanerogams in India (with which he included Malaya and Ceylon), 760 European species are indigenous in the Indian region, and of these some 400 are also in the British Isles. Of these 17,000 species, there are perhaps 4,000 Ceylon and Malayan plants which do not occur in India, and are of course also non-European. In other words, about 6 per cent. (760

* A Sketch of the Flora of British India. Imperial Gazetteer, 3rd Edition, 1906.

out of 13,000) of the Indian Phanerogams occur also in Europe. These percentage figures would not be much increased were one to add also the few score cultivated plants and weeds common to Europe and India. It seems safe to conclude that the fungi have an average distribution much wider than that of the vascular plants.

Comparison of the rusts of India and of other countries. Of the fungi recorded here for India, the Uredinales are best known not only in India but in the world as a whole. The total number of species hitherto found in India is 384, including unconnected forms, as compared with 252 species in the British Isles, 237 in California, 372 in Switzerland, 482 in Italy, about 1,200 known in the whole of North America, but only 162 in Australia.*

The 384 rusts in India include 102 unconnected forms in *Aecidium* and *Uredo*. The life cycles of many of the species in other genera are also not fully known, and, of course, many species still remain entirely undetected in India. In the case of the rusts, however, it is felt that in most cases the same name applied to a species in India and in England or in America really refers to the same fungus, and a different name to a different one. The following data are compiled, using distribution records in Sydow's Monographia, Grove's "British Rust Fungi" for occurrences in the British Isles, and Petch (369) for Ceylon.

The 102 unconnected forms are in the majority of cases known at present only in India. In some instances they will prove to be only stages of known rusts ; but at present they largely represent unknown quantities as regards distribution.

The remaining 282 rusts may be tabulated as follows :—

16 species, or 5·7 per cent., are listed otherwise only from Europe.
60 species, or 21·3 per cent., are listed otherwise from Europe and other countries.
9 species, or 3·2 per cent., are listed otherwise only from North America.
4 species, or 1·4 per cent., are listed otherwise from North and South America.
13 species, or 4·6 per cent., are listed otherwise only from Japan.
59 species, or 20·9 per cent., are listed also from the British Isles.
49 species, or 17·4 per cent., are listed also from Ceylon.

From the above it may be seen further that 76 species, or about 27 per cent., of these Indian Rusts have a distribution which includes Europe, and 13 species, or 4·6 per cent., are, so far as is now known, Indian-American in range. The 4·6 per cent. known only from India and Japan may well occur also in China or elsewhere in eastern Asia, which is as yet for the most part mycologically unexplored.

If it be considered that each of the 102 unconnected forms in India represents a definite species with a distribution limited to India, then 59 of the total of 384 rusts, or 15·3 per cent., would have a distribution including the British Isles and India, and nearly 20 per cent. would have a distribution including Europe as well as India, whereas only about 6 per cent. of indigenous Phanerogams are common.

The above figures for Uredinales and Phanerogams are not strictly comparable, because those for the former must include introduced species, since it is often

* Data from Arthur et al., Plant Rusts, 1929, and McAlpine, Rusts of Australia, 1906.

difficult or impossible to determine whether or not a fungus has been introduced. There are a considerable number of introduced flowering plants in India,* especially cultivated plants and weeds, and with some of these rusts have certainly immigrated. These introduced rusts are the most likely to be collected. If allowances be made for this element in the Indian flora both of hosts and rusts, the figures for distribution of the flowering plants and for the rusts become more similar, as indeed we should anticipate from the restricted and obligate parasitism of the latter.

Nevertheless, further study may indicate rather definitely that species of rusts have, on the average, wider ranges of distribution than species of Phanerogams. If we take again the rusts of the British Isles and of India, we may note that of the 252 species in the British Isles, 59, or 23·4 per cent. have been recorded also in India. And even more striking is the comparison between the rusts occurring in India and in Manitoba. Only a few cultivated plants and weeds, and a few species of indigenous Phanerogams, are common to these two areas which are so nearly antipodean, and yet 28 species of rusts are common to both : over 7 per cent. of the known Indian rusts occur in Manitoba, and some 21 per cent. of the 134 rusts known in Manitoba, occur also in India. Nine † of these 28 rusts were probably introduced into one or both areas, but the other 19 species ‡ appear to be native to both. It should be remarked that of these 19 species, three, (*Puccinia andropogonis*, *Uromyces proeminens*, and *U. scirpi*) are recorded with considerable doubt from India. All of these species apparently native to both India and Manitoba, with the exception of *Uromyces fabae*, have been found only in the colder, Himalayan region of India. *Puccinia circaeae*, *P. polygoni-amphibii*, and *Uromyces polygoni* utilize the same species of hosts in both countries, as does *U. fabae* in part ; the remainder apparently utilize different but related hosts in the two countries.

Species of certain non-parasitic groups of fungi certainly have a very wide distribution : thus, of the 69 names that have been applied to Discomycetes from India, over 40 per cent. are found in European lists, and 23 per cent. in the Manitoba list ; of the 75 recorded Gasteromycetes, 20, or 27 per cent. are known in England. It is to be expected that fungi which develop upon leaf mould or decaying wood may find a suitable substratum in almost any country, but we cannot pursue our comparisons far, because the data are too indefinite. Petch § has presented a discussion of the uncertainty involved in applying the names of European fungi to specimens from the tropics, in which he concludes that, in general, the

---

* Professor L. A. Kenoyer, Plant Life of British India, Sci. Monthly, XVIII, p. 58, 1924, states that "The present Indian flora contains many introduced plants, American ones being especially prominent."

† *Melampsora lini*, *Puccinia coronata*, *P. graminis*, *P. lolii*, *P. sorghi*, *P. anomala*, *P. taraxaci*, *P. triticina*, and *Uromyces trifolii*. The names used in the Manitoba list are in a few instances different from those used here.

‡ *Phragmidium disciflorum*, *P. potentillae*, *Puccinia absinthii*, *P. andropogonis*, *P. caricis*, *P. caricis-asteris*, *P. circaeae*, *P. hieracii*, *P. menthae*, *P. phragmitis*, *P. pimpinellae*, *P. polygoni-amphibii*, *P. punctata*, *P. violae*, *Pucciniastrum agrimoniae*, *Uromyces fabae*, *U. polygoni*, *U. proeminens*, and *U. scirpi*.

§ Petch, T. European fungi in the tropics.—Trans. Brit. Myc. Soc., III, pp. 340-347, 1912.

worse the material, the higher the percentage of European names applied to it. He was referring more particularly to the Agaricaceae of Ceylon, but there can be no doubt that the same rule would apply to the names given by earlier mycologists to many specimens of fungi from India. Bose (79) has, however, given some interesting data regarding the distribution of the Polypores in India and elsewhere.

Distribution of the fungi within India. The distribution of rusts in certain of the northern areas of India is indicated in Table III.

TABLE III.

Comparison of the Uredinales of three Northern Areas of India.

—	KASHMIR VALLEY		SIMLA		KUMAON	
	Number	Percentage	Number	Percentage	Number	Percentage
Genera exclusive of *Aecidium* and *Uredo*— Species recorded	49	..	88	..	52	..
Endemic*	2	4·1	15	17·0	6	11·5
Known only from India	15	30·6	52	59·1	33	63·5
Known also in Europe	28	57·1	26	29·5	13	25·0
Known also in North America	20	40·8	18	20·5	9	17·3
Form genera *Aecidium* and *Uredo*—						
Species recorded	3	..	15	..	11	..
Endemic *	..	..	6	40·0	5	45·5
Known only from India	2	66·7	5	33·3	4	36·4

It will be seen that Kashmir differs from Simla and Kumaon in having a low percentage of rusts which are apparently endemic or Indian, but a higher percentage with a distribution including Europe or North America. This may be due tc the fact that there is a similar distribution of the genera of the supporting higher plants, but it serves to confirm the statement made on an earlier page that the high mountain ranges to the north and north-west of India (especially the latter) are not a real barrier to the dissemination of plants.

The Erysiphaceae or white powdery mildews are generally stated to be rare in the hotter parts of the earth as compared with their numbers in temperate zones, and they have been contrasted in this respect with the sooty moulds, whose distribution is mainly tropical. So far as India is concerned, Erysiphaceae have been found throughout the country, but in the hotter parts they occur chiefly in the *Oidium* stage alone. Perithecial stages of Erysiphaceae have been found in India on some 59 different host plants, 15 species of mildews being represented but *Erysiphe polygoni* being on 25 of the 59 hosts. The 77 collections (excluding duplicate

* Endemic, so far as known, to Kashmir, Simla, or Kumaon, respectively.

collections from the same locality) in which perithecia have been found are distributed as follows :—

43 in the Himalaya, especially Kashmir.
3 in the foot-hills of the Himalaya.
1 in the Khasi Hills.
14 in the Indo-Gangetic Plain.
4 in Central India.
5 in Western India.
3 in South India.
4 in Burma.

It is evident, therefore, that a high proportion of the perithecial collections has been made in the north, either in the hills or in regions, such as the Indo-Gangetic Plain, in which there is a distinct cold season. Conidial stages are very prevalent in the Indo-Gangetic Plain : in Pusa, for instance, besides the 3 species bearing perithecia on 8 hosts, there are probably over 100 hosts on which *Oidium* alone occurs.

The sooty moulds or fumagines include a number of black, superficial leaf-inhabiting fungi, saprophytic in the sugary excretion of certain insects or haustorial parasites of the epidermis. They include, in the sense used by Arnaud,* representatives of several groups of the Ascomycetes and do not form a homogeneous whole. We have selected as typical genera, *Meliola* with 30 recorded Indian species, *Asterina* with 15, *Capnodium* with 7, *Asterinella* and *Dimerosporium* with 3 each, *Phaeosaccardinula* with 2, and *Parasterina*, *Phragmocapnias*, *Prillieuxina*, and *Pyrenocarpon* each with 1, or in all 64 species. The distribution of these, so far known, is as follows :—

	Species.
Eastern Bengal, Assam, and Lower Burma	22
Western Peninsula from Bassein to Travancore, including western Mysore and the Nilgiris	33
Eastern Peninsula from Orissa to the Godavari	6
Gangetic Plain to Calcutta	8
Madras, other than regions included above	4
Himalaya	3
" India "	1

Humidity as well as temperature is probably a factor in this distribution. The western side of peninsular India and the regions to the north and east of the Bay of Bengal represent the areas of highest rainfall †, and by far the largest number of species of the genera mentioned above occur in these areas. In the Himalaya and the whole of north-western India they seem to be rare. They appear also to be relatively scarce in the areas of extreme temperatures, both where the hot season is very hot and where the winters are relatively cold. Eight of the species are

* Arnaud, G. Les Astérinées. Thèse, Fac. des. Sci., Paris, 1918.
† Imperial Gazetteer of India, Atlas, Pl. 9, 1909.

common to the first two areas in the table, while four from the first and three from the second are also found in the Gangetic Plain.

Estimate of the number of fungi in India. The present list of 2,350 species of fungi recorded from India obviously includes but a small fraction of the total that occurs there. There are present some 15,000 known species of Phanerogams, or about one-tenth of those known in the world, and some 600 Pteridophytes. Hooker notes that " fungi abound in most parts of India ". More *genera* of Phanerogams are at present known in India, than *species* of fungi. It would be conservative to estimate that there are in India at least as many species of fungi as of vascular plants : indeed, in north temperate regions such as England and elsewhere in Europe, in Manitoba and, apparently, elsewhere in North America, the fungi have been found (owing especially to their wider average distribution) to outnumber the Phanerogams by at least 2 to 1.* If similar ratios exist in the tropics, it may well be that no more than 10 per cent. of the fungi of India are recorded, and many of those recorded are doubtful ; yet India is mycologically as well known as any other comparable area extending into the tropics.

IV.—CHARACTERS OF THE FUNGUS-FLORA.

The Myxomycetes have not been studied in India, but many of these widely distributed organisms occur there. A few species of bacteria parasitic upon cultivated plants have been observed, but are not listed here. The Phycomycetes have, however, received considerable attention. Several of that simple or primitive group, the Archimycetes, have been encountered, but only four of the Chytridiales have been identified. The Saprolegniales are undoubtedly present in considerable diversity, but only four species in four genera are named. The genera *Pythium* and *Phytophthora* have had comparatively extensive study in India, as the entries under these genera indicate.† Six of the very widely distributed species of *Cystopus* occur commonly in India. A number of the common downy mildews (Peronosporaceae) have been found in India, where they occur especially during the cool season, or in more northerly stations. Little study has been given to the Mucorales, but they apparently occur as moulds in the soil and on organic material, much as they do in more northern latitudes. Very few Entomophthoraceae are known as yet.

The Ascomycetes constitute a large assemblage in India, as elsewhere. Few Hemiascomycetes have been studied, but there are doubtless many Saccharomycetes and other lower Ascomycetes present. A few species of *Taphrina* (this name, rather than *Exoascus*, has been used) occur. The Discomycetes (including Helvellales,

---

* In Rabenhorst's Kryptogamenflora, covering north central Europe, nearly 15,000 fungi are listed. A figure for the vascular plants of an exactly similar area was not obtained, but Coste gives 4,263 species of Phanerogams and 91 of Pteridophytes for France and Corsica (Flore de la France, 1906).

† See also Ashby, S. F., Kew Bull., 1922, pp. 257-262, and Trans. Brit. Myc. Soc., XIII, pp. 86-95, 1928.

Pezizales, Phacidiales, and Tuberales) have never been found in great diversity in India; but they have never been diligently sought.

The Pyrenomycetes (here used to include all the Ascomycetes from Hysteriales on through the Laboulbeniales) are represented by some 400 entries. Only two or three each of the Hysteriales and Plectascales have been found in India. The Perisporiales are represented by a number of the widely distributed powdery mildews (Erysiphaceae) and by many of the Perisporiaceae. The genus *Meliola* has been divided by Theissen and Sydow* into three, and by von Hoehnel † into five genera, but since the Indian species could not be separated into these divisions without critical revision, they have been left in *Meliola*. The saprophytic sooty moulds, such as *Capnodium* and related fungi, are also rather common in India, frequently in their conidial and pycnidial forms alone. The Hemisphaeriales appear to be abundant throughout tropical regions, *Asterina* being the genus most commonly encountered in this order. In the Hypocreales there have been found several species of *Nectria*, and a few of *Hypocrea* and other genera. The genus *Cordyceps* has been reported only twice (both specimens collected by Hooker and Thomson in the Khasi Hills), and the genus *Claviceps* is represented only by *Sphacelia* stages, which are usually precluded from continued development by attacks of *Cerebella*. In the Dothidiales the genus *Phyllachora* predominates as one of the largest amongst the Ascomycetes.

The Sphaeriales occur in abundance in India, as they do everywhere. The Fimetariaceae have not been studied, and are represented only by three names in *Chaetomium*, two of which were applied to Indian material by Corda. The Sphaeriaceae have a few representatives known, but the Cucurbitariaceae and Coryneliaceae only one and two, respectively. The Xylariaceae is the only other family which requires mention: for the long list of species of *Xylaria*, and the considerable number of those of *Hypoxylon*, make this family dominant among the Sphaeriales.

The Laboulbeniales are doubtless present in India in large numbers, but the records in seven genera are made up principally from references in Thaxter's publications.

The Ustilaginales include a hundred species, many of which are widespread, and several of economic importance. Ciferri (121 : 28) agrees with Maire in restricting the genus *Cintractia* to those smuts on Juncaceae and Cyperaceae: we have indicated the two changes involved in this list, if Ciferri were to be followed. In India, as in other tropical countries, are to be found fungi classified under *Graphiola*, *Farysia*, and *Stylina*: while there is still some discussion as to the systematic position of these genera, they seem to fit best into the Ustilaginales. Certain other less common genera occur in India, but the genus *Ustilago* is the predominant one.

---

* See Ann. Myc., XV, pp. 461-463, 1917.
† Fragmente (264), XXIII, p. 22, 1919.

The Uredinales have already been discussed in part. As is usual in lists of tropical fungi, a considerable percentage of the species can be included only as form genera. India has its quota of those genera which develop rarely or not at all in temperate regions, such as *Blastospora*, *Chnoopsora*, *Crossopsora*, *Hamaspora*, *Hemileia*, *Masseeëlla*, *Monosporidium*, *Pucciniostele*, *Ravenelia*, *Schroeteriaster*, and *Sphaerophragmium*; and *Cystopsora* and *Gambleola* are known only from India. Nevertheless *Puccinia* with 140 species, and *Uromyces* with 47, are the common genera in their usual proportion to the total number, and to each other. About half the Indian rusts are known only from the northern part of the country, which has been most carefully surveyed. Some 53 (18·7 per cent.) of the 283 rusts for which a teleuto stage is known, are known only in this stage; but several of these certainly have other spore forms in their cycle. The genus *Melampsora* requires further study in India, but on the whole the rusts are comparatively well known.

Of the Hymenomycetes we can say but little. A few of the Auriculariales and Tremellales have been named, and several Thelephoraceae. The Polyporaceae have been collected by many in India, and despite our efforts to bring synonyms together, there still remains a total of 301, as follows: *Cyclomyces*, 1; *Daedalea*, 16; *Elmerina*, 1; *Favolus*, 9; *Fistulina*, 1; *Fomes*, 38; *Ganoderma*, 6; *Gloeoporus*, 2; *Hexagonia*, 12; *Lenzites*, 17; *Merulius*, 4; *Polyporus*, 79; *Polystictus*, 81; *Poria*, 9; and *Trametes*, 25. The Polypores are unquestionably abundant in variety in India; but one has a feeling that in this family there cannot be very many more than the 300 species which now purport to exist there.

The Boletaceae are represented by 12 entries in *Boletus*, each name applied by Berkeley as a new species based upon specimens and sketches from Hooker, and 5 species of *Strobilomyces*. None of the 12 Boleti has been reported from India since Hooker collected them about 1850, though fungi of this group are extremely prominent in the outer ranges of the Himalaya. Difficulties in preservation for subsequent study, no doubt, account for this hiatus in our knowledge.

For the Agaricaceae we have brought together the references, principally from Berkeley and P. Hennings. Thanks to Berkeley's efforts in particular, we can have some idea as to the mushroom flora of northern India, even though it may involve considerable labour to fit his names accurately to subsequent discoveries. As we have mentioned (p. ii), Berkeley found that the majority of the species required new names. He remarks (57, No. 291) that "Fries suspected that the Coprini would be found on a splendid scale in tropical countries. Later illustrations, whether of the pencil or herbarium, do not prove this, the species of *Coprinus* being, generally, either the same with our own, or obscure and uninteresting. The mushroom, on the contrary, assumes every conceivable luxuriance of form." We have now 6 entries under *Coprinus*, and 12 under *Agaricus*. *Amanita* is represented by one collection only, referred by Berkeley to the edible *A. caesarea*. *Armillaria mellea* occurs to some extent in the Himalaya as a root parasite: Miss Wakefield (see 266 : 437) found it practically identical with English examples. *Collybia* seems to be well represented in India. *Cortinarius*, so abundant in northern Europe, is known from India by only five collections, all made by Hooker. Only two species of *Crepidotus* are recorded, both collected by Gollan and determined by Hennings, and there are also two species of *Inocybe*. *Lactarius* and *Russula* have comparatively few entries in the Indian records, but *Lentinus*, *Lepiota*, and *Marasmius* are better represented. On the whole, the Agaricaceae of India are as yet scarcely known, and we cannot profitably discuss their features from the few and uncertain data available.

The short list of Gasteromycetes exhibits a number of the ubiquitous forms, and several, such as certain of the Phallales, which are known only in tropical regions.

The Fungi Imperfecti are, of course, present in India in enormous numbers, and include many important parasitic groups. Not much work has, as yet, been done upon them, and we present only comparatively few records gleaned from previous publications. Some special study has been given to *Cercospora* (457) with the result that 59 species are now known in India. This genus is commonly present as a parasite on cultivated and wild hosts in India and elsewhere in the tropics: about three-fourths of the Indian species are known in other countries also, and some of them are very widely distributed over the globe. *Diplodia*, *Phoma*, *Phyllosticta*, and *Septoria* are other common genera in India, as elsewhere.

V.—EXPLANATORY.

In the List of Species which follows, considerable grouping of orders has been made to facilitate reference, but a purely alphabetical arrangement of all genera and species has not been followed, because of the convenience involved in having separate lists of rusts, smuts, etc.

The nomenclature is, in most cases, that given in Saccardo's Sylloge Fungorum. A few of the species of fungi described from India long since, failed to obtain entry into Saccardo: these are usually indicated by "not in Sacc.". It is possible that in one or two of these cases we may have overlooked the Saccardo reference, for we have checked principally from the Indexes in that work, and in one case at least (*Cercospora solanacea*) the description is given, but the species is not indexed. Some would object, under the rules, to Saccardo's usage in compiling, for example, *Polyporus marginatus* Fr. as *Fomes marginatus* Fr., or *Agaricus* (*Collybia*) *camptopus* Berk. as *Collybia camptopoda* Berk., but it is conservative nomenclature, and we have not considered it necessary, in a case such as the latter, to mention that Berkeley's reference was to *Agaricus*. When no mention is made of Saccardo, the species is presumably one described too recently to have been included in the Sylloge. In cases in which a fungus was described by an author in another work than his own (e.g., *Diplodia citrina* Diedicke in Sydow & Butler), we have usually omitted the latter part of the reference, since the citation continues with the reference to the publication in which the name was used

Synonyms are listed with a number of the species. These are only cases in which different names have been applied to the Indian fungi, and are entered and indexed so that older references to these fungi can be traced. No effort has been made to give complete synonymies. Recent revisions of groups of fungi have been consulted and often followed. Bresadola (84, 85) has reported a considerable synonymy, particularly for Polypores ; we have followed him in these, and where Lloyd or others give views regarding synonymous species, we have indicated these without, however, altering the records as they stand.

A few records of Indian fungi have been shown definitely to require deletion ; these are given in square brackets, and not counted in the totals. Where the records are given as admittedly uncertain, the species name is preceded by a query (e.g., *Puccinia ?andropogonis* Schw.).

In a number of instances " India " is given in the published records of distribution of species of fungi when, upon investigation, it was found that the East or West Indies, Malay, Ceylon, or some other region should have been listed. These species have been left out of this list entirely, since no especial value would come from the attempted correction of all mistakes in geography, many of which are readily detected. In a very few cases in which Saccardo gives India as part of the distribution of a fungus, we have been unable to trace the original reference to India : these are included as " recorded by Saccardo from India ".

Host names are as tabulated in the Index Kewensis, except that in three or four cases where the name could not be found there, it is given in quotation marks, and in some others the names used are there listed as synonyms. Hosts infected by artificial inoculation only are not included in the Host Index. When a fungus is recorded, for example, on *Thea sinensis* and *Thea* sp., the latter is not included in the Host Index.

Reference to Saccardo's Sylloge Fungorum (given thus : " Sacc. VI : 120," referring to volume and page respectively) is made in every case in which such reference could be found. When the reference to Saccardo in given first, the fungus was not originally described from India, except in a few cases in which Saccardo transferred the species to the genus here used. In the Uredinales, a reference to Sydow's Monographia Uredinearum is similarly given. The other numbers following an entry (*e.g.* 240 : 130) refer to the number before the citation in the bibliography, and to the page. When an entry reads, for example, " Sacc. IV : 15 as *Oospora*," a synonym is involved which is not known to have been applied to an Indian record. The bibliography references following after a synonym applied to an Indian fungus, usually indicate that these referred to papers in which the synonym was used.

All specific names, whether of fungi or hosts, are given with lower case letters, for typographical convenience and because of the difficulty of using capitals correctly in every case.

No new species are described here, but five new combinations were necessary to transfer species recorded under *Vermicularia* into the genus *Colletotrichum*. Miss Duke (Trans. Brit. Myc. Soc., XIII, pp. 156-184, 1928) and others have shown that *Vermicularia* is untenable. One new name was required in *Sclerospora*.

Cross references are given, in the List of Species, to supposed synonyms, and to other cases where two entries need to be consulted to obtain the data regarding a species. An attempt has been made to give an approximately complete record of the literature pertaining to the occurrence of each species in India, particularly to the first record of its presence there, and also to monographic and other treatments which have brought the knowledge of a species down to date, or have revised it. The bibliography is not, however, entirely complete for references to Indian fungi, since we have omitted a few references of a plant pathological nature,* in which the mycological phase is not considered, and also a few older records in which references to fungi are too vague to permit entering the data under any species name. An effort has been made to include all the fungi recorded from India to the year 1930.

We have not undertaken to examine specimens of many of the fungi recorded by the older mycologists from India, but Montagne's and Léveillé's specimens at Paris have been looked up, as well as a very few of the specimens at Kew. Several Indian fungi have been issued as exsiccati, but we have listed very few of these references.

ACKNOWLEDGMENTS.

We are glad to acknowledge the help of all who have collected or studied Indian fungi, a number of whom are mentioned in the Historical Section above. Of the Indian mycologists, Munshi Inayat Khan, who will be repeatedly referred to as " Inayat " below, stands out as an indefatigable collector and field student of the fungi during his long association with one of us, terminated only by his death. His memory will be preserved as long as Indian fungi are studied, though he published nothing, having little gift for writing and no languages but his native tongue. Barclay deserves special mention for working out so thoroughly the rusts about Simla. The majority of the species he recorded have been re-collected by the Pusa staff, and have been examined, together with many Phycomycetes, Ascomycetes, and Ustilaginales, and a number of Fungi Imperfecti, by the Sydows. Mr. E. S. Salmon has kindly named specimens of most of the species of the Erysiphaceae recorded for India. The late C. G. Lloyd studied many of the Polyporaceae of India, which would be very poorly known except for his assistance. Though his work, and that of the Abbé Bresadola and others, have shown that many of the records of the early students of Indian fungi, such as Berkeley, Léveillé, and Cooke, require revision,

---

* In general the economic importance of the fungi recorded is not stressed. This aspect has been, in part, dealt with by one of us (111) in another publication.

we consider that these pioneer workers deserve more credit than they are sometimes given. Their labours have provided their successors with a starting point in many important groups, and have given at least a general view of some of the salient features of the mycological flora of India.

THE FUNGI OF INDIA

LIST OF SPECIES.

ARCHIMYCETES.

Olpidiopsis minor Fischer (Sacc. **XI**:247 ; **94**[1]:134, figs.). In hyphae of *Achlya polyandra*, Dehra Dun (Butler).

——**schenkiana** Zopf (Sacc. **XI**:247 ; **94**:135, figs.; **453**:509). In living cells of *Spirogyra* sp., Pusa (Butler).

Olpidium indicum Turner (**512**:164 ; Sacc. **XVI**:389). In *Oedogonium* sp., India (Wallich).

Plasmodiophora brassicae Woronin (Sacc. **VII**:464 ; **113** (20)). Recorded by McRae as occurring on *Brassica oleracea* (cabbage and cauliflower).

Pseudolpidium aphanomycis (Cornu) Fisch. (Sacc. **VII**:300 as *Olpidiopsis ;* **94**:132, figs.; **453**:510). In hyphae of *Aphanomyces laevis*, Pusa (Butler).

Rhinosporidium seeberi Wernicke, emend. Ashworth (**24**:301 ; **358**:505). This organism, considered by Ashworth (**24**) to be allied to the Olpidiaceae, was recorded by him and by Norrie in man in India, usually in nasal infections.

Spongospora subterranea (Wallr.) Johns. (**113** (9):66 ; **2**:14). On tubers of *Solanum tuberosum*, Mahableshwar, Bombay ;Himalayas ; Nilgiris.

Synchytrium collapsum Syd. (**453**:510 ; Sacc. **XXI**:839 ; **455**:247). In leaves of *Clerodendron infortunatum*, Comilla, and Wahjain, Assam (Butler); Pusa (Inayat).

——**rytzii** Syd. (**453**:510 ; Sacc. **XXI**:840 ; **445**:326 ; **455**:247). In leaves and petioles of *Anisomeles ovata*, Dehra Dun (Butler); Pusa (Chibber) ; in leaves, petioles, and stems of *Peristrophe* sp., Pusa (Butler, a); Kangra District, Punjab (Mitter); on *P. bicalyculata*, Pusa ; on leaves of *Justicia* sp., Nagpur (Pandit, b); Samalkota, Madras (Shaw); on *J. procumbens*, Pusa (Butler); on leaves of *Leucas* sp., Benares (Subramaniam, c); on *L. aspera*, Coimbatore (Sundararaman ; McRae); on *Dicliptera* sp., Dehra Dun (Butler); on *Lepidagathis* sp., Dehra Dun (Kar) ; on *Lagenaria vulgaris*, Pusa (Subramaniam). Sydow distinguished between the type on *Anisomelis* and forms a, b, and c on *Peristrophe, Justicia,* and *Leucas*, respectively, and was uncertain whether the differences observed were variations of one species or indicated distinct species. The fungus is widely distributed in India on several hosts.

---

[1] These numbers refer to the bibliography on pp. 174-209, the page of the cited work being given after the colon.

Woroninella aecidioides (Peck) Syd. (Sacc. **XXIV**:17 ; **445**:485). In leaves of *Amphicarpaea edgeworthii*, Simla, India. Gäumann (**245**:172) prefers the name *Synchytrium aecidioides* (Pk) Lagerh.

——**dolichi** (Cke) Syd. (Sacc. **XXIV**:17 ; **445**:484). In living leaves of *Dunbaria ferruginea*, Runnymede, Madras (McRae). (This fungus is known in other countries on *Dolichos gibbosus*, *Glycine javanica*, and *Vigna sinensis*). Gäumann (**245**:172) favours the name *Synchytrium dolichi* (Cke) Gm.

——**puerariae** (P. Henn.) Syd. (Sacc. **XXIV**:17 ; **445**:486). In leaves and stems of *Pueraria* sp., India.

PHYCOMYCETES.

Chytridiales.

Nowakowskiella ramosa Butler (**94**:141, figs. ; Sacc. **XXI**:847). On decayed leaves and culms of *Triticum*, Dehra Dun (Butler).

Physoderma schroeteri Krieger (Sacc. **XIV**:447 ; **453**:510). In radical leaves of *Scirpus supinus*, Pusa (Butler) ; in *S.* sp., Lahore (B. Das). The Indian specimens are very similar to the form on *S. maritimus* in Germany.

——**zeae-maydis** Shaw (**455**:245, figs. ; Sacc. **XXIV**:20 ; **3**:24 ; **113** (19)). (Perhaps a synonym of *P. maydis* Miyabe). In living leaves of *Zea mays*, Duars, Bengal (Shaw); Pusa (McRae). This fungus has been found to cause a serious disease of maize in the United States of America.

Urophlyctis alfalfae (Lagerh.) Magn. (Sacc. **XVII**:515). On *Medicago sativa* and *Vicia hirsuta*, Lahore (B. Das).

Oomycetes.

Achlya polyandra (Hildebr.) de Bary (Sacc. **VII**:275 ; **453**:511). On decomposing insects under water, Pusa (Butler).

Allomyces arbuscula Butler (**103**:1027 ; Sacc. **XXIV**:32). In still water and river water, on dead insects from a drain, and from garden soil cultures on boiled ants, Pusa (Butler). Also obtained from Poona.

Aphanomyces laevis de Bary (Sacc. **VII**:276 ; **453**:511). On decomposing insects under water, Pusa (Butler).

Bremia lactucae Regel (Sacc. **VII**:244 ; **453**:512). On *Lactuca dissecta*, Dehra Dun (Butler) ; *L. scariola*, Kashmir (Butler) ; *L.* sp., Lyallpur (Mitter) ; Mussoorie (Butler) ; Lahore (Mitra) ; *Launaea nudicaulis*, Lyallpur (Butler) ; *Sonchus oleraceus*, Punjab (B. Das) ; *Conyza* sp., Samalkota (Subramaniam).

Cystopus bliti (Biv.) de Bary (Sacc. **VII**:236 ; **453**:514). On leaves of the following Amaranthaceae : *Achyranthes aspera*, Coorg, Surat, Shillong, Sylhet, and Dehra Dun (Butler) ; Kumaon (Inayat) ; *A.* sp., Dehra Dun, Sylhet, and Pusa (Butler) ; Coimbatore (Subramaniam) ; Dacca (Som) ; *Alternanthera sessilis*, Pusa, Gauhati, and Sylhet (Butler) ; *A. triandra*, Bombay (Dastur) ; *A.* sp., Chittagong (R. Sen) ; Dacca (Som) ; *Amaranthus* *blitum*, Malda and Kashmir (Butler); Peshawar (Dastur); *A. paniculatus*, Kanaighat, Sylhet (Butler); *A. tristis*, Pusa (Butler); *A. viridis*, Pusa, Gauhati, Surat (Butler); *A.* sp., Burma (Dastur); *Digera arvensis*, Malda, Lyallpur, Pusa (Butler); Dohad Farm, Bombay (Kulkarni); Coimbatore (Subramaniam); Godavari (R. Sen). The form on *Amaranthus blitum* was considered by Zalewski (Bot. Centralbl. XV, p. 223, 1883) to be different from that on the other hosts, the latter being named *Cystopus amaranthacearum*. Subsequent mycologists have not followed Zalewski, and although the differences he mentioned are noticeable in the Indian forms, they are of doubtful specific value. *Coleosporium deeringiae* Pat. (**361**:123) is considered by Sydow (**442, III**:656) to be a *Cystopus*, apparently *C. bliti*. It was recorded on leaves of *Bosea* (*Deeringia*) *amherstiana* in India.

[Cystopus] candidus (Pers.) Léveillé (Sacc. **VII**:234; **453**:513; **455**:243; **445**:484 **111**:292, 295, figs.). On the following Cruciferae: *Brassica campestris*, Pusa and Kashmir (Butler); *B. campestris* var. *toria* (*B. napus* var. *dichotoma* Prain), Kangra, Punjab (Mitter); *B. nigra*, Gauhati (N. C. Das); *B.* sp., Kashmir (Butler); *Capsella bursa-pastoris*, Dehra Dun (Butler); *Cleome viscosa*, Dehra Dun (Butler); *Eruca sativa*, Lyallpur; Srinagar; and Nagina, United Provinces (Butler); *Gynandropsis pentaphylla*, Godavari (McRae); *Nasturtium palustre*, Kashmir (Butler); *Raphanus sativus*, Dehra Dun (Butler).

——**ipomoeae-panduratae** (Schw.) Stevens & Swingle (Sacc. **IX**:341; **453**:514). On *Ipomoea eriocarpa*, Achibal, Kashmir; Pusa; Dohad Farm, Bombay (Butler); *I. hederacea*, Dehra Dun (Inayat); Pusa (Butler); Poona (Ajrekar); Simla (Watt); *I. reniformis*, Samalkota (Barber); Surat, Dharwar, Akola Farm, and Hoshangabad (Butler); Gujrat (Burkill); Nagpur (Pandit); Bombay (Chibber); *I.* sp., Pusa (Butler); Kistna Dist. (Subramaniam). The form on *I. reniformis* causes marked deformity of the host plant, transforming it from a dorsiventral to a radial habit. That on *I. hederacea* does not produce this effect. Sydow (**442, IV**:569) considers it clear from the description that *Uredo ipomoeae* Barclay (**43**:228) is *C. ipomoeae-panduratae*.

——**platensis** Speg. (Sacc. **XI**:212; **453**:514). On leaves of *Boerhaavia repens*, Dehra Dun, Gauhati, Surat, Sibpur near Calcutta, and Ganjam (Butler); Godavari (S. N. Mitra). This species has been separated from *C. bliti* by American authors, because of apparent constant differences in the size of the areolae on the oospore wall. In the Indian material there is no constant difference, and it is doubtful that the two species can be kept separate.

——**portulacae** (DC.) Léveillé (Sacc. **VII**:235; **453**:513; **111**:319, fig.; **445**:485). On leaves of *Portulaca oleracea*, Dehra Dun (Inayat); Srinagar (Butler); on *P. quadrifida*, Malda, Bengal and Bassein, Burma (Butler); Coimbatore (McRae).

[Cystopus] tragopogonis (Pers.) Schroeter (Sacc. **VII**:234 ; **453**:513 ; **455**:243). On leaves of *Cirsium arvense*, Lyallpur (Butler) ; *Crepis glauca* and *Serratula pallida*, Harwan, Kashmir (Butler).

Dictyuchus monosporus Leitgeb (Sacc. **VII**:273 ; **453**:511). On dead plant fragments under water, Dehra Dun (Butler).

Peronospora affinis Rossm. (Sacc. **VII**:251). On *Fumaria parviflora*, Pusa (Butler).

——**arborescens** (Berk.) de Bary (Sacc. **VII**:251 ; **3**:26 ; **453**:512 ; **111**:344-5, figs. ; **191**:24, figs.). On leaves of *Papaver somniferum* throughout Bihar and United Provinces ; on *Argemone mexicana*, Pusa (Butler) ; Saharanpur (B. Das).

——**celsiae** Sydow (**455**:245 ; Sacc. **XXIV**:60). On leaves of *Celsia coromandelina*, Pusa (Butler).

——**chlorae** de Bary (Sacc. **VII**:247). On *Erythraea roxburghii*, Pusa (Butler).

——**effusa** (Grev.) Rabenh. (Sacc. **VII**:256 ; **453**:512 ; **111**:317, fig.). On leaves of *Chenopodium album*, Pusa, Lyallpur, Nagina, Cawnpore ; Harwan, Kashmir and Myingyan, Burma (Butler) ; Peshawar (Shaw) ; Jullundur (Mitter).

——**indica** Sydow (**443**, **XVI**:157, fig. ; Sacc. **XXIV**:59 ; **453**:513 as *P.? linariae* Fckl). On leaves of *Calceolaria scabiosaefolia*, Dehra Dun (Butler).

——**lamii** Braun (Sacc. **VII**:256 ; **455**:245). On leaves of *Plectranthus rugosus*, Verinag, Kashmir (Butler).

——**parasitica** (Pers.) de Bary (Sacc. **VII**:249 ; **453**:512 ; **455**:245 ; **111**:292, 299, figs.). On various Cruciferae: *Brassica campestris* and its cultivated varieties, Pusa and Kashmir (Butler) ; Lahore (Mitra) ; *B. juncea*, Dacca (Som) ; *B. napus*, Pusa (Butler) ; *Capsella bursa-pastoris*, Dehra Dun (Butler) ; *Eruca sativa*, Pusa and Lyallpur (Butler) ; *Raphanus sativus*, Dehra Dun (Butler ; *Sisymbrium irio*, Peshawar (Shaw).

——**rumicis** Corda (Sacc. **VII**:262 ; **445**:484). On leaves of *Rumex vesicarius*, Coimbatore (McRae) ; Bombay (Ajrekar).

——**trifoliorum** de Bary (Sacc. **VII**:252 ; **453**:513 ; **455**:245 ; **111**:266, fig.). On leaves of *Glycine hispida*, Larkipur, Kashmir (Butler) ; *Medicago denticulata*, Lyallpur (Butler) : *M. lupulina*, Pusa and Dehra Dun (Butler) ; *M. sativa*, Sargoda, Punjab (Goldon) ; *Melilotus alba*, Pusa (Butler) ; *M. indica*, Orai, United Provinces ; Lyallpur ; and Pusa (Butler) ; *M. parviflora*, Jullundur ; Gujranwala (Mitter). The form on the soy-bean in Manchuria and the United States has been regarded as a distinct species, *P. manshurica* (Naoum.) Gäumann, but the downy mildew of this crop in Kashmir agrees more nearly with *P. trifoliorum*.

——**viciae** (Berk.) de Bary (Sacc. **VII**:245 ; **453**:513 ; **455**:245 ; **111**:252, fig.). On leaves of *Vicia hirsuta*, Pusa (Butler) ; *Lathyrus sativus*, Lyallpur, Pusa, Cawnpore (Butler) ; *Pisum arvense*, Cawnpore, Pusa (Butler) ; *P. sativum*

Pusa (Butler); *Trigonella polycerata*, Jullundur (Mitter). R. Maire (Bull. Soc. Hist. Nat. de l'Afrique du Nord, VI, p. 151, 1915) records *Peronospora trifoliorum* as the downy mildew on *Trigonella foenum-graecum*.

Phytophthora arecae (Coleman) Pethybridge (**382**:555 ; Sacc. **XXIV**:35, name only ; **436**, figs. ; **333** ; **354** ; **353** ; **92**, and **453**:512 as *P. ? omnivora* ; **122**:620, figs., as *P. omnivora* var. *arecae* Colem. ; **124** ; **123**, figs. ; Sacc. **XXI**:860 as *P. cactorum* var. *arecae*). On leaves, fruits, and peduncles of *Areca catechu*, Mysore (Butler ; Narasimhan) ; Cochin ; and Karwar, Bombay; on nuts and inflorescences of *Cocos nucifera*, Malabar (Sundararaman, **436**), and reported on this host also by McRae (**113** (19)). Narasimhan (**354**) recently recorded the following hosts: leaves of *Colocasia antiquorum*, *Bryophyllum calycinum*, *Santalum album*, *Artocarpus integrifolia*, *Mangifera indica* ; fruits of *Ficus nitida*, *Jatropha glandulifera*, *Citrus medica*, *C. limonum*, and *Hevea brasiliensis*.

——**colocasiae** Raciborski (Sacc. **XVI**:396 ; **453**:512 ; **117** ; **111**:306, figs.). On leaves, petioles, flowers, and corms of *Colocasia antiquorum*, Burma (Dastur) ; Rangpur (A. Khan) ; Pusa (McRae) ; Comilla and Godagiri in Bengal, and Dehra Dun (Butler) ; on *C.* sp., Travancore and Burma (Butler) ; on wounded leaves and stems of seedlings of *Solanum tuberosum* (Pusa).

——**infestans** de Bary (Sacc. **VII**:237 ; **111**:277, figs. ; **87** ; **199**, figs.; **204** ; **113**(17): 53 ; **191**). On all parts of *Solanum tuberosum* and *Lycopersicum esculentum*, Himalayas, Khasi Hills, Plains of Bengal and Assam.

——**meadii** McRae (**328**:760 ; **327**:254, figs.; **329**, figs.; **200** ; **202**, figs.; **111**:494, figs.; **113**(8):49 ; **394**:6 ; **393**). On leaves, twigs, and fruits, and on the tapping cut, of *Hevea brasiliensis*, southwest India (McRae); Burma (Dastur ; Rhind).

——**palmivora** Butler (**113**(11):82 ; Sacc. **XXIV**:36 ; **332** ; **429** ; **271** ; **94**:82, figs., as *Pythium palmivorum* Butler ; **92** ; **96**:21 ; **99** ; **414**, figs.; **415**, figs.; **453**;511 ; **111**:491, figs., as *Phytophthora faberi* Maublanc ; **394**:14 ; **392** ; **202**:226 ; **123**:86 as *P. theobromae* Coleman ; discussion of nomenclature by Ashby, Trans. Brit. Myc. Soc., XIV, p. 33, 1929). On leaves and growing points of *Borassus flabellifer*, Godavari and Kistna (Butler) ; Hoogli District, Bengal (Kar ; Sundararaman) ; South Kanara ; On *Cocos nucifera*, Godavari, Kistna, and Travancore (Butler) ; Malabar (Shaw and Sundararaman) ; on *? Areca catechu*, Godavari and Kistna (Butler) ; on leaves, twigs, and fruit of *Hevea brasiliensis*, Burma (Rhind, **392**, **394**); on *Theobroma cacao*, and capable of infecting *Lycopersicum esculentum* (Coleman) ; perhaps this species on fruit of *Carica papaya*, Hmawbi, Burma. The cause of leaf-fall in *Hevea* rubber in Burma is identified with this species by Rhind, whereas in south India it is due to *P. meadii* ; but there is much confusion in the literature of the two species. According to Dastur (**202**) the Burma and south Indian fungi are probably the same species. The character which led to this species being first described as a *Pythium* (the discharge of zoospores into a transient bladder) has been found since in *Phytophthora parasitica, P. meadii*, and other species. The idea that *P. palmivora* and *P. faberi* were not distinct (**113**(12)), has received confirmation in the investigations of Ashby (loc. cit.).

[**Phytophthora**] **parasitica** Dastur (**198**:226, figs.; Sacc. **XXIV**:37 ; **113** (19); **111**:326, figs.; **203**, figs.; **345**, figs.). On leaves and young shoots of *Ricinus communis*, and on seedlings of various plants (*Solanum tuberosum, Sesamum*, etc.), Pusa ; on *Sesamum indicum*, Saharanpur (B. Das); on *Gossypium* and *Psidium*, Pusa (Mitra ; McRae). A physiologic form of this species was also found by Dastur (**203**) on *Vinca rosea* at Pusa, and inoculations produced infections on a number of other plants, chiefly seedlings. The *Phytophthora* found on *Piper betle*, Central Provinces (Dastur, **209**), may be this species.

——**pini** Leonian var. **antirrhini** Sundararaman and Ramakrishnan (**437** : 99, figs.). On *Antirrhinum majus*, causing collar-rot and wilt, Ootacamund (Sundararaman and Ramakrishnan).

Plasmopara obducens Schroeter (Sacc. **VII**:242 ; **453**:512). On leaves of *Impatiens balsamina*, Kasauli (Butler) ; Poona (Kulkarni) ; of *I.* sp., Dehra Dun (Butler).

——**viticola** Berkeley & Curtis (Sacc. **VII**:239 ; **455**:244 ; **3**:34 ; **86**). On leaves of *Vitis vinifera*, near Poona (Ajrekar) ; Madras (Sundararaman) ; Mandalay (Shroff).

——**wildemaniana** P. Henn. (Sacc. **XXI** : 861 ; **455**: 243, fig.). On leaves of *Justicia* sp., Pusa (Hafiz) ; Chittagong (R. Sen); on *Periotrophe bicalyculata*, Poona (Ajrekar).

Pseudoperonospora cubensis (Berk. & Curt.) Rost. (Sacc. **XVI** : 520 ; **113** (3) : 54 ; **455**:244 ; **111** : 311, figs.; **28, IV** :1, figs.). On leaves of *Luffa acutangula, L. aegyptiaca, Trichosanthes dioica*, and *T. cucumerina*, Pusa (Butler); on *Cucumis melo*, Lyallpur ; on *Trichosanthes dioica*, Bengal (Bal).

Pythium aphanidermatum (Edson) Fitzpatrick (**113**(19); **346** ; **432** ; **389** ; probably =*P. butleri* Subramanian, **422**, figs.; Sacc. **XXIV**:1332. Included at first under *P. gracile* Schenk: **94**:67, figs.; **111**:348, figs.; **453**:551 ; **325**). Parasitic on *Zingiber officinale*, Surat (Butler); Rangpur, Bengal (McRae); on *Nicotiana tabacum*, Pusa (Butler); on *Capsicum annuum*, Pusa (Butler); on *Carica papaya*, Pusa, Kathiawar, Dacca, Burma (Subramanian); inoculated with infection on *Solanum tuberosum* and *Ricinus communis* (Subramanian). The form on roots of *Ricinus* (**94**:70) is possibly the same. Other forms which are morphologically similar to this species have been more recently found on rotting Cucurbitaceae including *Cucumis sativus, Trichosanthes dioica, T. anguina, Luffa acutangula, L. aegyptiaca*, and *Lagenaria vulgaris*, at Pusa. Some physiologic specialization may occur. Recently recorded on *Amaranthus gangeticus* (**5**:3), and on *Opuntia dillenii* from which infections were obtained on the stems of *Cucumis*, *Trichosanthes*, *Datura*, *Solanum*, and *Physalis*, and the leaves of *Amorphophallus* and *Basella* (**113**(19)). *P. aphanidermatum* appears to be cosmopolitan and to have a very large number of host plants.

[Pythium] artotrogus (Mont.) de Bary (Sacc. **XI**:244 ; **94**:100, figs.; **453**:512 ; **113**(17):54). In rotting tubers of *Solanum tuberosum*, Calcutta (from Hooghly) (Butler); Khasi Hills (McRae).

——**de Baryanum** Hesse (Sacc. **VII**:270; **113**(5):66 ; **106**:262, figs.). On seedlings of *Lepidium sativum* and in garden soil, Pusa (Butler); on seedlings of *Clarkia* and of *Gilia*, Pusa (Dastur).

——**gracile** Schenk (Sacc. **VII**:272 ; **94**:67, figs., pro parte ; **453**:511, pro parte). In rotting plant débris, Calcutta (Butler). Parasitic forms first placed here are now considered to be *P. aphanidermatum*, q. v.

——**graminicolum** Subramaniam (**424**). On *Triticum vulgare*, associated with discoloration and rot of collar and roots, Dharwar (Subramaniam).

——**indigoferae** Butler (**94**:73, figs.; Sacc. **XXI**:884 ; **453**:511). On leaves of *Indigofera arrecta*, Calcutta (Butler).

——**monospermum** Pringsheim (Sacc. **VII**:271 ; **94**:71, figs.). A saprophytic form found growing on the rotted rhizomes of *Zingiber*, and decaying seedlings of *Lepidium sativum*, Pusa (Butler).

——**proliferum** de Bary (Sacc. **VII**:271 ; **94**:76, figs.; **453**:511). On decomposing insects under water, Calcutta, Dehra Dun, and Pusa (Butler); on seeds of *Ricinus*, Pusa.

Saprolegnia monoica (Prings.) de Bary (Sacc. **VII**:268 ; **453**:511). On decomposing insects under water, Dehra Dun (Butler).

Sclerospora graminicola (Sacc.) Schroeter (Sacc. **VII**:238 ; **95**, figs.; **278**, figs.; **453**:513;**445**:326 ; **111**:50, 89, 218, figs.; **455**:245 ; **515**). On leaves and inflorescenses of *Setaria italica*, Pusa, Kashmir, and Dharwar (Butler) ; Godavari (Mitra); Wynaad (McRae); on *Pennisetum typhoideum*, Trichinopoli, Dharwar, Poona, Surat, Orai (Butler); Bombay (**515**); and throughout northern India.

——**graminicola** var. **andropogonis-sorghi** Kulkarni (**278**:268, figs.; **111**:203, fig.) On leaves of *Andropogon sorghum*, Coimbatore (Butler ; Barber) ; Poona (Kulkarni) ; Bombay (Chibber) ; Dharwar (Butler); and Burdwan ; on leaves of *A. halepensis*, Kirkee, near Poona (Barber); on *Euchlaena mexicana* (*E. luxurians*), Kirkee (S. N. Mitra). W. H. Weston (Journ. Agric. Res., XXVII, p. 781, 1924) gives good reasons for considering this to be a distinct species.

——**indica** Butler, nom. nov. (**107**:279, figs., as *S. maydis* (Racib.) Butler ; Sacc. **XXIV**:66 ; **3**:22 ; **111**:191,fig.; **113**(19)). On leaves of *Zea mays*, Pusa (Butler), Poona (Ajrekar). The fungus described by Raciborski under the name *Peronospora maydis* (see Sacc. **XIV**:460), with which the Indian species was, at first, considered identical, has been shown by Palm (Meded. Laboratorium Plantenziekten, Buitenzorg, No. 32, 1918) to be a *Sclerospora* distinct from the Indian maize downy mildew. Palm gave it the new name, *S. javanica,* but under the International Rules Raciborski's specific name " maydis " should stand and the Javan fungus should have been called *S. maydis* (Racib.) Palm. A new name has, therefore, to be given to the Indian species.

ZYGOMYCETES.

Choanephora cucurbitarum (Berk. & Rav.) Thaxter (Sacc. **XVII**:507 ; **207**, figs.). On leaves, stems, and flowers of *Capsicum* spp., Pusa (Dastur).

——**infundibulifera** (Currey) Cunningham (**187**:417 ; Sacc. **IX**:339 ; **190**:163, figs.; **195**:333, figs., as *Cunninghamia infundibulifera* Currey ; **195**:578 as *Choanephora cunninghamiana* Currey; **455**:248). On flowers of *Hibiscus rosa-sinensis,* Calcutta (Cunningham); Pusa (Butler); on inflorescence of *Tabernaemontana coronaria,* Pusa (Butler).

——**simsoni** Cunningham (**190**:169, figs.; Sacc. **XIV**:432 ; **453**:515). Parasitic on *Ipomoea rubro-caerulea,* Calcutta (Cunningham) ; on living flowers of *Zinnia elegans,* Calcutta (Cunningham); Pusa (Butler); Poona (Chibber).

Cunninghamella elegans Lendner (Sacc. **XXI**:828 ; **113**(7):56). In soils, Pusa (Shaw).

Empusa lecanii Zimmermann (**125** ; **375**, XI). On *Lecanium viride, L. colemanii,* etc. on *Coffea* and *Psidium guajava,* South India (Coleman); Kotagiri, Nilgiris (Anstead). Petch (1. c.) has failed to find a true *Empusa* on any of the specimens (including Indian collections) examined by him, but found a mycelium of fine, hyaline, non-septate hyphae, bearing pyriform conidia which eventually became fuscous or pale brown and germinated either by one or two germ-tubes or by the protrusion of the spore contents into a globose vesicle which in some cases has a thick wall and appears to be persistent. It is suggested that this is the fungus described by Zimmermann, and that it is a *Pythium* or allied genus.

——**muscae** Cohn (Sacc. **VIII**:281). On house-flies, Lahore (B. Das).

Entomophthora aphidis Hoffm. (Sacc. **VII**:283). On *Aphis* sp., Lahore (B. Das); on *Aphis* on *Pisum sativum,* Pusa (Dutt).

Mucor glomerula Lendner (Sacc. **XXI**:821 ; **461**). From the soil in Madras (Thakur and Norris).

——**plumbeus** Bonorden (Sacc. **VII**:194 ; **461**). From the soil in Madras (Thakur and Norris).

——**praini** Nechitsch (**355**, figs.; Sacc. **XXI**:818 ; **267**:147 ; **113**(6):58 ; **111**:14, fig.). In fermenting rice, Sikkim (recorded by Nechitsch); Ranchi, Chaibassa, Balasore, Rajmahal, Dunka, Sambalpur (Hutchinson and Ayyar).

——**racemosus** Fresenius (Sacc. **VII**:192 ; **461**). From the soil, Madras (Thakur and Norris).

Rhizopus artocarpi Raciborski (Sacc. **XVI**:386 ; **344**). On *Artocarpus integrifolia*. Benares (Dastur); Pusa (Shaw); Andaman Islands (Mitra). This species is considered by Lendner (Les Mucorinées de la Suisse, 1908, p. 116) to be "une forme vigoreuse du *R. nigricans*."

——**cambodja** (Chrzas.) Vuillemin (Sacc. **XVII**:503 ; **355** ; **267**:147 ; **113**(6):58). In fermenting rice, Darjeeling (Hutchinson and Ayyer); Khasi Hills.

——**nigricans** Ehrenberg (Sacc. **VII**:212 ; **461**). From soil, Madras (Thakur and Norris).

ASCOMYCETES.

HEMIASCOMYCETES.

Blastocystis hominis Brumpt (**275**). Knowles and Das Gupta, who studied this organism from human patients at the Calcutta School of Tropical Medicine, consider it to be allied to the *Schizosaccharomycetes*.

Nematospora coryli Peglion (Sacc. **XVIII**:202 ; **396**). Recorded by Rhind [as "Nowell's form D," which Ashby and Nowell recently recorded (Ann. Bot., XL, pp. 69-83, 1926) as *N. coryli*] as an internal disease of unripe bolls of *Gossypium* in Burma, in which rotting of the contents of affected fruits ensued.

——**gossypii** Ashby & Nowell (**396**). Recorded by Rhind (as "Nowell's form C") as being more common than the preceding in affected bolls of *Gossypium* in Burma.

Protomyces macrosporus Unger (Sacc. **VII**:319 ; **454**:372 ; **111**:359, figs.). Common in living green parts of *Coriandrum sativum*, Pusa (Sen); Lahore (B. Das).

Pseudosaccharomyces indicus Kloecker (**272**:335, fig.; Sacc. **XXIV**:1310). From soil, Himalayas ; characters of the fungus described by Kloecker from cultures in beer-wort.

Saccharomyces cerevisiae Meyen (Sacc. **VIII**:916). In the usual fermentation processes.

Taphrina aurea (Pers.) Fr. (Sacc. **VIII**:812 ; **454**:373). On leaves of *Populus ciliata*, Murree (Butler).

——**cornu-cervi** Giesenhagen (**246**:135, figs.; Sacc. **XI**:437). In fronds of *Aspidium aristatum*, Nepal (Wallich); Coonoor (Gamble). "*Aspidium cornu-cervi*" is based on ferns affected by this fungus, and *Urobasidium rostratum* occurs on the galls produced by it.

——**deformans** (Berk.) Tul. (Sacc. **VIII**:816 as *Exoascus deformans* (Berk.) Fckl ; **454**:372 ; **113**(9):66; **113**(10):74 ; **2**:44 ; **230**). On leaves of *Prunus persica*, Shillong (Butler); Ranchi (Dobbs); Peshawar (Shaw); Mussoorie (Butler); Kashmir (**230**). The disease on *Nephelium litchi* at Calcutta attributed by Bal (**28**, II·4) to this species is an erineum caused by a mite of the genus *Eriophyes*.

[**Taphrina**] **laurencia** Giesenhagen (**246**:150, figs.; Sacc. **XI**:437 ; **111**:89, fig.). On fronds of *Pteris quadriaurita*, Assam (Simons).

——**maculans** Butler (**102**:36, figs.; Sacc. **XXIV**:1303 ; **111**:48, 347, figs.). On leaves of *Curcuma amada*, Saharanpur (Gollan); *C. angustifolia*, Kumaon (Inayat); Comilla (Butler), *C. longa*, Dehra Dun, Pusa, Poona, Surat (Butler); Samalkota (Subramaniam); Birbhum (Basu); Godavari (Shaw); Rangpur (Mitra); *Hedychium* sp., Mussoorie (Butler); *Zingiber cassumunar*, Rangpur, Bengal (Butler); *Z. zerumbet*, Bihar (Butler).

——**pruni** (Fckl) Tul. (Sacc. **VIII**:817 as *Exoascus pruni* Fckl ; **454**:373 ; **111**:86, fig.). On fruit of *Prunus padus*, Murree (Butler); Simla (Watt). *Cerasus cornuta* Wall. is based on the elongated fruit caused by this parasite (cf. Gardeners' Chronicle, 1850, p. 406) and not, as stated in Hooker's Flora of British India, II, p. 316, on deformities caused by insects.

——**rhomboidalis** Sydow & Butler (**454**:373, fig.; Sacc. **XXIV**:1303). On fronds of *Pteris quadriaurita*, Kumaon (Inayat).

DISCOMYCETES.

Aleurina orientalis (Patouill.) Sacc. & Sydow (Sacc. **XXII**:720 as *Sarcosoma ;* **454**:374). On cow dung, Pusa (Butler).

Ascobolus gollani P. Henn. (**263**:338 ; Sacc. **XVIII**:119). On manured ground under trees, Botanic Garden, Saharanpur (Gollan).

Aulographum pandani Cke (**133**:17, fig.; Sacc. **II**:729). On *Pandanus odoratissimus*, India (Hobson).

——**vagum** Desmaz. (Sacc. **II**:727 ; **130**:75). On coriaceous leaves, Himalayas (Fleming). " Without fruit, but probably this species," Cooke (**130**).

Bulgaria chalybea (Berk. in Herb.) Cke & Mass. (**173**:74 ; Sacc. **X**:41 ; **57**, after No. 359, as *B. inquinans* var. *chalybea* Berk). On trunks of trees, Jalla-pahar, Darjeeling, alt. 7500 ft. (Hooker f.).

Cenangium chrysoprasum Cke (**183**:74 ; Sacc. **XI**:423). On naked wood, India.

Chlorosplenium aurigeneum (Berk.) Sacc. (Sacc. **VIII**:318 ; **57**, No. 474 as *Peziza aurigenea* Berk.). On dead wood, Khasi Hills (Hooker f. & Thomson).

——**aeruginosum** (Oeder.) de Not. (Sacc. **VIII**:315 ; **319**, **III**:153). On fallen wood, Mandali, Jaunsar, 8-9,000 ft. (Duthie).

Coccomyces vilis Syd. & Butler (**454**:377 ; Sacc. **XXIV**:1267). On leaves of *Mangifera indica*, Malda (Butler).

Cryptomyces pongamiae (Berk. & Broome) Sacc. (Sacc. **VIII**:708 ; **454**:376). On leaves of *Pongamia glabra*, Islampur, Bombay (Chibber).

Dasyscypha aleurodes Cke (**183**:73 ; Sacc. **XI**:413). On palm petioles, Nilgiris.

——**clandestina** (Bull.) Fckl (Sacc. **VIII**:457 ; **57**, after No. 357, as *Peziza clandestina* Bull.). Over the upper side of the leaves of a species of *Pyrus*, Tonglo, Sikkim (Hooker f.).

[Dasyscypha] **emerici** (Berk. & Phill.) Sacc. (Sacc. **X**:22; **173**:74 as *Lachnella emerici* Berk. & Phill.). On twigs, Nilgiris.

Elaphomyces sapidus Massee (**319**, **XI**:252). United Provinces (Burkill). Edible.

Erinella corticola Massee (**319**, **I**:115 ; Sacc. **XVI**:756). On bark, Dehra Dun (Gamble).

Geoglossum?alveolatum Durand (Sacc. **XVIII**:8 as *Leptoglossum*). On the ground, Simla (Butler).

——**hirsutum** Pers. (Sacc. **VIII**:46 ; **301**:12, figs.). On bare earth, India (Gollan).

——**ophioglossoides** (L.) Sacc. (Sacc. **VIII**:43 ; **288**, No. 65:4 ; **57**, after No. 472, as *G. glabrum* Pers.). On the ground, Yeumtong, 13,000 ft. (Hooker f.); India (Kashyap).

Helotium citrinum (Hedw.) Fr. (Sacc. **VIII**:224 ; **57**, after No. 359, as *Peziza citrina* Pers.; **196**:128). On rotten wood, Sikkim, 8-9000 ft. (Hooker f.); Sikkim, 5-7,000 ft. (Kurz).

——**pusense** Syd. (**454**:374 ; Sacc. **XXIV**:1183). On stems of *Ricinus communis*, Pusa (Inayat).

Helvella crispa (Scop.) Fr. (Sacc. **VIII**:18 ; **145**:13 ; **454**:374 ; **288**, No. 65:4). On the ground, Punjab (Aitcheson) ; Achibal, Kashmir (Butler); Simla (Butler) ; India (Kashyap).

——?**fargesii** Patouill. (Sacc. **XI**:391 ; **288**, No. **65**, pp. 4, 8). India (Kashyap).

Humaria masseeana Sacc. & D. Sacc. (Sacc. **XVIII**:26; **319**, **III**:152 as *H. coccinea* Massee, not (Crouan) Quél). Garhwal (Gamble).

——**rutilans** (Fr.) Sacc. (Sacc. **VIII**:133 ; **196**:128). On mud banks, Toukyeghat River, Toungoo, Burma (Kurz).

Lachnea geneospora (Berk.) Sacc. (Sacc. **VIII**:178 ; **57**, No. 357, as *Peziza geneospora* Berk.). On rotten wood, Sinchul, Sikkim, 8,000 ft. (Hooker f.).

Lachnella nilgherrensis Cke (**173**:73 ; Sacc. **X**:20). On herbaceous stems, India (Herb. Berkeley).

Lagerheima carteri (Berk.) Sacc. (Sacc. **X**:55 ; **173**:75 as *Patellaria carteri* (Berk.) Phill.). On dead, decorticated wood, Bombay (Carter).

Leotia lubrica Pers. (Sacc. **VIII**:609 ; **57**, after No. 354). On clay banks, Sinchul, Sikkim, 8,600 ft. (Hooker f.).

Lophodermium hysterioides (Pers.) Sacc. (Sacc. **II**:791 ; **349**:23 as *Hysterium foliicolum* Fr.). The specimen in herb. Montagne, marked only " Neelgherri [Nilgiris] M. Perrottet " consists of a small coriaceous leaf bearing a very few hysterioid fruit bodies.

——**pinastri** (Schrad.) Chev. (Sacc. **II**:794 ; **454**:378). On needles of *Picea excelsa*, Achibal, Kashmir (Butler). The asci and spores are smaller than in European specimens (**454**:379).

Marchalia ustulata (Cke) Sacc. (Sacc. **VIII**:738 ; **133**:17 as *Rhytisma ustulatum* Cke ; **264**, No. 499, as *Phyllachora ustalata* (Cke) v. Hoehn.). " On dead leaves, probably of *Ficus*, Col. Hobson " (**133**). Cooke mentioned that this fungus occurred on leaves of the same tree which bore "*Trichobasis hobsoni*" (see *Cerotelium fici*). Sydow (**442**, **IV**:592) states that an examination of the original material showed that the host is undoubtedly *Ficus*, either *F. infectoria* or *F. religiosa*. Theissen and Sydow (**482**:450) consider this species to be a conidial form doubtless belonging to *Catacauma infectorium* (q. v.), and that the host is probably *F. infectoria*.

Midothis macrotis (Berk.) Sacc. (Sacc. **VIII**:547;**57**, No. 356, as *Peziza macrotis* Berk.; **59**:424 and **131**, fig. 94 as *Wynnea macrotis* Berk.; **287**:934). On rotten wood, Darjeeling, 7,500 ft. (Hooker f.).

Mitrula rosea Lloyd (**287**:885, figs., 955). On bare earth, India (Cave).

——**viridis** (Pers.) Karst. (Sacc. **VIII**:38 ; **57**, after No. 472, as *Geoglossum viride* Pers.). On the ground, Yeumtong, 12,000 ft.; Lachoong, 3,000 ft. (Hooker f.).

Morchella bohemica Krombholz (Sacc. **VIII**:14 ; **133**:16). On the ground, Kashmir (Herb. Berk.). Cooke notes that it is eaten by the natives, and dried and sent down into the Plains. The same is true for some of the other species, especially *M. deliciosa* which can be regularly bought in the bazaars of the Punjab cities.

——**conica** Pers. (Sacc. **VIII**:9 ; **467**:158 ; **263**:338). On the ground, Dehra Dun (Blatter) ; var. **acuminata** Kickx, Siwalik Hills (Gollan).

——**deliciosa** Fr. (Sacc. **VIII**:10 ; **128**:439 ; **454**:374). On the ground, Kashmir (Naudan); Amritsar and Lahore (Butler); Nepal (Smith); Kumaon (Mitra).

——**esculenta** (L.) Pers. (Sacc. **VIII**:3 ; **454**:374 ; **133**:16 ; **520**:259). On the ground, Amritsar, and Ranikhet, Kumaon Himalayas (Butler) ; Kashmir (Aitcheson).

——**gigaspora** Cke (**128**:442, fig.; Sacc. **VIII**:14 ; **131**:187). On the ground, Kashmir (Stewart). Recorded by Stewart in his list of Punjab plants as "*M. semilibera*."

Ombrophila indica Syd. (**454**:375 ; Sacc. **XXIV**:1237). On rotten wood and on the ground, Dehra Dun (Butler).

Otidea darjeelensis (Berk.) Sacc. (Sacc. **X**:4 ; **57**, No. 355, and **131**, pl. 215, as *Peziza darjeelensis* Berk.). On the ground, Darjeeling.

Peziza aurantia Pers. (Sacc. **VIII**:74 ; **57**, after No. 356 ; **196**:128). On earth, Sikkim (Kurz); on clay banks, Darjeeling, 7,000 ft., "so conspicuous that every one asks whether you have seen the scarlet fungus" (Hooker f.).

——**epispartia** Berk. & Broome (Sacc. **VIII**:89 ; **144**:96). Sent to Cooke from Belgaum (Hobson).

Phacidium symplocinum Syd. (**454**:376 ; Sacc. **XXIV**:1261). On living leaves of *Symplocos* sp., Darjeeling (McRae).

Phialea fructigena (Bull.) Gill. (Sacc. **VIII**:265 ; **57**, after No. 357, as *Peziza fructigena* Bull.). On stems of dead Umbelliferae, Sikkim, 8-9,000 ft. (Hooker f.).

Plicaria repanda (Wahl.) Rehm (Sacc. **VIII**:100 as *Discina* ; **28**, **III**:35-36, figs.). On rotten fruits of *Borassus flabellifer*, Calcutta (Bal).

Pseudopeziza medicaginis (Lib.) Sacc. (Sacc. **VIII**:724 ; **454**:375). On leaves of *Medicago sativa*, Poona (Ajrekar) ; of *M. lupulina*, Harwan, Kashmir (Butler).

——**repanda** (Fr.) Karst. (Sacc. **VIII**:727 ; **454**:375). On leaves of *Galium* sp., Kasauli (Butler).

——**trifolii** (Biv. Bernh.) Fckl (Sacc. **VIII**:723 ; **454**:375). On leaves of *Trifolium pratense*, Verinag, Kashmir (Butler).

Pseudophacidium indicum Syd. (**454**:375 ; Sacc. **XXIV**:1272). On dead branches, Dehra Dun (Butler).

Rhizina reticulata Berk. & Broome (**67**, No. 919 ; Sacc. **VIII**:58). On wood, Nilgiri Hills (E. S. Berkeley).

——**zonata** Berk. (**57**, No. 473 ; Sacc. **VIII**:59). Amongst leaves of *Pinus* sp., Darjeeling (Hooker f.).

Rhytisma acerinum (Pers.) Fr. (Sacc. **VIII**:753 ; **454**:377). On leaves of *Acer caesium*, Harwan, Kashmir (Butler); Darjeeling (McRae).

——**conoideum** Cke (**133**:16 ; Sacc. **VIII**:761). On leaves, India (Hobson). " Without fruit."

——**durissimum** Cke (**133**:16 ; Sacc. **VIII**:761). On coriaceous leaves, India (Hobson). " No fruit."

——**fuscum** Fr. (**238**: 551 ; Sacc. **VIII**:759). On leaves of Sapindaceae, India (Rudolphi).

——**himalense** Syd. & Butler (**454**:377, fig. ; Sacc. **XXIV**:1266). On leaves of *Ilex* (probably *I. dipyrena*), Ranikhet, Kumaon Himalaya (Butler).

——**lagerstroemiae** Rabenh. (**387**:31, name only ; Sacc. **VIII**:764 ; **445**:489). On upper surfaces of leaves of *Lagerstroemia* sp., Central India (Kurz) ; Belgaum, Bombay (Ajrekar) ; of *L. lanceolata*, Coimbatore.

——**piceum** Berk. (**57**, No. 475 ; Sacc. **VIII**:761 ; **454**:378). On living leaves of *Pieris*, Tambur Valley, Nepal (Hooker f.); Ranikhet, Kumaon (Butler) ; of *P. ovalifolia*, Nepal (Burkill).

Saccobolus kerverni (Crouan) Boud. (Sacc. **VIII**: 524 ; **454**:374). On horse-dung, Pusa (Butler).

Sarcoscypha macropus (Pers.) Lamb. (Sacc. **VIII**:28 as *Helvella* ; **288**, No. 65:4). India (Kashyap). A species of *Sarcoscypha* is also discussed in **467**:158.

Sclerotinia cinerea (Bon.) Schroet. (Sacc. **IV**:34 as *Monilia* ; **230**). On *Prunus persica*, Kashmir. The correctness of this determination is doubtful and the identity of the brown rotting fungi of fruits in the Himalaya has still to be worked out. They have not been found on orchard fruits in the Plains of India.

——**ricini** Godfrey (**3**:26). On inflorescence and fruits of *Ricinus communis*, Mysore (Narasimhan).

[Sclerotinia] sclerotiorum (Lib.) de Bary (Sacc. **XXII**:644 as *S. sclerotiorum* (Lib.) Sacc. & Trott. ; **270**, figs. ; **413**:177, figs., as *Rhizoctonia napi* Westend.) On *Brassica campestris* var. *sarson, Carthamus tinctorius, Amaranthus tristis, Argemone mexicana, Avena sativa, Beta bengalensis, Brassica campestris* var. *glauca, Calamintha* sp., *Cannabis sativa, Chenopodium album, Cicer arietinum, Cirsium arvensis, Fumaria parviflora, Hordeum vulgare, Lathyrus sativus, Lens esculenta, Leucas* sp., *Linum usitatissimum, Medicago lupulina, Pisum sativum, Scoparia dulcis, Triticum vulgare, Vicia hirsuta* (Shaw and Ajrekar ; Joshi). Joshi found, by inoculation, that several other plants could be infected.

Trichocoma paradoxa Junghuhn (Sacc. **X**:82 ; **232**:161, figs. ; **57**, No. 551, and after No. 455 ; **196**:128 ; **287**:1205, fig.). East Nepal (Hooker f.) ; on dry dead wood, Sikkim, 7-9,000 ft. (Hooker f.) ; on mossy dead sticks, Sikkim Himalaya (Kurz). This fungus is often considered to be allied to the Tuberaceae, but Fischer (**232** ; **233**:196) places it in the Plectascaceae, as does Dodge (**226**:151).

Triblidiella rufula (Spreng.) Sacc. (Sacc. **II**:757 ; **454**:375). On dead branches of *Citrus* sp., Sagaing, Burma (Butler) ; on dead branches, Puliyanur, Travancore (Butler) ; Dehra Dun (Inayat) ; Pusa (Butler) ; Bassein, Bombay (Kulkarni) ; and Tellicherry, Malabar (Butler).

Tuber indicum Cke & Massee (**180**:67 ; Sacc. **XI**:444). In the earth, Mussoorie (Duthie).

Vibrissea stilboidea (Berk.) Sacc. (Sacc. **VIII**:53 ; **57**, No. 359, as *Peziza stilboidea* Berk.). On the main nerves of leaves of *Pyrus*, Tonglo, Sikkim (Hooker f.).

——**turbinella** (Berk.) Sacc. (Sacc. **VIII**:52 ; **57**, No. 358, as *Peziza turbinella* Berk.). On the lower surfaces of leaves of *Pyrus*, Tonglo, Sikkim (Hooker f.).

PYRENOMYCETES.

Acanthostigma heterochaete Sydow & Butler (**454**:403 ; Sacc. **XXIV**:971). On living leaves of *Phaseolus mungo* var. *radiatus*, Pusa (Butler) ; of *Dumasia villosa*, Pusa (Butler) ; Nagpur (Pandit) ; of *D.* sp., Samalkota (Shaw) ; Dacca (Som) ; of *Alysicarpus vaginalis* var. *nummularifolia*, Samalkota (Shaw).

Acanthostoma wattii (Syd. & Butler) Theiss. (**469**:45 ; **472**:1285 ; **454**:383 as *Dimerium wattii* Syd. & Butl. ; Sacc. **XXIV**:256 ; **468**:188). On *Asterina camelliae* on leaves of *Thea sinensis*, Assam (G. Watt) ; Darjeeling (McRae).

Acrospermum parasiticum Syd. (**454**:379 ; Sacc. **XXIV**:1128). On living or fading leaves of *Heptapleurum venulosum*, Kumaon (Inayat).

Allescherina boehmeriae Syd. & Butler (**454**:413 ; Sacc. **XXIV**:734). On dead stems of *Boehmeria nivea*, Pusa (Inayat).

[Allescherina] cajani Syd. & Butler (**454**:413 ; Sacc. **XXIV**:733). In the bark of branches of *Cajanus indicus*, Pusa (Butler).

Amphisphaeria khandalensis Rehm (**467**:158 ; Sacc. **XXIV**:940). On *Bambusa*, Khandala, Bombay (Blatter).

Anthostoma carteri (Berk. & Cke) Berl. & Vogl. (Sacc. **IX**:520 ; **153**:51 as *Fuckelia carteri* Berk. & Cke). On bark, Bombay (Carter).

Anthostomella bambusae (Lév.) Sacc. (Sacc. **I**:289 ; **285**:50 as *Sphaeria bambusae* Lév. ; **57**, after No. 485, as *Hypopteris bambusae* (Lév.) Berk.). On dead stems of bamboo [*Bambusa ? arundinacea*], Nangki Hills, 6,000 ft. (Hooker f.).

——**pandani** (Rabenh.) Sacc. (Sacc. **I**:292 ; **387**:45 as *Sphaeria pandani* Rabenh.). On leaves of *Pandanus furcatus*, Calcutta. Pycnidia also present.

Apiospora camptospora Penzig & Sacc. (Sacc. **XIV**:534 ; **454**:402). On leaf sheaths of *Saccharum officinarum*, Bilin, Burma (Butler).

——**indica** Theiss. & Syd. (**481**:420 ; Sacc. **XXIV**:612 ; **454**:402 as *A. montagnei* Sacc. ; **57**, after No. 485, as *Hypopteris apiospora* (Mont.) Berk.) On culms of *Bambusa*, Nangki Hills (Hooker f.) ; Wynaad and Dehra Dun (Butler). The identity of the fungus recorded by Berkeley (**57**) is uncertain (**481**:419).

——**rhodophila** Sacc. (Sacc. **XIV**:534). On *Rosa macrophylla*, Garhwal, 11-12,000 ft. (Duthie).

Asterina balii Syd. (**443**, **XVII**:308). On leaves of *Alangium lamarckii*, Bhubaneshwar, Orissa (Bal).

——**camelliae** Syd. & Butler (**454**:389, figs. ; Sacc. **XXIV**:474 ; **111**:463, figs. ; **473**:83, fig.). On leaves of *Thea sinensis*, Dunmar Dulling, Sibsagar Dist., Assam (Watt).

——**cansjerae** Ryan (**401**:103). On *Cansjera rheedii*, Dharwar (Sedgwick).

——**carbonacea** Cke (**144**:96 ; Sacc. **I**:42 ; **473**:69). On coriaceous leaves, Belgaum (Hobson).

——**celtidicola** P. Henn., var. **capparidis** (Syd. & Butler) Theiss. (**473**:94 ; **454**:390 as *A. capparidis* Syd. & Butler ; Sacc. **XXIV**:446). On leaves ot *Capparis* sp., Madras (Butler).

[——**cincta** Berk. (**57**, No. 477 ; Sacc. **I**:43). On leaves ot *Camellia*, Khasi Hills (Hooker f. & Thomson). According to Theissen (**468**:188), this is a mixture of *Asterina camelliae* and *Acanthostoma wattii* (q.v.), and being wrongly described should be deleted.]

[——**concentrica** Cke (**161**:13 ; Sacc. **IX**:377 as *Asterula concentrica* (Cke) Sacc.). On culms of *Saccharum*, N. W. India. Theissen (**468**:13 ; **473**:24) found this to be an undeterminable, unripe member of the Dothidiaceae.]

——**congesta** Cke emend. Theiss. (**144**:95 ; Sacc. **I**:42 ; corrected diagnosis in **468**:198 ; **368**, **I**:62 ; **473**:99, fig. ; **445**:488). On leaves of *Santalum album*, Belgaum (Hobson) ; Malabar (McRae).

——**crebra** Syd. (**445**:327 ; Sacc. **XXIV**:463). On leaves of *Opilia amentacea*, Coimbatore (Fischer).

[**Asterina**] **delicatula** Syd. & Bal (**443, XVII**: 308). On leaves of *Aegle marmelos*, Hooghly, Bengal (Bal).

——**holarrhenae** Ryan (**401**:103). On upper surfaces of leaves of *Holarrhena antidysenterica*, Assam (Subramaniam).

——**indica** Syd. (**454**:390 ; Sacc. **XXIV**:473 ; **473**:86, fig.). On leaves of *Symplocos* sp., Darjeeling (McRae).

——**lawsoniae** P. Henn. & Nyman (Sacc. **XVI**:646 ; **454**:391 ; **18**:179 ; **473**:81, figs. ; **445**:488). On *Lawsonia alba* (*L. inermis*), Pusa (Butler) ; Bengal (Lafont) ; and various other localities. Arnaud (**18**:179), from a study of the pycnidial stage, transferred this to *Dimerosporium*. The pycnidia (**18**:214) (*Asterostomella* (*Hyphaster*) *balanseana*) occur commonly (**454**:391).

——**memecylonicae** Ryan (**401**:104). On leaves of *Memecylon edule*, Karwar, Bombay (Sedgwick).

——**nothopegiae** Ryan (**401**:104). Amphigenous on leaves of *Nothopegia colebrookiana*, Mysore (Kulkarni).

——**pleuriporus** Ryan (**401**:104). Hypophyllous on *Shorea talura*, Siddapur, Kanara (Sedgwick).

[——**scutellifera** Berk. (**57**, No. 478 ; Sacc. **I**:50). On leaves of *Antidesma*, Chittagong (Hooker f. & Thomson). Theissen (**468**:15) states that the specimen is sterile, and the species should be deleted.]

——**spissa** Syd. (**454**:392 ; Sacc. **XXIV**:464 ; **473**:61, figs.). On leaves of *Jasminum* sp., Chittagong (Sen) ; of *J. malabaricum*, Bombay (Ajrekar). Arnaud (**18**:178) has transferred this species to *Dimerosporium*, without having studied it, because he considers it evidently identical with an undetermined *Asterina*-like fungus on *Jasminum pubescens* in which Marshall Ward discovered the first haustoria known in the Microthyriaceae (Quart. Journ. Microsc. Sci., XXII, pp. 347-354, 2 pls., 1882).

——**travancorensis** Syd. (**443, XIII**:38). Epiphyllous on *Marsdenia* sp., Pulliyanur, Travancore (Butler).

Asterinella intensa (Cke & Mass.) Theiss. (**471**:120, figs. ; Sacc. **IX**:382 as *Asterina* ; **401**:105). On *Elaeodendron glaucum*, Madras.

——**malabarensis** (Syd.) Theiss. (**471**:106, fig. ; **454**:391 as *Asterina malabarensis* Syd. ; Sacc. **XXIV**:444). On leaves of *Pothos scandens*, Kanouth, Malabar (Butler).

——**stuhlmanni** (P. Henn.) Theiss. (**471**:120, fig. ; Sacc. **XVII** : 881 and **454**:392 as *Asterina stuhlmanni* P. Henn.). On leaves of *Ananas sativus*, Thurya Ghat, Sylhet (Butler).

Asterocystis mirabilis Berk. & Broome (Sacc. **I**:293 ; **454**:411). On culms of *Bambusa* sp., Bulsar, Bombay (Chibber) ; Tellicherry, Malabar (Butler).

Balansia andropogonis Syd. (**454**:395 ; Sacc. **XXIV**:697). On inflorescence of *Andropogon aciculatus*, Forbesganj, Purnea Dist., Bihar (Burkill) ; Chittagong (Sen); Sylhet and Noakhali (Butler); Assam (Burkill). Sydow states that *Ephelis pallida* Patouill., common in Tonkin and the Philippines, is doubtless the conidial stage.

[**Balansia**] **sclerotica** (Pat.) v. Hoehn. (**264**, No. 704; Sacc. **IX**:1002 as *Epichloe*; **454**:402 as *Ophiodothis sclerotica* (Pat.) P. Henn.). On inflorescence of *Andropogon nardus*, Erramacola, Wynaad (McRae); of *A. schoenanthus*, Vayitri, Wynaad; (Butler); of *A.* sp., Chatrapur, Ganjam (Butler); of *A. ?contortus*, Belgaum (Butler); on *Apluda aristata*, Dehra Dun (Butler); on *Panicum distachyum*, Coimbatore (McRae); on *P.* sp., Bangalore (Butler).

——**thanatophora** (Lév.) v. Hoehn. (**264**, after No. 630; **57**, No. 485, as *Dothidea vorax* Berk. & Curt.; Sacc. **II**:652 as *Ophiodothis vorax* (B. & C.) Sacc.). On deformed spikes of *Carex* sp., Churra, Khasi Hills (Hooker f.).

A variety (*Ophiodothis vorax* (B. & C.) Sacc. var. *pilulaeformis* (B. & C.) Sacc.) is recorded (**445**:489) on culms of *Panicum prostratum*, Kistna Dist., Madras (McRae).

Balladyna butleri Syd. (**454**:388, figs.; Sacc. **XXIV**:373). On culms of *Bambusa* sp., Khasi Hills, Assam (Butler).

——**gardeniae** Racib. (Sacc. **XVI**:411). On *Gardenia gummifera*, Siddapur, N. Kanara (Kulkarni). Von Hoehnel (**264**, No. 482) states that this is not specifically distinct from *Asterina velutina* Berk. & Curt., and makes the combination *B. velutina* (B. & C.) v. Hoehn.

Boerlagella effusa Syd. & Butler (**454**:403; Sacc. **XXIV**:1023). On wood or decorticated branches of *Populus ciliata*, Mussoorie (Inayat).

Botryosphaeria agaves (P. Henn.) Butler (**454**:415; Sacc. **XVII**:585 as *Physalospora*). On leaves of *Agave* sp., Pusa (Butler). Theissen (**479**:333) states that the Indian species differs from that of Hennings (which was collected in East Africa), and that neither belongs to *Botryosphaeria*.

——**egenula** Syd. & Butler (**454**:415; Sacc. **XXIV**:812). On leaves of *Cymbidium ?aloifolium*, Gauhati, Assam (Butler). Weese (Ber. deut. Bot. Ges., XXXVII, pp. 83-96, 1919) transfers this and many other species of *Botryosphaeria* to *Melanops*.

——**?pruni-spinosae** Delacr. (Sacc. **XI**:295). On *Prunus persica*, Pusa (Butler).

——**tamaricis** (Cke) Theiss. & Syd. (**481**:663; Sacc. **II**:590 as *Bagnisiella tamaricis* (Cke) Sacc.; **454**:401). On stems of *Tamarix gallica*, Pusa (Inayat).

——**xanthocephala** (Syd. & Butler) Theiss. (**479**:326; **454**:408 as *Physalospora xanthocephala* Syd. & Butler; Sacc. **XXIV**:800). On twigs of *Cajanus indicus* Pusa (Sen). Von Hoehnel (**264**, No. 1194) has made this the type of a new genus *Creomelanops*. He considers that *Botryosphaeria inflata* Cke & Massee (at least as found in Java) is probably the same.

Butleria inaghatahani Sacc. (**403**:303; Sacc. **XXIV**:1134). On leaves of *Vangueria* sp., Krishnapur, Comilla (Inayat Khan). Saccardo misread the locality and collector's names, and they are printed "Krishnapone" and "Inaghatahan" respectively. The species name should really be "*inayatkhani.*" Petrak and Sydow (**385**:99) have reported in detail on this genus and species.

Camillea bacillum Mont. (Sacc. **I**:347 ; **467**:159 ; **304**:3, fig.). On bark, Khandala, Bombay (Blatter).

Capnodium anonae Patouill. (Sacc. **XVII**:555 ; **454**:384). On leaves and twigs of *Ficus retusa, Agave vera-crucis,* and various weeds, Bilikere, Mysore, and on leaves of *Ficus retusa, F. glomerata,* and *F. bengalensis,* Hassan, Mysore (Butler) ; on *F. benjamina,* Khed, Bombay (Chibber).

——**brasiliense** Puttemans (Sacc. **XVII**:556 ; **454**:384 ; **111**:487, fig.). On leaves and twigs of *Coffea arabica,* Tuttapullem, Nilgiris (Brock) ; on leaves of *C. arabica* infested with *Lecanium hemisphaericum,* southern India (Lefroy).

——**eugeniarum** Cke (**144**:96 ; Sacc. **I**:78). On leaves of *Eugenia jambos* (*Jambosa vulgaris*), Belgaum (Hobson). Spores unknown.

——**lanosum** Cke (**144**:96 ; Sacc. **I**:77). On leaves of *Ficus* sp., Belgaum (Hobson).

——**pini** Berk. & Curt. (Sacc. **I**:75). Recorded on *Pinus excelsa,* Kashmir (Hole). Specimen at Pusa.

——**ramosum** Cke (**184**:76 ; Sacc. **XI**:271). On leaves of *Mangifera indica,* India (Herb. Berk.). It is probable that this fungus is really a *Phaeosaccardinula* (*Limacinula*), as the numerous collections of sooty mould of mango in the Pusa Herbarium, from various parts of India, seem to belong to *Phaeosaccardinula,* although they are all more or less immature.

——**salicinum** Mont. (Sacc. **I**:73 ; **196**:130 as *Fumago salicina* Tul.). On leaves of *Sphenodesme eryciboides,* Wa-choung, Yomah, Burma (Kurz). A doubtful record.

Carnostroma thyrsus (Berk.) Lloyd (**304**:27 ; **53**:384, fig., as *Sphaeria thyrsus* Berk. ; Sacc. **I**:320 as *Xylaria thyrsus* (Berk.) Sacc. ; Grev. XIII, pl. 170, fig. 128 ; **454**:418, fig., as *Xylaria excelsa* Syd.). In the Royal Botanic Garden, Calcutta ; ? on the ground, Narsinghpur Dist., Central Provinces (Burkill).

Catacauma acaciae Theiss. & Syd. (**481**:389 ; Sacc. **XXIV**:560). Epiphyllous on *Acacia leucophloea,* Coimbatore (McRae) ; Amritsar (Cheema).

——**aspidea** (Berk.) Theiss. & Syd. (**481**:380 ; Sacc. **II**:598 as *Phyllachora aspidea* (Berk.) Sacc. ; **454**:396). On leaves of *Ficus ?scandens,* Ranikhet, Kumaon (Butler).

——**infectorium** (Cke) Theiss. & Syd. (**481**:384 ; Sacc. **IX**:1013 as *Phyllachora infectoria* Cke ; **454**:396). Epiphyllous on *Ficus infectoria,* Bilikere, Mysore (Butler) ; Wynaad (McRae) ; on *F. religiosa,* Jessore (Bhattacharyia) ; Bandra, Bombay (Dastur) ; Insein, Burma (Butler). Theissen & Sydow (**482**:450) state that *Marchalia ustulata* (Cke) Sacc. (q.v.) is a conidial fungus doubtless belonging to this species.

[Catacauma] microcentrum (Berk. & Broome) Theiss. & Syd., var. **graphica** Theiss. & Syd. (**481**:385 ; Sacc. **XXIV**:562). Epiphyllous on *Ficus mysorensis*, Yelwal, Mysore (Butler). The species was recorded as *Phyllachora ficuum* Niessl (**454**:395).

——**repens** (Cda) Theiss. & Syd. (**481**:383 ; **185**, **IV**:42, figs., as *Sphaeria repens* Cda ; Sacc. **II**:597 as *Phyllachora repens* (Cda) Sacc. ; **454**:396 ; **133**:16, and **144**:95 as " *Dothidea repens* Corda "). On leaves of an unknown tree, India (Helfer) ; of *Ficus religiosa*, Belgaum (Hobson) ; Madras, Comilla, and Royal Botanic Garden, Calcutta (Butler) ; Bombay (Kulkarni) ; on *F. gossypina*, India (Wallich). Theissen and Sydow (**481**) think that Corda's type, which no longer exists, was perhaps upon *Ficus religiosa*.

Ceratomyces selinae Thaxt. (**462**, **V**:26 ; Sacc. **XVII**:918). On prothorax of *Selina westermanni*, India.

Ceratostomella adiposum (Butler) Sartoris (**93**:95, figs., as *Sphaeronema adiposum* Butler ; Sacc. **XXII**:926 ; **89**:47 ; **456**:190 ; **111**:383, fig.). On planted cuttings and old culms of *Saccharum officinarum* throughout India. Sartoris (Jour. Agr. Res., XXXV, pp. 577-585, 1927) found the supposed pycnospores to be produced in evanescent asci.

——**paradoxa** (de Seynes) Dade (Trans. Brit. Myc. Soc., XIII, p. 191, 1928 ; Sacc. **XXII**:1341 as *Thielaviopsis paradoxa* (de Seyn.) v. Hoehn. ; **426** ; **433** ; **344** ; **93**:32 as *T. ethaceticus* Went). On cuttings of *Saccharum officinarum*, Bengal (Butler) ; on *Cocos nucifera*, *Areca catechu*, *Borassus flabellifer*, and several other hosts infected by inoculation, Madras (Sundararaman) ; on *Cocos nucifera*, Andaman Islands (Mitra). The *Sphaeronema*-like form recorded by Butler (**93**:40) was doubtless the *Ceratostomella* stage.

Chaetomium amphitrichum Corda (**185**, **IV**:37, fig. ; Sacc. **I**:228 ; **454**:402). On rotting plants, Tenasserim (Helfer) ; at the base of rotting stems of *Gossypium indicum*, Nagpur (Evans) ; Pusa (Butler).

——**indicum** Corda (**185**, **IV**:37, figs. ; Sacc. **I**:222 ; **132**:117). On decayed paper, Moulmein (Helfer) and Tenasserim.

——?**setosum** Wint. (Sacc. **IX**:484 ; **384**:488). See entry under *Chaetomella furcata* in the *Sphaeropsidales*.

Chaetosphaeria indica Niessl (**356**:98 ; Sacc. **II**:96). On leaves of *Alangium lamarckii* (*A. decapetalum*), Calcutta (Kurz). Niessl states that the habit is like a *Capnodium*, and Saccardo observes that it may therefore be a *Meliola*.

Clematomyces pinophili Thaxt. (**462**, **II**:440 ; Sacc. **XVI**:692). On the inferior surface of *Pinophilus* sp., Burma.

Clypeosphaeria? crenulatum (Berk.) Sacc. (Sacc. **II**:91 ; **57**, No. 484, as *Hypoxylon crenulatum* Berk.). On dead stems of *Bambusa*, Nangki Mountains, 6000 ft. (Hooker f.).

Cordyceps falcata Berk. (**57**, No. 479, fig. ; Sacc. **II**:575 ; **375**, **IV**:36, fig.). On a dead caterpillar, Myrong, Khasi Hills (Hooker f. and Thomson).

[Cordyceps] racemosa Berk. (**57**, No. 480, fig.; Sacc. **II**:576). On a dead caterpillar, Myrong, Khasi Hills (Hooker f. and Thomson).

Corynelia clavata (L.) Sacc. (Sacc. **IX**:1073 ; **57**, after No. 487, as *Corynelia* "*uberiformis* Fr."). On *Podocarpus*, Khasi Hills (Hooker f.).

——**fructicola** (Pat.) v. Hoehn. (Sacc. **IX**:441 as *Capnodium* ; **454**:406 ; **235**:240). On fruits of *Myrsine africana*, Mussoorie (Butler).

Cryptovalsa indica Syd. (**454**:412 ; Sacc. **XXIV**:735). On dead branches, Dehra Dun (Inayat).

——**planiuscula** Syd. & Butler (**454**:412 ; Sacc. **XXIV**:735). On dead branches, Pusa (Sen).

——**rabenhorstii** (Nits.) Sacc. (Sacc. **I**:190 ; **454**:412). On branches of cultivated *Morus* sp., Pusa (Hafiz).

Cucurbitaria agaves Syd. & Butler (**454**:406 ; Sacc. **XXIV**:1055). On leaves of *Agave* sp., Dehra Dun (Butler). Associated with *Microdiplodia agaves*, q. v.

Daldinia concentrica (Bolt.) Ces. & de Not. (Sacc. **I**:393 ; **263**:339 ; **467**:159 **454**:417 ; **196**:130 ; **349**:23 as "*Hypoxylon concentricum* Grev." ; **263**:339 as *D. gollani* P. Henn. ; Sacc. **XVII**:617 ; **304**:24, figs.). On wood of *Citrus aurantium*, Nagpur (Pandit) ; on trunks of *Dalbergia sissoo*, Pusa (Dastur) ; of *Excoecaria acerifolia*, Dehra Dun (Inayat) ; on dead branches of *Ficus carica*, Siwalik Hills (Gollan) ; Bengal (Hutchings) ; and elsewhere ; on *Ixora undulata*, Pusa (Butler) ; and on old wood in many parts of India. A specimen is in the Montagne Herbarium at Paris, marked "ad truncos emortuos prope Nedoubatta Nelligheri cl. Perrottet".

——**vernicosa** (Schw.) Ces. & de Not. (Sacc. **I**:394 ; **57**, after No. 483, as *Hypoxylon vernicosum* (Schw.) ; **304**:25, fig.). Darjeeling (Hooker f.).

Diaporthe (Chorostate) curvatispora Wakef. (**519**, **XXVI**:164, fig.). On bark of *Mesua ferrea*, Sibsagar (Hole).

——**(Chorostate) taxicola** Sacc. & Syd. (Sacc. **XVI**:493). A specimen so determined is at Pusa on *Taxus baccata*, Darjeeling.

Diatrype chlorosarca Berk. & Broome (Sacc. **I**:195 ; **454**:415). On dead culms of *Bambusa*, Tellicherry, Malabar (Butler) ; Palghar, Bombay (Ajrekar) ; on dead branches, Pulliyanur in Travancore, and Dehra Dun (Butler) ; Palghat, Madras (Subramaniam). The form from Palghat has larger stromata and spores, and that from Dehra Dun smaller stromata but larger spores, than the type. It is somewhat doubtful if all belong to the same species (**454**:415).

——**rugosa** Currey (**196**:130). On hard wood, Yomah, Burma (Kurz).

Dichomyces hybridus Thaxt. (**462**, **II**:422 ; Sacc. **XVI**:678). On *Philonthus* sp. Sylhet.

Didymella kariana Sacc. (**403**:301 ; Sacc. **XXIV**:909). On dying leaves of *Polygonum* sp. with broad leaves, accompanied by *Puccinia ?polygoni* and *Darluca filum*, Mussoorie (Kar).

[**Dimerosporium aterrimum** Cke & Wint. (**181**:83 ; Sacc. **XI**:259). On coriaceous leaves, Manipur (Watt). Theissen (**468**:194) states that the fungus is unripe and not describable, and should be deleted.]

——**erysiphoides** Ell. & Ev. (Sacc. **IX**:407 ; **454**:383). On leaves of *Cynodon dactylon*, Pusa (Butler) ; Bassein, Bombay (Burkill) ; of *Paspalum scrobiculatum*, Bassein, Bombay (Burkill).

——**mangiferum** (Cke & Broome) Sacc. (Sacc. **I**:53 and 77 ; **132**:117, figs., and **134**:5, figs., as *Capnodium mangiferum* Cke & Br.). On leaves of *Mangifera indica*, Mysore.

Diplocarpon rosae Wolf (Sacc. **XXIV**:911 ; **5**:3 ; **456**:220 as *Marssonina rosae* (Lib.) Died.). On leaves of *Rosa* spp., Maymyo (Butler) ; India (Anstead). The conidial stage alone has been reported from India.

Diplochorella indica (Sacc.) Theiss. & Syd. (**481**:622 ; Sacc. **XXIV**:632 ; **403**:302 as *Dothidella indica* Sacc.). On living leaves and petioles of Meliaceae (?*Turraea*), Lonavla, Bombay (Saxton).

Dothidella trifolii (Pers.) Bayliss-Elliott & Stansfield (Sacc. **II**:613 as *Phyllachora trifolii* (Pers.) Fckl ; **113**(8):54 ; **2**:27). On leaves of *Trifolium resupinatum*, Peshawar (Shaw).

Echinosphaeria profusa Syd. (**443**, **VIII**:407 ; Sacc. **XXIV**:974). On the stem of *Jasminum malabaricum*, Matheran, Bombay (Ajrekar).

Enarthromyces indicus Thaxt. (**463**:276, figs. ; Sacc. **XIV**:726). On *Pheropsophus* sp., Booloo Valley, N. W. India.

Endodothella bambusae (Rabenh.) Theiss. & Syd. (**481**:585 ; Sacc. **XXIV**:614 ; **387**:45 as *Sphaeria bambusae* Rabenh. ; Sacc. **I**:446 as *Physalospora bambusae* (Rab.) Sacc. ; **454**:401 as *Dothidella bambusicola* Syd. & Butler). On leaves of *Bambusa spinosa*, Royal Botanic Garden, Calcutta ; of *Bambusa* sp., Moulmein, Burma (Butler).

—— **dispar** Syd. (**481**:584 ; Sacc. **XXIV**:614 ; **454**:401 as *Dothidella dispar* Syd.). On leaves of *Andropogon contortus*, Tellicherry, Malabar (Butler).

?**Endothia hypocreoides** (Berk. & Cke) v. Hoehn. (Ann. Myc. **VIII**:466 ; **155**.81 as *Nectria hypocreoides* Berk. & Cke ; Sacc. **IX**:958). On bark, Bombay (Carter).

Epichloe bambusae Pat. (Sacc. **XIV**:655 ; **454**:395). On culms of *Bambusa* sp., South India (Barber).

——**cinerea** Berk. & Broome (Sacc. **II**:579 ; **454**:394 ; **111**:91, fig.). On inflorescence of *Eragrostis tenuifolia*, Hunsur and Bilikere, Mysore (Butler). Petch (Ann. Peradeniya, vii, p. 88, 1920) states that it seems probable that the Indian fungus is not identical with *E. cinerea* B. & Br. described from Ceylon.

——**typhina** (Pers.) Tul. (Sacc. **II**:578). On *Stipa* sp., Harwan, Kashmir (Butler).

Eriosphaeria calospora Speg. (Sacc. **IX**:698). On *Bambusa* sp., Pusa (Mitra) ; Dehra Dun (Butler).

Erysiphe acaciae Blumer (**69**:182). On *Acacia catechu*, Poona (Ajrekar ; Sydow, Fungi Exotici No. 75). This species has been separated from the allied *E. polygoni* by Blumer, on account of the great size of its perithecia.

——**cichoracearum** DC. Sacc. **I**, under several names ; **4** ; **518** ; **111**:314 and 338, figs.). On the following :—*Coccinia* (*Cephalandra*) *indica*, Cawnpore and Pusa (Butler) ; *Galium verum*, Harwan, Kashmir (Butler) ; *Mangifera indica*, inflorescence, Bombay (Wagle, **518**:4) [the name *Oidium mangiferae* Berthet has been applied to a mildew on mango in Brazil] ; *Momordica balsamina*, Pusa (Butler) ; *Nicotiana tabacum*, Koilpatti, Madras (Sampson) ; Rangpur ; and elsewhere (*Oidium* only) ; *Plantago brachyphylla*, Srinagar (Butler) ; *P. major*, Harwan, Kashmir (Butler) ; *Trichosanthes dioica*, Pusa (Butler).

——**galeopsidis** DC. (Sacc. **I**:16). On *Phlomis stewartii*, Verinag, Kashmir (Butler).

——**graminis** Lév. (Sacc. **I**:19 ; **406**:476 ; **111**:173, figs.). On *Hordeum vulgare*, Pusa ; *Stipa sibirica*, Verinag, Kashmir ; *Triticum vulgare*, Dehra Dun (Butler) ; Lahore (B. Das) ; Peshawar (Shaw) ; Gilgit. Salmon (**406**) found that the form on wheat in India affected wheat and *Hordeum silvaticum*, but not barley or oats.

——**polygoni** DC. (Sacc. **I**:19 ; **111**:253, figs. ; **113**(10):81 ; **404**:174). On the following :—*Berberis* sp., Achibal and Verinag, Kashmir (Butler) ; *Brassica campestris*, Chenab, Punjab (Milligan) ; *Chenopodium botrys*, Arrah, Kumaon, and Larukhpur (Butler) ; *Coriandrum sativum*, Coimbatore (Butler) ; *Eryngium billardieri*, Harwan, Kashmir (Butler) ; *Ipomoea cymosa*, Lahore (B. Das); *I.* sp., Uri, Kashmir (Butler) ; *Lathyrus* sp., Ranikhet, Kumaon (Butler) ; *Lespedeza bicolor*, Harwan, Kashmir (Butler) ; *Ligusticum thomsoni*, Harwan, Kashmir (Butler) ; *Melilotus alba*, Harwan, Kashmir (Butler) ; *M. indica*, Pusa (Butler) ; *Papaver somniferum*, Jaora State, Central India (Coventry) ; *Pisum sativum*, Dehra Dun, Pusa, and Verinag, Kashmir (Butler) ; Maymyo, (Shroff) ; *Plectranthus rugosus*, Verinag, Kashmir (Butler) ; *Polygonum aviculare*, Harwan, Kashmir (Butler) ; *P.* sp., Mussoorie (Butler) ; *Ranunculus diffusus*, Baramulla, Kashmir (Butler) ; *R. laetus*, Dubgaon and Harwan, Kashmir (Butler) ; *R.* sp., Kumaon (Butler) ; *Rumex nepalensis* and *R. orientalis*, Harwan, Kashmir (Butler) ; *Thalictrum minus*, Harwan, Kashmir (Butler) ; *Trigonella foenum-graecum*, Pusa and Dehra Dun (Butler) ; *Vicia faba*, Dehra Dun (Butler).

Eurotium herbariorum (Wigg.) Link (Sacc. **I**:26 ; **132**:117 ; **340**:76 as *Aspergillus herbariorum* Wigg.). " Common on various substances. Bengal, Chittagong, Burmah " ; common as a laboratory mould, Goa (de Mello) and Pusa.

Eutypa phaselina (Mont.) Sacc. (Sacc. **I**:179 ; **196**:130 as *Sphaeria phaselina* Mont.). Evergreen forests, Nakawa, Toukyeghat, Burma (Kurz).

Eutypella vitis (Schw.) Ell. & Ev. (Sacc. **I**:180 as *Eutypa* ; **454**:413). On wood of *Vitis vinifera*, Poona (Chibber). Spores only 6-9 $\times$ 2μ instead of 12-14 $\times$ 2½μ.

Collections from North America contain similar small-spored forms, which Sydow and Butler suggest may perhaps constitute a var. *microspora* (**454**:413-14).

[**Eutypella**] **zizyphi** Syd. & Butler (**454**:413 ; Sacc. **XXIV**:721). On dead branches of *Zizyphus jujuba*, Pusa (Sen). An allied or perhaps identical species occurs on *Indigofera arrecta* at Pusa.

Glomerella cingulata (Stonem.) Spaulding & v. Schrenk (Sacc. **XVII**:573 and **XVI**:452 ; **507** ; **1**:11-12 ; **2**:53, 56 ; **487**:703 ; **392**:6 ; **393**:4 ; **395** ; **205**, figs.; **111**:355, 448, 512, figs.). On *Capsicum* spp., Burma and Bihar (Dastur). Dastur regards the conidial forms *Gloeosporium piperatum* Ell. & Ev. and *Colletotrichum nigrum* Ell. & Hals. as belonging to this species but Higgins [Anthracnose of pepper (*Capsicum annuum* L.). Phytopathology, XVI, pp. 333-345, 1926] considers them to be as yet unattached to any perithecial form. Higgins regards the perithecial form found on *Capsicum* as *Glomerella piperata* (Stonem.) Spauld. & v. Schrenk. The brown blight and the bark disease of coffee in south India and Burma is caused by *Colletotrichum coffeanum* Noack (**113**:(4):62 ; **111**:483, figs. ; **392** ; **487**:703), which is generally considered to be a conidial stage of *G. cingulata* ; and the brown blight of tea in Assam, Burma, and Madras is caused by *C. camelliae* Massee, the perfect stage of which was found by Tunstall (**2**:56 ; **507**:37 ; **503** ; **504**) to agree with *G. cingulata*.

——**gossypii** (Southw.) Edgerton (Sacc. **XXII**:77 ; **434** ; **111**:365, figs.). Found only in the imperfect stage, *Colletotrichum gossypii* Southw., on *Gossypium* sp., Pusa (Subramaniam).

——**lindemuthiana** (Sacc. & Mag.) Shear. See entry under *Colletotrichum lindemuthianum*.

Hypocrea flavo-virens Berk. in Herb. (**155**:100 ; Sacc. **IX**:976). On bark, Nilgiris.

——**neilgherrensis** Berk. & Cke (**155**:79 ; Sacc. **IX**:979). On bark, Nilgiris (E. Berkeley).

—— **peltata** (Jungh.) Berk. (**57**, after No. 359 ; Sacc. **I**:536 as *H. peltata* (Jungh.) Sacc.). On dead bark, Darjeeling, 7,500 ft. (Hooker f.).

——**rugulosa** Berk. & Cke (**155**:79 ; Sacc. **IX**:973). On rotten wood, Nilgiris (E. Berkeley).

——var. **major** Cke (**155** : 79; Sacc. **IX**:973). On bark of trees, Nilgiris (E. Berkeley).

——**subrufa** Berk. & Cke (**155**:79 ; Sacc. **IX** :972). On branches, Nilgiris (E. Berkeley).

[——**variabilis** Currey (**196**:130 ; Sacc. **IX**:536). On living leaves of *Bambusa* sp., Yomah, Burma (Kurz). This is a mixture of *Aschersonia badia* and *Hypocrella mollii* (**375**,**11**:250).]

Hypocrella discoidea (Berk. & Broome) Sacc. (Sacc. **II**:580 ; **375**, II:234, synonymy and literature ; **456**:216 as *Aschersonia cinnabarina* P. Henn.). On

Aleyrodes on *Tectona grandis*, Rangoon (Hole) ; on *Aleyrodes* on *Spondias mangifera*, Chittagong (Sen) ; on *Aleyrodes* on *?Millettia*, Travancore (Butler).

[**Hypocrella**] **javanica** (Penz. & Sacc.) Petch (**367**, **II**:431 ; **375**, **II**:220, synonymy and literature ; Sacc. **XVII**:820 as *Fleischeria*). On *Lecanium hemisphaericum* var. *coffeae* on *Thea sinensis*, Assam ; on *Lecanium* sp. on *Thea sinensis*, Kurseong and Darjeeling ; on *Eriochiton theae* on *Thea sinensis*, Assam. (The records of *H. ceramichroa* (Berk. & Br.) Petch from India in **367**, **II**:428, refer in part to *H. javanica* and in part to *H. reineckeana* ; see **375**, **II**:215, 221).

——**mollii** Koorders (**375**, **II**:238, synonymy and literature ; Sacc. **XVI**:985 as *Aschersonia confluens* P. Henn. ; **456**:216 ; **196**:130 as *Hypocrea variabilis* Currey, in part). On *Aleyrodes* on *Tectona grandis*, Rangoon ; on *Aleyrodes* on *Castanopsis* sp., Shan States, Burma (I. H. Burkill). See *Aschersonia badia.*

——**oxystoma** (Berk.) Petch (**375**, **II**:232 ; **57**, No. 463, as *Aschersonia oxystoma* Berk. ; Sacc. **III**:620 as *A. oxyspora* Berk.). On [some insect on] Myrsineae, lower part of India (Hooker f. and Thomson).

——**raciborskii** Zimmermann (Sacc. **XVII**:818 ; **375**, **II**:242, synonymy and literature ; **67**:89 as *Aschersonia placenta* Berk. ; Sacc. **III**:620 ; **456**:216). On *Aleyrodes* on leaves of *Citrus medica*, Kumaon (Inayat) ; of *Morus indica*, Manipur.

——**reineckeana** P. Henn. (Sacc. **XIV**:654 ; **375**, **II**:209 (misprinted " reineckiana "), synonymy and literature). On *Lecanium marsupiale*, South India (Barber).

——**semiamplexa** (Berk.) Sacc. (Sacc. **II**:581; **57**, No. 483, as *Hypocrea semiamplexa* Berk.). On the spikes of *Bambusa* sp., Darjeeling (Hooker f.). Petch (**375**, **II**:266) excludes this species, which he states is apparently a *Balansia*.

Hypocreopsis carteri (Berk. & Cke) Sacc. (Sacc. **IX**:981 ; **155**:79 as *Hypocrea carteri* Berk. & Cke ; **287**:1256, fig.). On bark, Bombay (Carter).

——**undulata** (Berk. & Cke) Sacc. (Sacc. **IX**:981 ; **155**:79 as *Hypocrea undulata* Berk. & Cke). On rotten wood, Nilgiris (E. Berkeley).

Hypomyces chrysospermus Tul. (Sacc. **II**:467 ; **454**:394). On *?Boletus* sp., Khasi Hills (E. M. Coventry).

——**floccosus** Fr. (Sacc. **II**:472 ; **57**, after No. 483, as *Hypocrea floccosa* Fr.). On *Lactarius* sp., Pomrang (Hooker f.).

Hypoxylon atropurpureum Fr. (Sacc. **I**:375 ; **454**:415). On decorticated wood of *Tamarindus indica*, Surat (Gleadow).

——**coccineum** Bulliard (Sacc. **I**:353 ; **57**, after No. 483, as " *H. fragiforme* P."). On dead timber, Sone River (Hooker f.).

——**fusco-purpureum** (Schw.) Berk. & Curt. (Sacc. **I**:378 ; **454**:416). On dead culms of *Bambusa* sp., Bulsar, Bombay (Chibber).

——**hookeri** Berk. in Herb. (**151**:129 ; Sacc. **II** (appendix), p. XXVI, and **IX**:548 On wood, India (Hooker f.).

[Hypoxylon] ?hypomiltum Mont. (Sacc. **I**:354 ; **454**:416 ; **376**:156). On branches of *Zizyphus jujuba* and stems of *Celastrus* sp., Pusa (Butler).

——**indicum** Syd. (**454**:416 ; Sacc. **XXIV**:1083). On fallen branches, Pusa (Butler).

——?**investiens** (Schw.) Berk. (Sacc. **I**:383 ; **454**:416). On wood, Pusa (Inayat) ; on bark of *Dalbergia sissoo*, Pusa (McRae).

——**marginatum** (Schw.) Berk. (Sacc. **I**:371 ; **196**:130 ; **376**:151). On wood, Nakawa, Toukyeghat, Burma (Kurz).

——**multiforme** Fr. (Sacc. **I**:363 ; **57**, after No. 483). On *Betula* sp., Yangma Valley, E. Nepal, 11,000 ft. (Hooker f.).

——**ochraceo-fulvum** Berk. & Cke (**151**:133 ; Sacc. **II** (appendix), p. XXVIII, and **IX**:554). On bark, Nirwab jungle, India.

——**perforatum** (Schw.) Fr. (Sacc. **I**:375 ; **57**, after No. 483 ; **376**:160). On *Bambusa* sp., Nangki Mountains, 6,000 ft. (Hooker f.).

—— **pistillare** P. Henn. & E. Nym. (Sacc. **XVI**:446 ; **454**:416). On bark, Barguai, Mysore (Butler).

——**rubiginosum** (Pers.) Fr. (Sacc. **I**:376 ; **454**:416). On branches, Kasauli (Butler); Dehra Dun (Basu) ; on culms of *Bambusa* sp., Chittagong (Sen) ; on trunks of a palm, Poona (Chibber) ; on bark of *Dalbergia sissoo*, Pusa (Dastur) ; on twigs of *Rosa* sp., Pusa (Butler).

——**stygium** (Lév.) Sacc. (Sacc. **I**:379 ; **454**:416). On bark of *Ficus glomerata*, Pusa (Butler).

——**udum** (Pers.) Fr. (Sacc. **I**:386). Recorded by Saccardo as occurring in India ; no other reference noted.

——**vividum** Berk. & Broome (Sacc. **I**:359 ; **454**:417; **376**:154 ; **467**:159, fig.). On bark, Bombay (Blatter) ; Pusa and Royal Botanic Garden, Calcutta (Butler) ; on decorticated wood, Chittagong (Sen) ; on bark of *Tamarindus indica*, Surat (Gleadow).

Konradia bambusina Racib. (Sacc. **XVI** :606 ; **403**:302). On withering branches of *Bambusa* sp., Chaumuhani, Noakhali, Bengal (Inayat).

Kretzschmaria kurziana (Currey) Sacc. (Sacc. **IX**:567 ; **196**:129, fig., as *Xylaria kurziana* Currey ; **150**:94 and **152**:3 as *Rhopalopsis kurziana* (Curr.) Cke). On brick-laid soil upon which fires had been burnt, Royal Botanic Garden, Calcutta (Kurz). Lloyd (**304**:21) states that this may be the same as *K. cetrarioides* (Welw. & Curr.) Sacc. ; and also states (**287**:939, 1033) that it is a *Poronia* rather than a *Kretzschmaria*.

——**micropus** (Fr.) Sacc. (Sacc. **IX**:565 ; **494**:117 ; **5**:4 ; **376**:136 ; **505**). On roots of *Thea sinensis*, Assam (Tunstall).

Laboulbenia assamensis Thaxt. (**462**, **I**:159 ; Sacc. **XVI**:683). On *Catascopus?*, Assam.

——**coarctata** Thaxt. (**462**, **I**:165 ; Sacc. **XVI**:683). Along median depression of elytrae of *Orechtochilus?*, Bengal (Hope Coll.).

[**Laboulbenia**] **crassipes** Speg. (**417**:469, fig. ; Sacc. **XXIV**:173). On elytrae of *Orechtochilus lucidus*, Tenasserim (L. Fea).

——**dineutis** Thaxt. (**462**:171 ; Sacc. **XVI**:684). On *Dineutes* spp., Bengal ; Nilgiris.

——**euschizomeri** Speg. (**417**:463, fig. ; Sacc. **XXIV**:203, under *L. brachyonychi* Thaxt.). On *Euschizomerus aeneus*, Tikelae, Burma (L. Fea).

——**orechtochili** Thaxt. (Sacc. **XVI**:687 ; **417**:476, fig.). On the margin of the elytrae of *Orechtochilus typus*, Tenasserim (L. Fea).

——**orechtochilicola** Speg. (**417**:477, fig. ; Sacc. **XXIV**:173). On *Orechtochilus feae*, Tenasserim (L. Fea).

——**pheropsophi** Thaxt. (Sacc. **XI**:452 ; **417**:504). On *Pheropsophus ?africanus*, Calcutta ; on *P.* sp., Deccan.

——**podontiae** Thaxt. (**464** :36 ; Sacc. **XXIV**:166). On elytrae of *Podontia 14-punctata*, Himalayas.

——**tenuis** Thaxt. (**462, I**:204 ; Sacc. **XVI**:689). On ?*Catoscopus* sp., Assam.

Laestadia buxifolia (Cke) Sacc. (Sacc. **IX**:584 ; **154**:69 as *Sphaerella buxifolia* Cke). On leaves of *Buxus sempervirens* var. *himalayensis*, Botanic Garden, Saharanpur.

——**camelliae** (Cke) Berl. & Vogl. (Sacc. **IX**:583 as *L.* " *camilleae* "; **157**:4 as *Sphaerella* "*camilleae*" Cke ; **111**:443, figs.; **454**:406 as *L. theae* Racib. ; **113** (3):52 ; **408**:78 ; **2**:56 ; **509** ; **503**). On leaves of *Thea sinensis*, Dooars, Bengal (Hope) ; and elsewhere in north eastern India. Petch (**368**:192) considers this to be probably the same as *Glomerella cingulata*, but the two fungi as they occur in India are quite distinct.

——**perusta** (Berk. & Broome) Sacc. (Sacc. **I**:430 ; **454**:406). On leaves of *Dioscorea* sp., Chittagong (Sen).

Lasiobotrys butleri Theiss. & Syd. (**480**:177 ; Sacc. **XXIV**:252 ; **454**:384 as *L. lonicerae* Kunze). On branches and leaves of *Lonicera* sp., Harwan, Kashmir, and Bowali, Kumaon (Butler).

——**elegans** (Syd.) Theiss. (**475**:13 ; Sacc. **XXII** :38 as *Dimerium elegans* Syd.). Stated by Theissen (**475**) to occur on *? Quercus* in Sikkim and to bear *Trichothyriella quercigena* (Berk.) Theiss.

Lembosia caespitosa (Cke) Sacc. (Sacc. **II**:742 ; **144**:95 as *Ailographium caespitosum* Cke ; **474**:448 ; **316**:35). On coriaceous leaves, Belgaum (Hobson).

——**incisa** (Syd.) Theiss. (**474**:443, fig. ; **454**:390 as *Asterina incisa* Syd. ; **473**:28). On leaves of *Webera corymbosa*, Balehonnur, Mysore (Butler).

Leptosphaeria agaves Syd. & Butler (**454**:409 ; Sacc. **XXIV**:979). On fading leaves of *Agave rigida* var. *sisalana*, Dauracherra, Sylhet (Butler).

——**culmifraga** (Fr.) Ces. & de Not. (Sacc. **II**:95). On *Oryza sativa*, Noakhali (Som).

——**eriobotryae** Syd. & Butler (**454**:409 ; Sacc. **XXIV**:999). On leaves of *Eriobotrya japonica*, Saharanpur (Inayat).

[Leptosphaeria] indica Syd. & Butler (**454**:409 ; Sacc. **XXIV**:987). On fading leaves and stems of *Asparagus* sp., Wynaad, Madras (McRae).

——**sacchari** Breda de Haan (Sacc. **XI**:324 ; **454**:409 ; **93**, fig. ; **111**:382, fig.). On leaves of *Saccharum officinarum*, Dehra Dun ; Bilin, Burma ; Tellicherry ; Coimbatore ; Samalkota ; Champaran ; and Comilla, Bengal (Butler) ; Assam (Som).

Leucoconis erysiphina Syd. (**484**:456 ; **443**, **XIV**:259 as *Zukalia erysiphina* Syd. ; Sacc. **XXIV**:380 ; **18** : 51). On leaves of *Quercus* sp., Kumaon (Butler's collector).

Linospora populina (Pers.) Schr. (Sacc. **II**:357 ; **57**, after No. 475, as "*Phacidium ceuthocarpa* Fr.." [a new combination for *Sphaeria ceuthocarpa* Fr.] with the statement "This is no *Sphaeria* "). On leaves of *Populus ciliata*, Khabili River, 6000 ft. (Hooker f.).

Massaria marginata Fckl (Sacc. **II**:9 ; **454**:411). On twigs of *Rosa* sp., Harwan, Kashmir (Butler).

Massarina usambarensis (P. Henn.) v. Hoehn. (Sacc. **XIV**:594 as *Holstiella* ; **454**:410). On the bark of *Feronia elephantum*, Chittagong (Sen) ; on *Citrus aurantium*, Moulmein, Burma (Butler) ; on *Mangifera indica*, Chittagong (Sen) ; on a dead stem, Noakhali (Butler).

Melanomma citricola Syd. and Butler (**454**:405 ; Sacc. **XXIV** : 1011). On the bark of *Citrus medica*, Chittagong (Sen).

——**glumarum** Miyake (Sacc. **XXII**:244 ; **454**:406). On culms and glumes of *Oryza sativa*, Chittagong (Sen.).

Melanospora parasitica Tul. (Sacc. **I**:464 ; **454**:392). On *Cephalosporium lecanii* parasitic on *Lecanium* infesting *Coffea arabica*, Mysore (Lefroy).

——**zamiae** Corda (Sacc. **I**:463; **454**:392). On leaves and culms of *Oryza sativa*, Noakhali (Som).

Meliola ? ambigua Pat. (Sacc. **IX**:424 ; **403**:303). See entry under *Spegazzinia meliolae* in the Hyphomycetes.

——**amphitricha** Fr. (Sacc. **I**:63 ; **144**:96 ; **241**:76, figs. ; **454**:379). On an unknown host, Belgaum (Hobson) ; on leaves of *Terminalia catappa*, Wynaad (McRae) ; on *Olea* sp. Tellicherry, Malabar (Butler) ; on *Zanthoxylum ovalifolium*, Talagupha, Mysore (Kulkarni). The form on *Terminalia catappa* has larger spores than the normal for the species. A *Meliola* found on *Phaenix sylvestris* and *Citrus medica* var. *acida* is attributed by Bal (**28**, **I**:3) to *M. amphitricha*.

——**arundinis** Pat. (Sacc. **XIV**:473 ; **420** : 109). On *Phragmites karka*, Kamrup and Puttimari, Assam (Taslim).

——**asterinoides** Wint., var. **major** Gaill. (Sacc. **XI**:262 ; **28**, **V**:4 ; **454**:379). On leaves of *Webera corymbosa*, Bilikere, Mysore (Butler) ; on *Plectronia* (*Canthium*) *umbellata*, Matheran, Bombay (Ajrekar) ; on *P. parviflora*, Bhubaneswar, Orissa (Bal).

[Meliola] bicornis Wint. (Sacc. **IX**:422). On *Desmodium* sp., Bassein, Burma (Butler).

——**butleri** Syd. (**454**:389, figs. ; Sacc. **XXIV**:338 ; **28, V**:1, fig.). On leaves of *Citrus medica* var. *acida*, Chittagong (Sen); of *C. decumana*, Kya-in, Amherst District, Burma (Butler).

——**cadigensis** Yates (Sacc. **XXIV**:339 ; **28, V**:2). On leaves of *Glycosmis pentaphylla*, Bengal (Bal).

——**cameliae** (Catt.) Sacc. (Sacc. **I**:62). On *Citrus* sp., Dehra Dun (Kar) ; Pusa (Butler).

——**cladotricha** Lév. (Sacc. **I**:63, and **IX**:419 ; **454**:380). On leaves of *Eugenia jambolana*, Cottamunda, Wynaad (McRae) ; on coriaceous leaves of an unknown host, Kya-in, Amherst District, Burma (Butler).

——**clavulata** Wint. (Sacc. **IX**:426 ; **454**:380). On leaves of *Ipomoea* sp., Tellicherry, Malabar ; and Pulliyanur, Travancore (Butler) ; on leaves of *Argyreia hirsuta*, Balehonnur, Mysore (Butler).

——**clerodendricola** P. Henn. (Sacc. **XVI**:413). On *Clerodendron infortunatum*, Chaumuhani, Noakhali Dist. (Butler).

——**crescentiae** Stevens in lit. (**420**:109). On *Heterophragma roxburghii*, Dharwar, Bombay (Sedgwick).

——**densa** Corda (Sacc. **IX**:419 ; **156**:85). On leaves of ?*Ilex* sp., Khasi Hills.

——**diospyri** Syd. (**454**:381 ; Sacc. **XXIV**:284). On leaves of *Diospyros montana*, Sidrabunna, Koppa District, Mysore (Butler).

——**elmeri** Syd. (Sacc. **XXIV**:331 ; **420**:109). On *Pittosporum dasycaulon*, Ghat Forests, N. Kanara (Sedgwick).

——**eugeniicola** Stevens (**420** : 107, figs.). On *Eugenia eucalyptoides*, Pachanadi, Mangalore (Subramaniam).

——**geniculata** Syd. & Butler (**454** : 381, figs. ; Sacc. **XXIV** : 265). On leaves of *Odina wodier*, Pulliyanur, Travancore (Butler).

——**holigarnae** Stevens (**420**:108, figs.). On leaves of *Holigarna grahamii*, Aumod and Ekambi, N. Kanara (Sedgwick).

——**indica** Syd. (**454**:382; Sacc. **XXIV**:300). On leaves of *Barringtonia acutangula*, (Som).

—— —— var. **car[e]yae** Stevens (**420**:109 as var. "*caryae*") on *Careya arborea*, Gairsoppa Falls, N. Kanara (Sedgwick).

——? **jasminicola** P. Henn. (Sacc. **XI**:265 ; **28, V**:3). On *Jasminum malabaricum*, Sirsi, N. Kanara (Kulkarni) ; on *J. auriculatum* and J. *sambac*, Bengal (Bal) ; on *J.* sp., Insein, Burma (Inayat).

——**mangiferae** Earle (Sacc. **XXII**:48 ; **454**:382). On leaves of *Mangifera indica*, Pulliyanur, Travancore (Butler) ; Sirsi, N. Kanara (Kulkarni) ; Insein, Burma (Inayat).

——**memecyli** Syd. (**443, XII**:198 ; Sacc. **XXIV**:314). On leaves of *Memecylon edule*, Dapoli, Bombay (Ajrekar). The species identified by Bal (**28, V**:4) as *M. heudeloti* Gaill. on the same host, Bhubaneshwar, Orissa, agrees well with this species, except for some discrepancies in the description of the setae (see Stevens, **419**:459).

[**Meliola**] **opiliae** Syd. (**445**:327 ; Sacc. **XXIV**:327). On leaves of *Opilia amentacca*, Coimbatore, Madras (Fischer).

——**palmicola** Wint. (Sacc. **XI**:267 ; **241** ; **454**:382 ; **28, V**: 2). On leaves of *Phoenix* sp., Mudigere and Munsur, Mysore (Butler) ; of *P. sylvestris*, Chittagong (Sen) ; Burdwan (Battachariya) ; Godavari (Sundararaman) ; Calcutta (Bal) ; of *P. humilis*, Talagupha, Mysore (Kulkarni).

——**pterospermi** Stev. (**420**:108, figs.). On *Pterospermum* sp., Burma (Butler).

——**sakawensis** P. Henn. (Sacc. **XVII**:548 ; **420**:109). On *Vitex leucoxylon*, N. Kanara (Sedgwick).

——**simillima** Ell. & Ev. (Sacc. **XVI**: 414 ; **420**:109). On *Holarrhena antidysenterica*, Dacca (Som).

——**stenospora** Wint. (Sacc. **IX**: 423 ; **28, V**:5 ; **419**:281). On *Strychnos nux-vomica*, Bhubaneshwar, Orissa (Bal).

——**tamarindi** Syd. (Sacc. **XXIV** : 308). On *Tamarindus indica*, Mangalore (Subramaniam).

——**zigzag** Berk. & Curt. (Sacc. **I**:67 ; **144**:96). On an unknown host, Belgaum (Hobson). An immature specimen of what seems to be this species has been collected (**454**:383) on *Cinnamomum* sp., Wahjain, Assam (Butler).

Metasphaeria albescens Thuem. (Sacc. **IX**:843 ; **105**: 35). On *Oryza sativa*, Burma (Butler).

——**boehmeriae** (Rabenh.) Sacc. (Sacc. **II**:156 ; **387**:46 as *Sphaeria*). On stems of *Boehmeria nivea*, Royal Botanic Garden, Calcutta.

——**celastrina** Syd. & Butler (**454**:408 ; Sacc. **XXIV**:952). In the bark of branches of dead *Celastrus* sp., Pusa (Butler).

Micropeltis applanata Mont. (Sacc. **II**:669 ; **196**:130). On leaves of *Gironniera*, South Andaman Island (Kurz).

Microsphaera alni (Wallr.) Salm. (Sacc., under several synonyms). On *Corylus colurna*, Mussoorie (Butler).

Microthyrium annuliforme Syd. (**445**:488 ; Sacc. **XXIV**:423). On leaves of *Capparis sepiaria*, Coimbatore (McRae).

——**microscopicum** Desm. (Sacc. **II**:662 ; **144**:96). Host not stated, Belgaum (Hobson).

Morenoella shoreae Ryan (**401**:104). On leaves of *Shorea talura*, Siddapur, N. Kanara (Sedgwick).

Munkiodothis melastomata (v. Hoehn.) Theiss. & Syd. (Sacc. **XXIV**:609). On *Melastoma* sp., Cherrapunji, Assam (Subramaniam).

?**Myiocopron gironnierae** Hariot & Karst. (**256**:129 ; Sacc. **IX**:1053). On leaves of *Gironniera* sp., Andaman Islands.

[——**oribiculare** (Cke) Sacc. (Sacc. **II**:661 ; **140**:118 as *Micropeltis orbicularis* Cke). Epiphyllous on *Symplocos spicata* Dinagepore [Dinajpur], India. According to v. Hoehnel (**264**, No. 426), this is a lichen of the genus *Raciborskiella* (*R. orbicularis* (Cke) v. Hoehn.).]

Myriangium cinchonae Rehm (**390**:325 ; Sacc. **XXII**:580). On the bark of *Cinchona* "*regia*," India.

——**duriaei** Mont. & Berk. (Sacc. **I**:198 as *Diatrype* ; **454**:395 ; **375**,**V**:65, figs.). On the bark of fallen branches, Pusa (Butler).

Nectria alutacea Berk. & Cke (**155**:81 ; Sacc. **IX**:958). On bark, Nilgiris (E. Berkeley).

——**bolbophylli** P. Henn. (Sacc. **XVII**:790 ; **454**:392, with synonymy of forms ; **91**:25, figs. ; **111**:17, fig. ; **105**:35). On stems of *Cajanus indicus*, Dehra Dun (Butler) ; on *Capsicum annuum*, Dacca (Som) ; on the trunk of *Cocos nucifera*, Pulliyanur, Travancore (Butler) ; on dead culms and glumes of *Oryza sativa*, Chittagong (Sen) ; Burma (Inayat) ; on the trunk of *Piper betle*, Shiggaon, Bombay (Chibber) ; on bark of *P. nigrum*, Vayitri, Wynaad (Butler) ; Hassan, Mysore (Lamb). The Indian specimens resemble the description of the form of this species called *N. coffeicola* by Zimmermann, that on *Oryza* resembling *N. coffeicola* var. *ochroleuca* Zimm.

——**cinnabarina** (Tode) Fr. (Sacc. **II**:479 ; **454**:393 ; **492**, figs. ; **495** ; **497**:37 ; **504**:54 ; **522**:408, figs. as "*Nectria ?ditissima*.") On branches of *Populus ciliata*, Verinag, Kashmir, and of *Prunus armeniaca*, Achibal, Kashmir (Butler) ; of *Pyrus communis*, Lansdowne, United Provinces ; on stems of *Thea sinensis*, Darjeeling and Cachar (Tunstall).

——**collabens** Berk. & Cke (**155**:81 ; Sacc. **IX**:958). On bark, Bombay (Carter).

——**diploa** Berk. & Curt. (Sacc. **II**:504 ; **375** ; **I**:105). The conidial stage, *Pseudomicrocera henningsii* (Koord.) Petch, on *Aspidiotus* on *Citrus aurantium*, Jorhat, Assam (Butler) ; on a scale insect on *Indigofera*, Bassein, Burma.

——**diversispora** Petch (Sacc. **XXII**:456 ; **454**:393). On fruits of *Hevea brasiliensis*, Mergui, Burma.

——**eugeniae** Currey (**196**:130 ; Sacc. **II**:511). On dead leaves of *Eugenia* sp., Yomah, Burma (Kurz).

——**heterosperma** Kalchbr. & Cke (Sacc. **II**:485 ; **454**:393). On dead branches, Dehra Dun (Mitra) ; Pusa (Butler). A variety of this species distinguished by somewhat smaller spores occurs on branches of *Citrus aurantium*, Sagaing, Burma (Butler ; **454**:393).

——**laetifulva** (Berk. & Cke) Sacc. (Sacc. **IX**:961 ; **155**:82 as *Dialonectria laetifulva* Berk. & Cke). On bark, Nilgiris (E. Berkeley).

——**tjibodensis** Penzig & Sacc. (Sacc. **XIV**:636 ; **454**:393). On the bark of *Guazuma tomentosa*, Pusa (Subramaniam).

——**vilis** (Syd.) Petch (**378**:191 ; Sacc. **XXIV**:693 as *Hypocrella vilis* Syd. ; **375**, **I**:157 as *N. tuberculariae* Petch). Associated with *Asterolecanium* sp. on stem of *Bambusa*, south India.

[Nectria] xanthostigma (Berk. & Cke) Sacc. (Sacc. **IX**:962 ; **155**:82 as *Dialonectria xanthostigma* Berk. & Cke). On herbaceous stems, Nilgiris (E. Berkeley).

Neocosmospora vasinfecta Smith (Sacc. **XVI**:562 ; **101**:6, figs. ; **454**:394 ; **111**:21, fig.) On roots of *Cajanus indicus*, Pusa (Butler) ; on roots and cotyledons of *Cicer arietinum*, Pusa (Hafiz) ; Coimbatore (McRae) ; Bannu ; on roots of *Crotalaria juncea*, Pusa and Samalkota (Butler) ; of *Cyamopsis psoraloides*, Pusa (Butler) ; of *Dolichos biflorus*, Pusa (Subramaniam) ; of *D. lablab*, Pusa (Butler) ; of *Gossypium* sp. Mirpurkhas and Nagpur ; of *Indigofera arrecta* and *I. sumatrana*, Pusa (Sen) ; of *Saccharum officinarum*, Pusa (Subramaniam); of *Vigna catjang*, Rajshahi Farm ; of *Zingiber officinale*, Rangpur (McRae).

Nummularia cinnabarina P. Henn. (Sacc. **XIV**:514). On *Dalbergia sissoo*, Pusa (Butler).

—**suborbicularis** (Welw. & Curr.) Sacc. (Sacc. **I**:399 ; **196**:130 as *Hypoxylon suborbiculare* Welw. & Curr.). On old logs, Royal Botanic Garden, Calcutta (Kurz).

Oidiopsis taurica (Lév.) Salm. (Sacc. **I**:16 as *Erysiphe taurica* Lév. ; **111**:271, fig.; **404**:220 ; **2**:28 ; **3**:26). On *Saussurea* (*Aplotaxis*), Himalaya (Jacquemont) ; on *Capparis* sp. (as *Erysiphe intertexta* Berk. mss. in Herb. Kew.), between Lama Yara and the Photu Pass (Thomson) ; on *Medicago sativa*, Madras (Sundararaman) ; on *Capsicum* sp., Bombay (Kulkarni) ; on *Cyamopsis psoraloides*, Nadiad, Bombay ; on *Vinca pusilla*, Coimbatore (Butler).

Ophiobolus butleri Syd. (**443**, **XIV**:260 ; Sacc. **XXIV**:1062). On stems of *Capparis* sp., Pusa (Butler).

——**cajani** Syd. (**454**:409 ; Sacc. **XXIV**:1064). On dead stems of *Cajanus indicus*, Pusa (Butler).

——**manihotis** Syd. (**454**:410 ; Sacc. **XXIV**:1063). On petioles of *Manihot utilissima*, Pulliyanur, Travancore (Butler).

——**oryzae** Miyake (Sacc. **XXII**:312). On *Oryza sativa*, Noakhali (Som).

——**porphyrogonus** (Tode) Sacc. (Sacc. **II**:338 ; **454**:410). On stems of *Vigna catjang*, Pulliyanur, Travancore (Butler).

Palawania grandis (Niessl) Syd. (Sacc. **XXIV**:424). On dead leaves of *Calamus* sp., Calcutta (Kurz). von Hoehnel remarks (**264**, No. 350) that this is perhaps only a strongly developed *Seynesia*.

Paranthostomella capparidis Syd. & Butler (**454**:408 ; Sacc. **XXIV**:825). On living or languishing leaves of *Capparis spinosa* var. *leucophylla*, Pusa (Inayat) ; on *C.* sp., Kulna, Bengal (Battachariya) ; Poona (Chibber) ; Pusa (Butler).

Parasterina pemphidioides (Cke) Theiss. (**484**:421 ; **133**:16, fig., as *Asterina pemphidioides* Cke ; Sacc. **I**:40 ; **184**:76 as *A. crustosa* Berk. & Cke ; **454**:390 ; *A. hobsoni* Berk. in Herb. ; **468**:18, 186 ; **473**:40, figs. ; **474**:426, figs.; **356**:99 as *Meliola fumago* Niessl ; **241**:119). On leaves of *Eugenia heyneana* (erroneously given as *E. jambolana* in **454**:390), Godavari (Sundararaman). On the upper surface of leathery, red-brown leaves, India (Hobson) ; on leaves of *Celastrus* sp., Calcutta (Kurz).

Parodiella paraguayensis Speg. (Sacc. **IX**:410 ; **483**:130). On leaves of *Crotalaria filipes*, Koondapur, S. Kanara (McRae).

——**perisporioides** (Berk. & Curt.) Speg. (Sacc. **I**:717 ; **454**:383 ; **483**:126 ; **132**:117 as *Dothidea perisporioides* B. & C.). On leaves of some leguminous plant, Bombay (Hobson) ; on leaves of *Medicago lupulina*, Dehra Dun (Butler) ; of *Desmodium rufescens*, Wahjain, Assam (Butler) ; of *D. triflorum*, Taliparamba, S. Kanara (McRae) ; of *D.* sp., Dehra Dun (Kar) ; Ranchi (Mitra) ; of *Indigofera trifoliata*, Kistna, Madras (Barber) ; Bassein, Bombay (Kulkarni) ; of *I.* sp., Deolali (Vakil).

——**spegazzinii** Theiss. & Syd. (**483**:131 ; included, with preceding two species, as *P. grammodes* (Kze) Cke in **445**:327). On leaves of *Crotalaria leschenaultii*, Coimbatore (McRae). Immature, but apparently this species.

Peroneutypella ambiens Syd. (**454**:414 ; Sacc. **XXIV**:732). On fallen twigs, Dehra Dun (Butler).

——**indica** Syd. & Butler (**454**:414 ; Sacc. **XXIV**:730). On dead twigs of *Dalbergia sissoo*, and of an unknown host, Pusa (Butler).

——**pusilla** Syd. (**454**:414 ; Sacc. **XXIV** :732). On dead branches of *Citrus* sp., Sagaing, Burma (Butler).

Phaeosaccardinula butleri (Syd.) Theiss. & Syd. (**484**:481 ; Sacc. **XXIV**:386 ; **454**:385 as *Limacinula butleri* Syd.). On leaves of "*Artocarpus* [? *Ficus*] *mysorensis*", Vayitri, Wynaad (Butler).

——**theae** (Syd. & Butler) Theiss. & Syd. (**484**:481 ; Sacc. **XXIV** : 386 ; **454**:386 as *Limacinula theae* Syd. & Butl. ; **111**:461, fig.; **264**, No. 1089, as *Capnites theae* (Syd. & Butl.) v. Hoehn.). On leaves and twigs of *Thea sinensis*, Rungmook, Darjeeling (Watt).

Phaeosphaeria oryzae Miyake (Sacc. **XXII**:214). On *Oryza sativa*, Jorhat, Assam.

Phragmocapnias betle (Syd. & Butler) Theiss. & Syd. (**484**:480 ; Sacc. **XXIV**:385 ; **454**:384 as *Capnodium betle* Syd. & Butler). On leaves of *Piper betle*, Dacca (Som) ; Mudon, Burma (Butler).

Phragmocauma viventis (Cke) Theiss. & Syd. (**481**:411 ; **133**:16, fig., as *Dothidea viventis* Cke ; Sacc. **II**:601 as *Phyllachora viventis* (Cke) Sacc.). On living leaves of Leguminosae, India (Hobson).

Phyllachora ajrekari Syd. (**443**, **VIII**:408 ; Sacc. **XXIV**:566). On leaves and petioles of *Ceropegia ?hirsuta*, Matheran, Bombay (Ajrekar).

——**aliena** Syd. (**443**, **VIII**:407 ; Sacc. **XXIV**:596). On leaves of *Memecylon edule*, Matheran (Ajrekar).

——**ambigua** Syd. (**481**:537 ; Sacc. **XXIV**:600 ; **454**:411 as *Trabutia ambigua* Syd.). On leaves of *Eugenia jambolana*, Koppa, Mysore (Butler).

[——**annulata** (Cke) Sacc. (Sacc. **II**:610 ; **144**:95, as *Dothidea annulata* Cke). On leaves of an unknown tree, Belgaum (Hobson). This imperfectly described species was re-examined by Theissen and Sydow (**481**:571), and it should be deleted.]

[Phyllachora] assimilis Syd. (**481**:439 ; Sacc. **XXIV**:573). On leaves of *Andropogon assimilis*, Kumaon (Inayat) ; Dehra Dun (Butler) ; of *A. micranthus*, Maymyo, Burma (Butler). (See also *P. graminis*, below).

——**bambusae** Syd. & Butler (**481**:441 ; Sacc. **XXIV**:576 ; **454**:400 as *Metachora bambusae* Syd. & Butler). On living leaves of *Bambusa*, Kanouth, Malabar (Butler).

——**bauhiniae** (Wint.) Theiss. & Syd. (**481**:491 ; **523**:27 as *Trabutia bauhiniae* (Wint.) ; Sacc. **IX**:601). On leaves of *Bauhinia vahlii*, Royal Botanic Garden, Calcutta (Kurz) ; Kirkee, Bombay (Chibber). This fungus was identified doubtfully as *P. tenuis* (Berk.) Sacc. in **454**:398. Theissen (**480**:183) stated that Kurz's collection includes both this fungus and *Pseudothis bauhiniae* (v. Hoehn.) Theiss., and Petrak & Sydow (**384**:305) found *Lasmeniella globulifera* also.

——**bischofiae** Syd. (**454**:396 ; Sacc. **XXIV**:571). On leaves of *Bischofia javanica*, Panora, Wynaad (McRae).

——**catervaria** (Berk.) Sacc. (Sacc. **II**:598 ; **481**:469 ; **454**:396 as *P. topographica* Sacc.). On leaves of *Ficus hispida*, Chittagong (Sen) ; on *F.* sp., Wynaad (Butler) ; Comilla (Inayat).

——**centothecae** Syd. (**443**, **XIII**:39 ; Sacc. **XXIV**:577). On leaves of *Centhotheca lappacea*, Moulmein, Burma. (See also *P. graminis*, below).

——**chionachnes** Syd. (**443**, **I**:164 ; Sacc. **XVII**:839). On leaves of *Chionachne barbata*, Punjab (Hooker f. & Thomson).

——**coicis** P. Henn. (Sacc.**XI**:373 ; **454**:399 ; **111**:243, fig.). On leaves of *Coix lachryma-jobi*, Wynaad, Madras (Butler).

——**cynodontis** (Sacc.) Niessl (Sacc. **II**:602 ; **454**:399 ; **445**:328). On leaves of *Cynodon dactylon* ; common in many parts of India.

——**cyperi** Rehm var. **donacis** Berl. & F. Sacc. (Sacc. **IX**:1029 ; **454**:399). On leaves of *Arundo* sp., Wahjain, Assam (Butler) ; of *Andropogon muricatus*, Pusa (Inayat) ; of *Imperata arundinacea*, Khasi Hills (Butler).

——**dalbergiae** Niessl (**356**:99 ; Sacc. **II**:594 ; **454**:397). On leaves of *Dalbergia* sp., Calcutta (Kurz) ; of *D. sissoo*, Pusa (Butler) ; of *D. lanceolata*, Bassein, Burma and of *D.* sp., Bilin, Burma (Butler). The form on *D.* sp. from Burma has shorter spores (15-18μ) than that on *D. sissoo* (19-22μ), but is otherwise identical.

——**desmodii** P. Henn. (Sacc. **XIV**:664 ; **454**:397). On leaves of *Desmodium* sp., Harwan, Kashmir, and Mussoorie (Butler).

——**dolichospora** Syd. (**454**:396 ; Sacc. **XXIV**:598). On leaves of *Tinospora cordifolia*, Solebile, Mysore (Butler).

——**erebia** Syd. (**454**:396 ; Sacc. **XXIV**:590). On leaves of *Caragana* sp., Harwan, Kashmir (Butler).

[**Phyllachora**] ?**fimbristylicola** Speg. (Sacc. **XXII**:423 ; **454**:398). On leaves of *Fimbristylis* sp., Dauracherra, Sylhet, and *F. dichotoma*, Kanaighat, Sylhet (Butler). Theissen and Sydow (**481**:373) make this species a synonym of *P. gracillimum*.

——**graminis** (Pers.) Fckl (Sacc. **II**:602 ; **454**:399-400). On leaves of the following grasses : *Andropogon serratus*, Dehra Dun (Butler) ; *A. triticeus*, Matheran, Bombay (Ajrekar) ; *A.* sp., Koppa, Mysore (Butler) ; *Anthistiria* sp., Wynaad (McRae) ; Mussoorie (Inayat) ; Kumaon (Butler) ; *Chloris barbata*, Coimbatore (McRae) ; *Elionurus hirsutus*, Hoshangabad (Butler) ; *Oryzopsis* sp., Wynaad (McRae) ; Kumaon (Inayat) ; *Panicum patens*, Kanaighat, Sylhet (Butler, erroneously given as *Isachne* sp. in **454**:399) ; *P. ramosum*, Wahjain, Assam (Butler) ; *P. ?colonum*, Champaran (Burkill) ; *P.* sp., Panora, Wynaad (McRae) and Wahjain, Assam (Butler) ; *Pollinia grata*, Moulmein, Burma (Butler). The mass of forms included under this species will doubtless be subdivided into distinct varieties and species as has already been done with the following forms, which were given as *P. graminis* in **454**:399-400 : on *Pogonatherum saccharoideum* : *P. pogonatheri* Syd. ; on *Ischaemum laxum* : *P. ischaemi* Syd. ; on *Centotheca lappacea* : *P. centothecae* Syd. ; on *Andropogon assimilis* and *A. micranthus* : *P. assimilis* Syd.

——**indica** Theiss. & Syd. (**481**:488 ; Sacc. **XXIV**:589). On leaves of *Acacia penninervis*, India.

——**ischaemi** Syd. (**481**:449 ; Sacc. **XXIV**:580). On leaves of *Ischaemum laxum*, Bilin, Burma (Butler).

——**ixorae** Theiss. & Syd. (**481**:553 ; Sacc. **XXIV**:602). On leaves of *Ixora parviflora*, Coimbatore (McRae) ; of *I. polyantha* and *I.* sp., Matheran, Bombay (Ajrekar) ; of *?I.* sp. (Hobson). Theissen and Sydow (**481**:572) consider that *P. demersa* (Cda) Sacc. (**185**, **IV**:fig. 121, as *Sphaeria* ; Sacc. **II**:595 ; **133**:16 ; **445**:488) should be deleted, as Corda's type does not exist, and specimens subsequently referred to *P. demersa* are in part *P. ixorae*.

——**malabarensis** Syd. & Butler (**454**:398 ; Sacc. **XXIV**:576). On leaves of *Bambusa* sp., Wynaad (Butler).

——**permixta** Syd. (**454**:397 ; Sacc. **XXIV**:605). On leaves of *Schima wallichii*, Maymyo, Burma (Butler).

——**pogonatheri** Syd. (**443**, **XIII**:40 ; Sacc. **XXIV**:584). On leaves of *Pogonatherum saccharoideum*, Kumaon (Inayat).

——**pongamiae** (Berk. & Broome) P. Henn. (Sacc. **VIII**:708 as *Cryptomyces pongamiae* (B. & Br.) Sacc. ; **454**:376, figs., 397 ; **445**:328 ; **481**:508 ; **367**, **III**:291). On leaves of *Pongamia glabra*, Islampur, Bombay (Chibber) ; Madras (Butler).

——**rhytismoides** (Cda) Sacc. (Sacc. **II**:594 ; **481**:506 ; **185**, **IV**:42, figs., as *Dothidea rhytismoides* Cda). On leaves of *Mimosa* (" *Acacia* " in Sacc.), Tenasserim, Burma (Helfer).

[**Phyllachora**] **rottboelliae** Syd. & Butler (**454**:400 ; Sacc. **XXIV**:585). On leaves of *Rottboellia exaltata*, The Droog, Nilgiris (Butler).

——**sacchari** P. Henn. (Sacc. **XVII**:838 ; **445**:489). On leaves of *Saccharum spontaneum*, South Malabar (McRae) ; on *S.* sp., Khasi Hills (Subramaniam).

——**sacchari-spontanei** Syd. (Sacc. **XXIV**:586 ; **481**:459). On leaves of *Saccharum spontaneum*, Godavari (Subramaniam) ; Suri, Birbhum (Basu) ; Pusa (Inayat) ; Bassein, Burma (Butler).

——**shiraiana** Syd. (Sacc. **XVI**:622 ; **454**:398). A form on leaves of *Arundinaria* sp., Wahjain, Assam (Butler).

——**sorghi** v. Hoehn. (Sacc. **XXII**:426). On leaves of *Sorghum vulgare* [*Andropogon sorghum*], Sabour (Dutt).

——**spissa** Syd. (**454**:397 ; Sacc. **XXIV**:592). On leaves of *Dalbergia sissoo*, Wynaad (McRae) ; of *D.* sp., Khasi and Jaintia Hills (Subramaniam).

——**transiens** Syd. & Butler (**454**:397 ; Sacc. **XXIV**:605). On leaves of *Eurya acuminata*, Kumaon (Inayat).

Phyllachorella micheliae Syd. (**445**:489 ; Sacc. **XXIV**:607 ; **481**:576). On leaves of *Michelia nilagirica*, Ootacamund (McRae).

Phyllactinia corylea (Pers.) Karst. (Sacc. **I**:5 ; **98**:12, figs ; **405**:498 ; **113** (16):4 ; **113** (17):2 ; **392**). On leaves of the following plants : *Indigofera gerardiana* and *Juglans regia*, Verinag, Kashmir (Butler) ; *Morus alba*, Pusa, and Dubgaon in Kashmir (Butler) ; *M.* sp., Kashmir (Butler) ; Ootacamund (McRae) ; Maymyo (Rhind) ; *Pyrus communis*, Shillong (Som) ; *P. pashia*, Mussoorie (Butler).

—— —var. **subspiralis** Salmon (**405**:501 ; Sacc. **XXII**:20 as *P. suffulta* var. *subspiralis*). On leaves of *Dalbergia sissoo*, Dehra Dun and Pusa (Butler) ; Poona (Woodrow) ; Bombay (Chibber) ; Nagpur (Pandit).

Physalospora calami Syd. (**454**:407 ; Sacc. **XXIV**:804). On leaves of *Calamus tenuis*, Chittagong (Sen).

——**piperina** Syd. (**443**, **XIII**:38 ; Sacc. **XXIV**:1334). On stems of *Piper nigrum*, Vayitri, Wynaad (Butler).

——**rhodina** (Berk. & Curt.) Cke (Sacc. **IX**:592 ; **421**:212). Specimens agreeing with the conidial stage of this fungus in morphological and cultural characters have been recorded by N. E. Stevens on *Mangifera indica* and *Albizzia* sp., Poona (Ajrekar) ; on tubers of *Ipomoea batatas*, Pusa (Shaw) ; and on various other hosts including tea, cacao, and rubber from Ceylon and the Federated Malay States. The possibility that the species generally known as *Diplodia cacaoicola* [*Botryodiplodia theobromae*, q. v.], so frequent as a parasite of *Hevea* rubber and tea in India as elsewhere, is really this fungus is suggested by Stevens (**421**:217), but Tunstall refers the latter fungus to *Thyridaria tarda* (q. v.).

——**transversalis** Syd. (**454**:407 ; Sacc. **XXIV**:804). On living leaves of *Cocos nucifera*, Bilin, Burma (Butler).

[**Physalospora**] **ventricosa** (Durieu & Mont.) Cke (Sacc. **XI**:292 ; **454**:408). On dead stems of *Ricinus communis*, Pusa (Butler).

Placostroma elettariae (Berk. & Broome) Theiss. & Syd. (**481**:408 ; Sacc. **II**:633 as *Dothidella*). On *Elettaria cardamomum*, Bangalore (Anstead).

Pleogibberella calamia (Cke) Berl. & Vogl. (Sacc. **IX**:992 ; **159**:8 as *Gibberella calamia* Cke). Surrounding fruits of *Calamus fasciculatus*, Vizagapatam, S. India.

Pleospora spinarum Syd. (Sacc. **XVI**:545 ; **454**:410). On spines of *Astragalus* sp., Achibal, Kashmir (Butler). The Indian specimens have somewhat longer asci and spores than those of the type from France.

Podocrea grossa (Berk.) Lloyd (**287**:1259, fig. ; **57**, No. 360, as *Hypocrea grossa* Berk. ; Sacc. **II**:528). On rotten wood, Darjeeling, 7-8,000 ft. (Hooker f.).

Podonectria coccicola (Ell. & Ev.) Petch (**375**, **I**:161 ; **378**:196 ; Sacc. **IX**:996 as *Ophionectria coccicola* (Ell. & Ev.) Berl. & Vogl.). On *Chionaspis manni* on *Thea sinensis*, Darjeeling (conidial stage only).

Podosphaera leucotricha (Ell. & Ev.) Salmon (Sacc. **IX**:365 as *Sphaerotheca* ; **113**(11):74). On *Pyrus malus*, Lahore (Mitra) ; Ramgarh, Kumaon (Shaw) ; Srinagar, Kashmir (Butler).

Polyrhizon terminaliae Syd. (**481**:234 ; Sacc. **XXIV**:412 ; **454**:401 as *?Dothidea terminaliae* Syd.). On leaves of *Terminalia catappa*, Wynaad (McRae).

Polystigma ochraceum (Wahlenb.) Sacc. (Sacc. **II**:458 ; **140**:117 as *P. fulvum* DC.). On leaves of *Prunus padus*, Jubal State, N. W. Himalayas, 18,000 ft. (Gamble).

Poronia arenaria Syd. & Butler (**454**:420, fig. ; Sacc. **XXIV**:1094). On sand dunes near *Casuarina* trees, Chatrapur, Ganjam (Butler).

——**gigantea** Sacc. (**403**:302 ; Sacc. **XXIV**:1094). On dung of elephant, Mondhomuzhi, Travancore (Ramaswami).

——**oedipus** Mont. (Sacc. **I**:349 ; **133**:16 ; **263**:340 ; **454**:420). On dung, India (Hobson) ; on horse dung, Pusa (Butler) ; Botanic Garden, Saharanpur (Gollan).

——**polyporoides** P. Henn. (**263**:340 ; Sacc. **XVII**:620). On dead twigs and on the ground, Botanic Garden, Saharanpur (Gollan).

Prillieuxina winteriana (Pazschke) Arnaud (**18**:162 ; Sacc. **XI**:255 as *Asterina* ; **401**:103 as *Asterinella winteriana* (Pazsch.) Theiss. ; **471**:122, fig.). Epiphyllous on *Castanopsis* sp., Mangalore (Subramaniam).

Pseudothis bauhiniae (v. Hoehn.) Theiss. (**480**:183 ; **264**, No. 774, as *Roussoella bauhiniae* v. Hoehn., in part ; see **481**:491 and *Phyllachora bauhiniae* above). On living leaves of *Bauhinia vahlii*, Royal Botanic Garden, Calcutta (Kurz).

Pyrenocarpon magnificum (Syd. & Butler) Theiss. (**473**:31 ; Sacc. **XXIV**:509 ; **478**:639, figs. ; **454**:391 as *Asterina magnifica* Syd. & Butler). On leaves of *Terminalia* sp., Moulmein, Burma (Butler).

Rickia coleopterophagi Paoli (**359**:286, fig.; Sacc. **XXII**:8). On the acarus *Coleopterophagus procerus*, India.

Rosellinia andurnensis Ces. & de Not. (Sacc. **I**:253; **454**:404). On bark, Dehra Dun (Inayat).

——**arcuata** Petch (Sacc. **XXIV**:834; **111**:437, fig.; **505**; **493**:9 as *R. bothrina* (B. & Br.) Sacc.; **1**:12; **2**:57; **494**:117). On roots of *Thea sinensis*, The Droog, Nilgiris (Butler); Assam (Tunstall).

——**bunodes** (Berk. & Broome) Sacc. (Sacc. **I**:254; **454**:404; **111**:357, fig.; **367**, **II**:434; **317**:178). At the base of the trunks of *Litsea angustifolia*, *L. wightiana*, *Schleichera trijuga*, *Holigarna longifolia*, *Grevillea robusta*, and *Piper nigrum*, Hassan, Mysore (Lamb). Specimens of this disease on pepper were sent from Mysore to Massee in 1895 (**317**) by Middleton, but Massee identified them no farther than to say the fungus was allied to *Dematophora* [*Rosellinia*] *necatrix*. Petch (**367**, **II**:434) states that *R. bunodes* is recorded as parasitic on *Coffea* in southern India.

——**cocoes** P. Henn. (Sacc. **XXII**:110; **5**:4). On *Areca catechu*, India (Shaw).

——**leprantha** (Fr.) Sacc. (Sacc. **I**:255; **240**:131 as *Sphaeria leprantha* Fr.). On fallen bark, Nicobar Islands (Didrichsen).

——**necatrix** (Hartig) Berl. (Sacc. **XVII**:595; **243**:205). Recorded by Gandhi (as *Dematophora necatrix* Hartig) from Poona in 1928, on *Vitis vinifera*.

——**picta** (Berk.) Cke (**164**:81; Sacc. **IX**:426). On decorticated wood, Nilgiris.

——**spadicea** Ces. (Sacc. **I**:267; **242**:438). Recorded from India, on stems of *Arundinaria spathiflora*.

——**sublimbata** (Durieu & Mont.) Pass. (Sacc. **I**:259; **196**:130 as *Sphaeria sublimbata* Dur. & Mont.). "Apparently on stems of *Thysanolaena acarifera* [*T. agrostis*], Kambala toung", Burma (Kurz).

Sarcoxylon compunctum (Jungh.) Cke (**159**:107; Sacc. **I**:325 as *Xylaria compuncta* (Jungh.) Berk.; **57**, after No. 482; **240**:130 as *Hypoxylon compunctum* Fr.). Khasi Hills (Hooker f.); Nicobar Islands, rare (**240**). Theissen (**472**:1301) considered that this fungus probably belongs to the Hypocreaceae; Petch (**372**:145) thinks it probable that the Ceylon specimen belongs to *S. aurantiacum* Pat.; and Lloyd (**304**:29, and **287**:1203-4) states that the Indian record is really *S. aurantiacum*, which had better go in Moeller's genus *Entonaema* as *E. aurantiaca*.

Schizothyrium annuliforme Syd. & Butler (**454**:376; Sacc. **XXIV**:527). On living leaves of *Acer oblongum*, Mussoorie (Butler). It is questionable if the fungus is correctly to be considered a *Schizothyrium* (**454**:377).

Sirrhodothis seriata Syd. & Butler (**481**:416; Sacc. **XXIV**:611; **454**:402 as *Scirrhia seriata* Syd. & Butler). On leaves of *Bambusa* sp., Moulmein (Butler).

Sphaerella* bambusina Syd. & Butler (**454**:407 ; Sacc. **XXIV**:866). On living leaves of *Bambusa* sp., Wahjain, Assam (Butler).

——**bhauria** Cke (**140**:118 ; Sacc. **I**:489). On leaves of "Bhauri" (*Symplocos spicata*), Dinajpur.

——**citrullina** C. O. Smith (Sacc. **XXII**:123). On leaves of *Lagenaria vulgaris*, Pusa (Sen).

——**coffeicola** Cke (Sacc. **I**:498 ; **111**:481, fig.). On *Coffea* sp., Wynaad (Watt).

——**fragariae** (Tul.) Sacc. (Sacc. **I**:505). On *Fragaria* sp., Harwan, Kashmir (Butler).

——**heveae** Petch (Sacc. **XXIV**:863 ; **344**). Recorded by Mitra on *Hevea brasiliensis*, Andaman Islands.

——**malinverniana** Cattan. (Sacc. **I**:527 ; **454**:407). On leaves of *Oryza sativa*, Nadia, Bengal (Mukerji) ; Murshidabad, Bengal (Mazumdar).

——**roseigena** Ell. & Ev. (Sacc. **IX**:643 ; **454**:407). On leaves of *Rosa* sp. cult., Darjeeling (McRae).

Sphaeria constellatio Berk. (**57**, No. 486 ; Sacc. **II**:398). On the upper surface of green leaves, Khasi Hills (Hooker f.). Several species of "*Sphaeria*" were recorded by early workers on Indian fungi, but only this one remains in that uncertain genus.

Sphaerostilbe aurantiicola (Berk. & Broome) Petch (**375**: **I**:158 ; Sacc. II:487 as *Nectria*). On *Mytilaspis piperis* on *Piper nigrum*, Wynaad (Barber) (given in **375**, **I**:127 as on *Lepidosaphes* sp. on *Piper nigrum*, South India (Barber, 1905) ; given in **454**:393 as *Nectria coccidophthora* Zimm. ; in **472**:1293 as *N. subcoccinea* Sacc. & Ell.).

——**coccidophthora** (Zimm.) Petch (**375**, **I**:160, 130 ; Sacc. **XVII**:784 as *Nectria*). On *Chionaspis* on an undetermined host, India (Butler).

——**gracilipes** Tul. (Sacc. **II**:513 ; **454**:394). On dead stems, Pulliyanur, Travancore ; and Noakhali (Butler).

——**lateritia** Berk. & Curt. (Sacc. **II**:516). On stems and branches, Nepal.

——**repens** Berk. & Broome (Sacc. **II**:516 ; **113**(9):67 ; **111**:432, figs. ; **1**:12 ; **2**:57 ; **493**:13, fig.; **494**:117 ; **394** ;**502** ; **505** ; **386**). On roots of *Thea sinensis*, Assam (Tunstall) ; of *Hevea brasiliensis*, Burma (Pinching ; Rhind).

Sphaerotheca euphorbiae (Cast.) Salmon (Sacc. **XIV**:462 as *S. tomentosa* Otth.) On *Euphorbia pilosa*, Harwan, Kashmir (Butler).

——**humuli** (DC.) Burr. (Sacc. **I**:3-4, older names). On *Agrimonia eupatorium*, Harwan, Kashmir (Butler).

——var. **fuliginea** (Schlecht.) Salm. (Sacc. **XXII**:20 ; **111**:514). On the following: *Bidens pilosa*, Poona (Ajrekar) ; *Lagenaria vulgaris*, Orai, United Provinces (Inayat) ; *Phaseolus aconitifolius*, Mandalay (Butler) ; *Siegesbeckia orientalis*, Verinag, Kashmir (Butler) ; *Taraxacum officinale* and *Cucurbita moschata*, Harwan, Kashmir (Butler).

---

* *Mycosphaerella* was proposed by Johanson in 1884 for this genus, because the name *Sphaerella* had been used earlier for a genus of Algae, but *Sphaerella* is still used by many mycologists.

[Sphaerotheca] pannosa (Wallr.) Lév. (Sacc. **I**:3 ; **143**:95). On *Rosa* sp. cult., Belgaum (Hobson) ; Srinagar, Achibal, and Harwan in Kashmir, Ranikhet and Dehra Dun (Butler) ; Simla (Coventry) ; Nagpur (Pandit) ; Lonavla, Bombay (K'rtikar). On *Prunus amygdalus*, Harwan, Kashmir (Butler) ; on *P. persica*, Darjeeling (McRae) ; Harwan, Verinag, and Srinagar in Kashmir (Butler).

Sphaleromyces indicus Thaxt. (**462**, **IV**:41 ; Sacc. **XVI**:692). On *Pinophilus* sp., near " *P. rufipennis* ", Malabar.

Starbaeckiella mangiferae Syd. (**447**:37 ; **454**:405 as *Rosellinia mangiferae* Syd. ; Sacc. **XXIV**:826). On bark of *Mangifera indica*, Chittagong (Sen).

Stomiopeltis aspersa (Berk.) Theiss. (**477**:86 ; **484**:432 ; **470** :219 as *Calothyrium aspersum* (Berk.) Theiss. ; **264**, No. 517, as *Microthyrium aspersum* (Berk.) v. Hoehn.; **57**, No. 476, as *Asterina aspersa* Berk. ; Sacc. **I**:45). On leaves of *Laurus* sp., Khasi Hills (Hooker f.).

Tettigomyces indicus Thaxt. (**465**:24 ; Sacc. **XXIV**:86). On bristles of *Gryllotalpa* sp., North India (Scudder Collection).

Thielavia basicola Zopf (Sacc. **I**:39 ; **454**:379, conidial stage only). On roots of *Viola odorata*, Lahore. McCormick [48th Ann. Rept. Conn. Agri. Exper. Stat., pp. 539-554, 1925] states that the conidial " stage " is a distinct fungus, *Thielaviopsis basicola* (Berk.) Ferraris.

Thyridaria tarda Bancroft (Sacc. **XXIV**:770 ; **493**:15, fig. ; **484**:118 ; **506**:68). On roots of *Thea sinensis*, Doom Dooma district and Surma Valley, Assam (Tunstall). Tunstall (**506**) reports obtaining immature pycnidia typical of *Botryodiplodia theobromae* (q. v.) from cultures from ascospores of *T. tarda.*

Titanella ilicina (Syd. & Butler) Syd. (**447**:36 ; Sacc. **XXIV**:1047 ; **454**:411 as *Pleomassaria ilicina* Syd. & Butler). On the bark of *Ilex* (probably *I. dipyrena*), Ranikhet, Kumaon (Inayat).

Trabutia butleri Theiss. & Syd. (**481**:354 ; Sacc. **XXIV**:557). On leaves of *Ficus* sp., Wahjain, Assam (Som). This fungus was identified as *Phyllachora abyssinica* P. Henn. in **454**:396. Petrak (**383**:386) has transferred *Trabutia butleri* to *Phyllachora* as *P. butleri* (Theiss. & Syd.) Petrak.

——? **cayennensis** (DC.) Sacc. (Sacc. **I**:449 ; **57**, after No. 485, as *Sphaeria cayennensis* Fr.). On leaves of *Ficus roxburghii*, Sikkim (Hooker f.).

——**ficuum** (Niessl) Theiss. & Syd. (**481**:352 ; **356**:99 as *Phyllachora ficuum* Niessl (pro parte) ; Sacc. **II**:598). On leaves of *Ficus infectoria*, Royal Botanic Garden, Calcutta (Kurz). Petrak (**383**:386) considers that this species should remain in *Phyllachora*.

Trichosphaeria macularis Syd. & Butler (**454**:402 ; Sacc. **XXIV**:809). On living leaves of an undetermined host, Pulliyanur, Travancore (Butler).

Trichothyiella quercigena (Berk.) Theiss. (**475**:12 ; **468**:180 as *Trichothyrium quercigenum* (Berk.) Theiss. ; **159**:67 as *Asterina quercigena* (Berk.) Cke (*Dothidea quercigena* Berk. in Herb.) ; Sacc. **IX**:376 as *Asterula quercigena*

(Berk.) Sacc.). On leaves of *Quercus* (?or *Pasania*), Sikkim, " auf dem Stroma von *Lasiobotrys ?elegans* (Syd.) Theiss."(**475**:13).

Uncinula necator (Schwein.) Burr. (**89**:48 ; **86** ; **1**:9 ; **2**:41 ; **3**:33 ; Sacc. **I**:22 as *Erysiphe*). Conidial stage on *Vitis vinifera*, Srinagar, Kashmir (Butler) ; Nasik and Poona (Kulkarni) ; Deolali (Vaz) ; Poona (Keatinge).

——**polychaeta** (Berk. & Curt.) ex Ellis (Sacc. **IX**:367). On *Celtis australis* (*C. caucasica*), Harwan, Kashmir (Butler).

——**salicis** (DC.) Wint. (**404**:87 ; **140**:117 as *Erysiphe martii*, in error ; Sacc. **I**:7 as *U. adunca*). On leaves of *Populus ciliata*, Simla (Gamble) ; on *Salix* sp., Mussoorie (Kar).

——**tectonae** Salmon (**406**:479 ; Sacc. **XXII**:22). On leaves of *Tectona grandis*, Jubbulpur (Forest Officer) ; Dohad, Bombay (Chibber) ; Burma (Hole) ; Nagpur (Pandit) ; on *Cordia macleodii*, Jubbulpur (Forest Officer).

Ustilaginoidea virens (Cke) Takahashi (Sacc. **XIV**:431 ; **262**:25 ; **191**:96, figs. ; **105**:30, figs. ; **111**:228, figs. ; **452**:425 ; **392**:1 ; **393** ; **141**:15 as *Ustilago virens* Cke ; **319**, **II**:167 as *Sphacelia oryzae* Massee). In the inflorescence of *Oryza sativa*, Tinnevelli (Western) ; Assam (Watt) ; Samalkota, Madras (Barber) ; and throughout India.

Ustulina tessulata (Berk.) Cke (**152**:3, *Hypoxylon tessulatum* Berk. in Herb. ; Sacc. **IX**:542). On wood, Bombay.

— **vulgaris** Tul. (Sacc. **I**:351 ; **263**:339 as *U. maxima* (Hall.) Schroet.). On dead trunks, Botanic Garden, Saharanpur (Gollan).

——**zonata** (Lév.) Sacc. (Sacc. **I**:352 ; **493**:11, figs. ; **494**:116 ; **393** ; **394**:5 ; **505** ; **23** ; **111**:348, figs.). On *Thea sinensis*, Assam (Butler) ; Darjeeling and elsewhere in N. E. India (Tunstall) ; on *Hevea brasiliensis*, Burma (Rhind); south India (Ashplant).

Valsa ceratophora Tul. var. **rosarum** de Not. (Sacc. **I**:109). On *Rosa* sp., Pusa (Butler).

——(**Euvalsa**) **corchori** Syd. & Butler (**454**:412 ; Sacc. **XXIV**:20). On stems of *Corchorus* sp., Poona (Chibber).

——**nepalensis** (Berk.) Sacc. (Sacc. **I**:125 ; **57**, No. 487, as *Sphaeria nepalensis* Berk.). On dead twigs of *Betula* sp., East Nepal, 9,000 ft. (Hooker f.).

Vizella conferta (Cke) Sacc. (Sacc. **II**:662 ; **140**:118 as *Micropeltis conferta* Cke ; **476**:14 ; **264**, No. 1007). Epiphyllous on *Symplocos spicata*, Dinagepore [Dinajpur]. Stated by Theissen (**476**) to belong to the Sphaeriaceae, but regarded by von Hoehnel (**264**) as a true subcuticular Hypodermiae, and recently discussed in detail by Petrak & Sydow (**385**:102).

Winteria profusa (Syd. & Butler) Sacc. (Sacc. **XXIV**:970 ; **454**:404 as *Rehmiomyces profusus* Syd. & Butler). On dead twigs of *Cajanus indicus*, Dehra Dun (Basu).

Xylaria aemulans Starb. (Sacc. **XVII**:630 ; **78**:256 ; **287**:1030, fig., 1069). " Very common at Barkuda [Chilka Lake] on prostrate logs " (Bose, **78**).

[**Xylaria**] **allantoidea** Berk. (Sacc. **I**:314; **454**:417; **472**:1301; **376**:121; **263**:340 as ?*Xylaria obtusissima* Berk.). On old wood, Wahjain, Assam (Basu); on dead trunks, Botanic Garden, Saharanpur (Gollan).

——**aristata** Mont. (Sacc. **I**:333; **144**:96). Recorded by Cooke from Belgaum (Hobson).

——**aspera** Massee (Sacc. **XVI**:444; **454**:417). On old wood, Pusa (Inayat); Godavari (Barber); on trunks, Royal Botanic Garden, Calcutta (Butler); on wood of *Dalbergia sissoo,* Pusa and Dehra Dun (Butler). Lloyd (**305**:14) states that this species is probably the same as *X. melisii* Berk.

——**assamensis** Lloyd (**287**:1294, fig.). Assam (Hole).

——**dealbata** Berk. & Curt. (Sacc. **I**:323; **467**:159 as *X. dealbata* B. & Br.). On wood, Khandala, Bombay (Blatter). Lloyd (**305**:8) states that Theissen (**466**:65) is in error in referring this species to *X. obovata* Berk.

——**delitschii** Auersw. (Sacc. **I**:336; **263**:340). On the ground, apparently on rotten fruits, Botanic Garden, Saharanpur (Gollan). Theissen (**472**:1302) states: " It is very doubtful whether the identification of this species, which is only known from Germany, is correct."

——**deserticola** Speg. (Sacc. **XVIII**:629; **454**:417). On old roots of *Ficus* sp., Fraserpet, Coorg (Butler); on old wood on the ground, Pusa (Butler). Theissen (see **454**:418) considered this species to be only a form of *X. cornu-damae* (Schw.) Berk.

——**digitata** (L.) Grev. (Sacc. **I**:339; **196**:129; **263**:341; **376**:119). Arnigadh, Mussoorie (Gollan); on moist logs, Royal Botanic Garden, Calcutta (Kurz). Lloyd (**288**, No. **62**:6) considered this name a synonym of *X. cornu-damae* (Schw.) Berk.

——**emerici** Berk. in Herb. (**150**:86; Sacc. **I**, Addenda:XIX, and **IX**:528). ?On the ground, Nilgiris (Berkeley f.). Lloyd considers that this species, known only from the one collection, is probably the same as *X. titan* Berk.

——**euglossa** Fr. (Sacc. **I**:324; **454**:418; **472**:1302; **466**:59). On dead wood of *Tamarindus indica,* Surat District, Bombay (Gleadow); on wood, Royal Botanic Garden, Calcutta (Butler); as *X. turgida* Fr. from the Nicobar Islands (**472**).

[——**fimbriata** Lloyd (**287**:726, figs., 1069). India (Bose). Lloyd stated later (**287**:1254) that the specimen is conidial only and should not have been named.]

[——**furcellata** Berk. in Herb. (**150**:88; Sacc. **1**, Addenda: XXI, and **IX**:537), Nilgiris. Theissen (**472**:1301) remarks " This species cannot be maintained, because quite undeveloped", and Lloyd (**287**:1254, 727, fig.) also excludes it.]

——**gomphus** Fr. (Sacc. **I**:316; **263**:340). In an underground cellar, Botanic Garden, Saharanpur (Gollan). Theissen (**466**: 63; **472**:1301) follows Bresadola (Ann. Myc., V:241, 1907) in considering this species identical with *X.*

involuta Klotz. (see *X. tabacina* below), but Lloyd (**305**:15) regards it as a good species.

[Xylaria] **guyanensis** Mont. (Sacc. **I**:312 ; **196**:129). In evergreen forests, Toukyeghat, Burma (Kurz). Lloyd (**287**:650) notes that *X. neilgherries* Mss. is the same.

——**heloidea** Penzig & Sacc. (Sacc. **XIV**:506 ; **454**:418). On fallen twigs, leaves, and capsules of *Cedrela toona*, Dehra Dun (Butler).

——**hispidissima** (Fr.) Sacc. (Sacc. **I**:345 ; **240**:129 as *Rhizomorpha hispidissima* Fr. ; **304**:14). Nicobar Islands (Didrichsen).

——**hypoxylon** (L.) Grev. (Sacc. **I**:333 ; **196**:129 ; **391**:146 ; **263**:340 ; **454**:418 ; **57**, after No. 480, and after No. 359, as " *Hypoxylon vulgare* (*Sphaeria hypoxylon* Ehr.)". Sone River, Bihar ; and on old wood, Darjeeling, 7,500 ft. (Hooker f.) ; on old tree stumps, Baronga Island, Arracan, Burma ; and on old logs, Royal Botanic Garden, Calcutta (Kurz) ; on buried wood, Botanic Garden, Saharanpur (Gollan) ; on branches of *Casuarina*, Nicobar Islands.

—— —form **tropica** Syd. & Butler (**454**:418 ; Sacc. **XXIV**:1099). On old wood, Solebili, Balehonnur, and Barguai, Mysore (Butler) ; Kallor Hills, South India (Calder). Some of the earlier collections called *X. hypoxylon* may also be this form.

——**nigripes** (Klotzsch) Sacc. (Sacc. **IX**:527 ; **273**:203, figs., as *Sphaeria* (*Cordyceps*) *nigripes* Klotzsch ; **67**:118 as *X. gardneri* Berk. ; **144**:96 as *X. escharoidea* Berk. ; **57**, No. 481, as *X. piperiformis* Berk. ; **467**:159 ; **263**:340, as *X. peperomioides* P. Henn. ; **196**:129, figs., as *X. flagelliformis* Currey and *X. mutabilis* Currey ; **454**:419 ; **150**:85, 89 ; **364**:242). On the ground, India (D. Wight) ; Belgaum (Hobson) ; Botanic Garden, Saharanpur, (Gollan) ; on earth, Yomah, Burma (Kurz) ; on brick-laid paths, Royal Botanic Garden, Calcutta (Kurz) ; rooting in the soil, Sikkim, 5,000 ft. (Hooker f.) ; on wood, Bombay (Blatter) ; on the ground, Dehra Dun (Rao) ; Koppa in Mysore, and Pusa (Butler) ; Jullundur (Dobbs) ; Dhulia, Bombay (Kulkarni) ; and elsewhere. Three forms (a, b, and c : **454**:419) are recognisable in the Pusa herbarium.

The large sclerotium of termite nests described by Berkeley & Currey as *Sclerotium stipitatum* (**58**:91 ; **197**:93 ; **111**:12, figs.), has been shown by Petch (**365**:401) to belong to this fungus. Bose (**78**:254) gives figures and a description of *X. nigripes* grown from termite nests.

——**obovata** Berk. (Sacc. **I**:317 ; **150**:82, 88, as *X. carteri* Berk. in Herb.; **466**:65). On wood, Bombay (Carter).

——**pistillaris** (P. Henn. & Nyman) Theiss. (Sacc. **XVI**:446 as *Hypoxylon pistillare* P. Henn. & E. Nym. ; **472**:1302 ; **454**:416). On bark, Barguai, Mysore (Butler).

——**plebeja** Cesati (Sacc. **I**:318 ; **454**:420 ; **472**:1301 ; **376**:124 ; **57**, No. 482, as *X. fistuca* Berk.). East Nepal, 8,000 ft., and woods, Myrong, Khasi Hills

(Hooker f.) ; on rotten wood, Pusa (Inayat) ; on *Areca catechu*, Sirsi, Bombay (Kulkarni).

[**Xylaria**] **polymorpha** (Pers.) Grev. (Sacc. **I**:309 ; **196**:129 ; **57**, after No. 359, as *Hypoxylon polymorphum* Ehr.). On old logs in the shade, rainy season, Royal Botanic Garden, Calcutta (Kurz) ; Sinchul, 8,000 ft. (Hooker f.).

——**regalis** Cke (**150**:86 ; Sacc. **I,** Addenda:XIX, and **IX**:530). On wood, Royal Botanic Garden, Calcutta (Kurz). "Perhaps merely a form of *X. plebeia* Ces." (Theissen, **472**:1301 and **466**:58). Lloyd (**288,** No. **64**:2) thought it might be the same as *X. titan* Berk.).

——**salmonicolor** Berk. in Herb. (**150**:87 ; Sacc.**I,** Addenda : XX, and **IX**:529). On wood, Nilgiris.

——**sanchezii** Lloyd (**287**:1283, fig., 1294, 1310, figs.). On the ground, India (Dr. H. Chaudhuri).

——**tabacina** (Kickx) Berk. (**57**, after No. 482 ; Sacc. **I**:324 ; **196**:129 ; **150**:82 ; **376**:119 ; **454**:418 ; **305**:3 ; **57**, after No. 359, as *Hypoxylon tabacinum* Kickx ; **454**:418 as *X. involuta* Kl.). On dead wood, Darjeeling, 7,500 ft., and Kali Pani, Khasi Hills (Hooker f.) ; on old fallen trees, Sikkim, 7-8,000 ft. (Kurz) ; on old wood, Dehra Dun (Butler). According to Bresadola (Ann. Myc., V:241, 1907), *X. gigantea*, *X. wrightii*, *X. involuta*, *X. portoricensis*, and *X. gomphus* are all forms of this species.

—— **timorensis** Lloyd (**287**:1056, figs., 1125). On bamboo, India (Gollan); India (D. Maruda Rajan).

——**trichopoda** Penzig & Sacc. (Sacc. **XVII**:633 ; **454**:420). On fallen twigs chiefly of Rosaceae, Dehra Dun (Butler) ; perhaps this species also on seeds of *Terminalia arjuna*, Dehra Dun (Parker).

——**tricolor** Fr. (Sacc. **I**:342 ; **400** :134). On trunks, near Darjeeling (Remy).

——**tuberosa** (Pers.) Cke (Sacc. **IX**:537 ; **144**:96 as *X. scopiformis* Mont. ; **466**:54 ; **454**:420). Belgaum (Hobson) ; on rotten wood, Barguai, Mysore (Butler).

——**vagans** Petch (Sacc. **XXIV**:1101 ; **287**:1069). India (Bose).

BASIDIOMYCETES.

USTILAGINALES.

Cintractia axicola (Berk.) Cornu (Sacc. **VII**:480 ; **452**:427 ; **455**:253). In the peduncles of *Fimbristylis* spp., Yelwal, Mysore (Butler) ; Godavari (Mitra) ; Samalkota (Shaw) ; Nagpur (Pandit) ; Wynaad (McRae) ; Amritsar (Hafiz Khan) ; Bassein, Burma (Butler) ; of *F. complanata*, Chatrapur, Ganjam (Butler). At least three different types of spores occur in the above collections, and the species is perhaps a composite one,

[Cintractia] cryptica Cke & Massee (**171**:34 ; Sacc. **IX**:285). In ovaries of *Pollinia argentea*, Munepore (C. B. Clarke). Ciferri (**121**:33) transfers this species doubtfully to *Ustilago* as *U.* (*?*) *cryptica* (Cke & Massee) Ciferri.

——**peribebuyensis** Speg. (Sacc. **VII**:458 ; **452**:427). In the peduncles of *Cyperus* sp., Bilikere, Mysore (Butler) ; Mozufferpore (Butler) ; Cocanada (Sundararaman) ; Chittagong (Sen) ; Samalkota (Shaw) ; Godavari (Subramaniam) ; Nagpur (Pandit) ; Bombay (Saxton).

——**pulverulenta** Cke & Massee (**171**:34 ; Sacc. **IX**:285). In ovaries of *Erianthus* sp., Nungklo, Khasi Hills (C. B. Clarke) ; of *Saccharum arundinaceum*, Coimbatore, Madras. Ciferri (**121**:33) transfers this species doubtfully to *Ustilago* as U. (*?*) *pulverulenta* (Cke & Massee) Ciferri.

Doassansia alismatis (Nees) Cornu (Sacc. **VII**:503 ; **455**:255). In the leaves of *Alisma plantago*, Achibal, Kashmir (Butler).

——**martianoffiana** (Thuem.) Schroet. (Sacc. **VII**:504 ; **455**:255). In the leaves of *Potamogeton* sp., Wular Lake, Kashmir (Butler).

——**nymphaeae** Syd. (**443**, **VIII**:406 ; Sacc. **XXIII**:630). In the petioles of *Nymphaea stellata*, Bassein, Bombay (Chibber).

Entyloma eugeniarum Cke & Massee (Sacc. **XI**:233). On *Eugenia ?tetragona*, Maymyo (Butler).

——**fumariae** Schroet. (Sacc. **VII**:494). On *Fumaria parviflora*, Pusa (Butler).

——**nymphaeae** (Cunningham) Setch. (**407**:189 ; **188**:32, figs., as *Rhamphospora nymphaeae* Cunn. ; Sacc. **IX**:287). In the leaves of *Nymphaea stellata* and *N. lotus* (*N. rubra*), Royal Botanic Garden, Calcutta (Cunningham) ; on *N.* sp., Chaumuhani and Begumganj, Noakhali District (Butler).

——**obesum** Syd. (**443**, **VI**:145 ; Sacc. **XXIII**:625). In the leaves of *Andropogon annulatus*, Nagpur (Pandit) ; Dehra Dun, Hoshangabad, and Dohad Farm in Bombay (Butler). Zundel (**526**:157) transferred this to *Tolyposporella.*

——**oryzae** Syd. (Sacc. **XXIII**:625 ; **105**:35 as *Entyloma* sp.). In the leaves of *Oryza sativa*, Lower Burma and Pusa (Butler) ; Bassein, Burma (Inayat) ; Alibag, Poona (R. K. Bhide).

——**physalidis** (Kalchbr. & Cke) Wint. (Sacc. **VII**:494 ; **452**:427). In the leaves of *Physalis minima* var. *indica*, Dehra Dun (Butler).

——**ranunculi** (Bonord.) Schroet. (Sacc. **VII**:488). In the leaves of *Ranunculus* sp., Ranikhet, Kumaon (Butler).

——**speciosum** Schroet. & P. Henn. (Sacc. **XIV**:424). In the leaves of *Panicum* sp., Bangalore (Butler).

Farysia butleri Syd. (**447**:42 ; Sacc. **XXIII**:631 ; **452**:424 as *Ustilago butleri* Syd.). In the inflorescence of *Scleria elata*, Thurya, Cherrapunji, and Syndai, Assam (Butler). The systematic position of the genus *Farysia* is uncertain (see Fischer, **233**:196).

—— **emodensis** (Berk.) Syd. (**447**:42 ; Sacc. **XXIII**:631 ; **57**, No. 354, as *Ustilago emodensis* Berk. ; **445**:486 ; **369**:223). Destroying the inflorescence of *Polygonum* sp., Tonglo, Sikkim and Nangki, E. Nepal, 10,000 ft. (Hooker f.); on *P. chinense*, Coimbatore (C. E. C. Fischer).

Graphiola applanata Syd. & Butler (**452**:428; Sacc. **XXI**:526). In living leaves of *Phoenix sylvestris*, Dehra Gopipur, Kangra, Punjab Himalaya (Burkill); Surat (Inayat). Notes are given on this and other species of *Graphiola* by Fischer (**233**:188; **234**:228).

——**borassi** Syd. & Butler (**453**:489, figs.; Sacc. **XXI**:527). On living leaves of *Borassus flabellifer*, Pusa; Godagiri, Bengal; Cocanada, Madras (Butler); Arrah (Battachariya); Suri, Birbhum (Basu); Bombay (Burkill); Bezwada.

——**phoenicis** (Moug.) Poiteau (Sacc. **VII**:522; **133**:15; **144**:96; **196**:130; **452**:428; **194**:225). In living leaves of *Phoenix paludosa*, Calcutta (Kurz); of *P. sylvestris*, Arnigadh, Mussoorie (Butler), and in very many stations in the plains; of *P. dactylifera*, Saharanpur (Gollan); Deolali (Butler); Muzaffargarh, and Hazaribagh; of *P. acaulis*, Kumaon Himalaya (Inayat).

Melanopsichium austro-americanum (Speg.) Beck (Sacc. **XVII**:484; **453**:486, figs.; **111**:21, figs.). In all parts of *Polygonum glabrum*, Pusa (Butler); Darjeeling (McRae); Lahore (G. S. Cheema).

Mykosyrinx arabica (P. Henn.) Penzig (Sacc. **XVII**:484; **445**:487). In peduncles of *Vitis quadrangularis*, Mattipalayam, Coimbatore (McRae).

Neovossia barclayana Bref. (**82**:170, figs.; Sacc. **XVI**:375, and **XIV**:422 as *Tilletia barclayana* (Bref.) Sacc. & Syd.). In ovaries of *Pennisetum orientale* (*P. triflorum*), Simla (Barclay). Sydow (**451**:422) has recently published notes on this fungus.

Sorosporium contortum Griff. (Sacc. **XVII**:485; **526**:154). On *Andropogon contortus*, Makla, Amraoti District, Central Provinces (Burkill).

——**flagellatum** Syd. & Butler (**453**:489; Sacc. **XXI**:512). In the unopened spikes of *Ischaemum timorense* (*Spodiopogon byronis*), Wynaad (Butler); Kurseong, near Darjeeling (McRae); Koppa, Mysore (Butler).

——**furcatum** Syd. & Butler (**455**:254; Sacc. **XXIII**:619). In ovaries of *Ischaemum aristatum*, Lendru, Nagpur (Pandit); Insein, Burma (Butler).

——**geminellum** Syd. & Butler (**455**:253; Sacc. **XXIII**:618; **526**:152). In the inflorescence of *Andropogon* sp., Maoryngkneng, Khasi Hills (Burkill).

——**paspali** McAlpine (Sacc. **XXI**:513; **455**:253; **111**:240, fig.; **113** (13):37; **113**(19); **113** (21):63). In inflorescences of *Paspalum scrobiculatum*, Chamta Ghat, Monghyr (Butler's collector); Pusa (Mitra; Butler).

——**pseudanthistiriae** Syd. & Butler (**455**:254; Sacc. **XXIII**:619). In the spikes of *Pseudanthistiria hispida*, Bassein, Bombay (Bhide).

——**reilianum** (Kuehn) McAlpine (Sacc. **VII**:471 as *Ustilago reiliana* Kuehn; **231**:219; **29**:275; **452**:425; **455**:249; **111**:199, 214, figs.; **279**, figs.; **363**; **526**: 151; **132**:115, fig., as *Ustilago pulveracea* Cke). On male florets of *Zea mays*, Lahore (Stewart); on male and female inflorescences of *Zea mays*,

Dohad Farm, Bombay (Gupta); Dubgaon, Kashmir (Butler); in inflorescence of *Andropogon sorghum*, Bombay (Chibber); United Provinces (Hayman); Coimbatore (Butler); Nagpur (Pandit); Jhelum, Punjab; in inflorescence of *A. halepensis*, Srinagar, Kashmir (Butler); Kolkhar,Central Provinces (Burkill).

[**Sorosporium**] **wildemanianum** P. Henn. (Sacc. **XXI**:513; **455**:253; **526**:146). In the inflorescence of *Andropogon* (*Cymbopogon*) *martini*, Kohana, Amraoti District, Central Provinces (Burkill).

Sphacelotheca cruenta (Kuehn) Potter (Sacc. **VII**:455 as *Ustilago cruenta* Kuehn; **279**, figs.; **363**; **111**:212, fig.; **4**; **526**:131). In the ovaries and rachis of *Andropogon halepensis*, Botanic Gardens, Saharanpur (Hafiz Khan); Asirgarh, Central Provinces (Burkill); of *A. sorghum*, Poona, Sholapur, Dharwar, and Surat (Kulkarni); Chindwara and Seoni Districts, Central Provinces (Evans); Bombay (Burns).

——**fagopyri** Syd. & Butler (**453**:486; Sacc. **XXI**:508). In ovaries of *Fagopyrum esculentum*, Kulu, Himalaya (Watt).

——**hydropiperis** (Schum.) de Bary (Sacc. **VII**:499; **455**:253). In ovaries of *Polygonum serrulatum*, Martand, Kashmir (Butler); of *P. glabrum*, Pusa (Butler); of *P. persicaria*, Kangra District, Punjab (Mitter); of *P. posumbu*, Dharmkot (Mitter); of *P.* sp., Shillong (Subramaniam); Harwan, Kashmir (Butler).

——**sorghi** (Link) Clinton (Sacc. **VII**:456 as *Ustilago sorghi* (Link) Pass.; **452**:425, 427; **281**; **111**:209, fig.; **29**:275; **1**:8; **2**:16-18; **392**:2; **393**:2; **526**: 132; **279**; **363**; **82**:120 as *U. tulasnei* Kuehn). In ovaries of *Andropogon sorghum* (*Sorghum vulgare*) throughout India and Burma; of *A. halepensis*, Botanic Garden, Saharanpur (Inayat).

Stylina disticha (Ehrenb.) Syd. (**233**:192, figs.; **237**, **II**:434 as *Sphaeria disticha* Ehrenb.; Sacc. **VII**:523 as *Graphiola ?*). Reported in the leaves of *Dracaena draco* from India. Fischer (**233**) states that the host is not *Dracaena*, but a palm. Subsequent collections in China were on *Livistona chinensis*. Petch (**366**) suggests that this fungus may be near the genus *Endocalyx*, but Fischer considers that it is near *Graphiola*.

Tilletia caries (DC.) Tul. (Sacc. **VII**:481 as *T. tritici* (Bjerk.) Wint.; **455**:254; **111**:167, fig.). In ovaries of *Triticum vulgare*, Sopor, Kashmir (Butler); Gilgit (Carrison); Simla and elsewhere in north-western India.

——**foetens** (Berk. & Curt.) Trel. (Sacc. **VII**:485 as *T. laevis* Kuehn; **455**:254; **111**: 166). In ovaries of *Triticum vulgare*, Sopor, Kashmir (Butler); Gilgit (Carrison); and elsewhere in north-western India.

——**horrida** Takahashi (Sacc. **XIV**:422; **105**, figs.; **111**:227, fig.; **393**:1; **395**). In the grains of *Oryza sativa* throughout India and Burma (Butler, Rhind, et al.).

[**Tilletia**] **tumefaciens** Syd. (**455**:255, fig.; Sacc. **XXIII**:621). In the leaves and culms of *Panicum antidotale*, Lyallpur (Milne).

Tolyposporium ehrenbergii (Kuehn) Pat. (Sacc. **XXI**:516 as *T. filiferum* W. Busse; **29**:275; **452**:427; **111**:215, figs.; **279**, figs.). In ovaries of *Andropogon sorghum*, Kistna, Madras (Barber); Sind (Kulkarni); Talaganj, Jhelum District, Punjab; Coimbatore and Koilpatti, Madras. Mason (Trans. Brit. Myc. Soc., XI, pp. 284-285, 1926) has reported on the nomenclature of this smut.

——**penicillariae** Brefeld (**82**:154, figs.; Sacc. **XIV**:426; **453**:489; **113** (14):46; **113** (15):54; **111**:225, fig.). In ovaries of *Pennisetum typhoideum*, Simla (Barclay); Poona and Pusa (Butler); Nadiad in Bombay, and Coimbatore (Subramaniam); Sialkot, Punjab (Cheema); Baroda. The form on cultivated bulrush millet in the plains and Deccan does not agree very closely with Brefeld's description of the form from Simla, and in particular the size of the spores is recorded by Brefeld as being larger. The original specimens collected by Barclay have not been traced.

Urocystis coralloides Rostrup (Sacc. **VII**:521; **5**:5; **113** (13):38; **113** (14):48; **113** (18); **343**). On roots of *Brassica campestris* var. *sarson*, Pusa (Shaw; Mitra).

——**tritici** Koern. (Sacc. **XXI**:526; **452**:427; **111**:172, figs.). In the leaves of *Triticum vulgare*, Lyallpur (Butler). The form on wheat has been identified by many authors with *U. occulta* (Wallr.) Rabenh., but one will not infect the host plant of the other, and McAlpine (**321**:63, 199) considered it better to keep them distinct.

Ustilago amadelpha Syd. & Butler (**455**:249; Sacc. **XXIII**:608; **526**:127). In peduncles and apices of culms of *Andropogon* sp., Mozufferpore District (Butler's collector).

——**andropogonis-annulati** Bref. (**82**:109, figs.; Sacc. **XIV**:419; **452**:425). In ovaries of *Andropogon annulatus*, Calcutta (Cunningham); Dehra Dun, Chatrapur, and Poona (Butler); Bassein, Bombay (Chibber); Samalkota, Madras (Barber); Punjab (Main); Nagpur. Zundel (**526**:132) has transferred this to *Sphacelotheca*.

——**andropogonis-tuberculati** Bref. (**82**:109, figs.; Sacc. **XIV**:419). In the ovary of *Andropogon tuberculatus*, Simla (Barclay). There appears to be no record of this grass as a Himalayan species, the Calcutta herbarium having only two sheets, both from the Central Provinces. The species most likely to be confused with *A. tuberculatus* Hack. is *A. tristis*.

——**aristidae-cyananthae** Bref. (**82**:102, figs.; Sacc. **XIV**:415; **452**:426). In fruits of *Aristida cyanantha*, India (Cunningham); in inflorescence of *A.* sp., Siwalik Hills near Dehra Dun (Hole). Magnus (**309**:434) states that this fungus or a near ally received through Massee from Gamble in Dehra Dun as *U. aristidae* Peck is really a *Sorosporium*.

[Ustilago] arundinellae Bref. (**82**:108, figs.; Sacc. **XIV**:415; **453**:486). In caryopses of *Arundinella* sp., Calcutta (Cunningham); in inflorescence of *A. setosa*, Burma Gori, Kumaon (Butler).

——**avenae** (Pers.) Jensen (Sacc. **IX**:283; **452**:424; **393**:5; **111**:179, figs.). In ovaries of *Avena sativa*, Dehra Dun and Pusa (Butler); Yawnghwe, Burma (Rhind); Lahore (Mitra & Das); Jagannathpur (Cave); Delhi (Burkill); Sind; Arrah; and elsewhere in India.

——**bengalensis** Syd. & Butler (**455**:250; Sacc. **XXIII**:609). In inflorescences of *Cymbopogon pendulus*, Banarhat, Duars, Bengal (Burkill).

——**burkilli** Syd. & Butler (**455**:248; Sacc. **XXIII**:607). In ovaries of *Aneilema nudiflorum*, Gauripur, Mymensingh (Burkill).

——**burmanica** Syd. & Butler (**455**:250; Sacc. **XXIII**:611). In inflorescences of *Ischaemum* sp., Kya-in, near Moulmein, Burma (Butler); of *I. spathiflorum*, Bombay (Kulkarni).

——**bursa** Berk. (**57**, No. 466; Sacc. **VII**:473; **455**:250, fig.). In ovaries of *Anthistiria arundinacea*, Sikkim (Hooker f.); Nimar District, Central Provinces (Burkill); of *A.* sp., Hoshangabad and Surat (Butler); between Labada and Ghatang, Amraoti District, Central Provinces (Burkill); Bassein, Bombay (Bhide). McAlpine (**321**:186) united this species with *Tolyposporium anthistiriae* Cobb as *T. bursa* (Berk.) McAlpine, but the Australian species is clearly different. *U. anthistiriae* Petch from Ceylon has smaller spores.

——**coicis** Bref. (**82**:110; Sacc. **XIV**:412; **111**:243, fig.). In ovaries of *Coix lachryma-jobi*, Simla (Barclay); Bombay (Kulkarni; Ajrekar).

——**consimilis** Syd. (**450**:281; **526**:127; **455**:249 as *U. sacchari* Rabenh.). In culms of *Saccharum fuscum*, Sibsagar, Assam (Basu). Eaten by cowherds in Assam. There is an allied fungus on *S. spontaneum* at Pusa (coll. Butler) which was referred (**455** :249) to "*U. sacchari*", but seems to be intermediate between *U. scitaminea* and the present species. *U. consimilis* is confused by Zundel (**526**), who uses this name instead of *U. scitaminea* for the common smut of sugar-cane.

——**cornuti** Syd. & Butler (**452**:426; Sacc. **XXI**:561). In the unopened spikes, rarely in the flowers, of *Ophiurus corymbosus*, Surat (Butler).

——**crameri** Koernicke (Sacc. **VII**:455; **82**:122; **455**:249; **111**:234, figs.; **425**, figs.; **1**:8). In ovaries of *Setaria italica*, Himalaya (Barclay); Manjri, near Poona (Chibber); throughout Madras (Sundararaman).

——**cynodontis** P. Henn. (Sacc. **XIV**:416; **82**:105, figs.; **452**:427; **336**, figs.). In inflorescences of *Cynodon dactylon*, Simla (Barclay); Calcutta, Dehra Dun, Pusa, Mysore, Kashmir, and Bombay (Butler); Peshawar (Shaw); Lucknow, Nepal, and Semaria Ghat, Monghyr (Burkill); Madras (Subramaniam); Agra (Mitra).

——**digitariae** (Kze) Rabenh. (Sacc. **VII**:454; **452**:426). In the spikes of *Panicum repens*, Yelwal, Mysore (Butler).

[Ustilago] duthiei Ricker (**398**:111 ; Sacc. **XXI**:505). In inflorescences of *Andropogon bladhii*, Dehra Dun (Duthie). Zundel (**526**:134) has transferred this to *Sphacelotheca*.

——**effusa** Syd. (**452**:425 ; Sacc. **XXI**:506 ; **526**:127). In leaf-sheaths, leaves, and culms of *Andropogon muricatus*, Kanaighat, Sylhet and of *Arundinella wallichii*, Wahjain, Assam (Butler).

——**egenula** Syd. & Butler (**455**:251, figs. ; Sacc. **XXIII**:609). In ovaries of *Eragrostis nutans*, near Pusa (Butler).

——**eleusinis** Kulkarni (**280**:186, figs. ; **333**). In ovaries of *Eleusine coracana*, Malkapur and various other parts of Bombay (Kulkarni). McRae (**333**) states that field experiments indicate that this smut is not seed-borne.

——**endotricha** Berk. (Sacc. **VII**:467 ; **57**, after No. 467). In ovaries of *Carex* "*baccata*" [? *baccans*] or some allied species, Khasi Hills (Hooker f. & Thomson) ; Tambur River, E. Nepal (Hooker f.). Berkeley notes that the spores are considerably smaller than those of the Ceylon and New Zealand specimens.

——**erythraeensis** Syd. (Sacc. **XXIII**:611 ; **455**:251). In inflorescences of *Manisuris granularis*, Pusa and Dharwar (Butler) ; Nepal and Amraoti District (Burkill).

——**goeppertiana** Schroet. (Sacc. **VII**:478). On *Rumex* sp., Mussoorie (Kar). Causing hypertrophy of mid-rib and petiole ; spores 12-16·5μ.

——**hordei** (Pers.) Kellerm. & Swingle (Sacc. **IX**:283 ; **452**:424 ; **111**:184, figs.). In ovaries of *Hordeum vulgare*, not uncommon in northern India.

——**inayati** Syd. & Butler (**453**:486 ; Sacc. **XXI**:500). In ovaries of *Iseilema laxa*, Orai, United Provinces (Inayat) ; Hoshangabad, Central Provinces (Butler) ; Nagpur (Pandit).

——**indica** Syd. & Butler (**455**:250 ; Sacc. **XXIII**:610). In culms and panicles of *Ischaemum angustifolium* (=*Pollinia eriopoda*), Pathankot, Punjab (Mitter).

——**iseilematis** Syd. & Butler (**452**:426 ; Sacc. **XXI**:500). In ovaries of *Iseilema laxa*, Samalkota, Madras (Barber) ; Nagpur (Butler) ; Godavari (Subramaniam).

——**microchloae** Syd. & Butler (**452**:427 ; Sacc. **XXI**:500). In inflorescences of *Microchloa setacea*, Bilikere, Mysore (Butler ; Barber).

——**nardi** Syd. (**452**:425 ; Sacc. **XXI**:505). In ovaries of *Andropogon nardus*, Wynaad (Butler). Zundel (**526**:137) transferred this to *Sphacelotheca*.

——**nuda** (Jensen) Kellerm. & Swingle (Sacc. **IX**:283). In inflorescences of *Hordeum vulgare*, Pusa (Butler).

——**ocrearum** Berk. (**57**, No. 468 ; Sacc. **VII**:458). In the ocreae of *Polygonum* sp., Nangki, East Nepal, 10,000 ft. (Hooker f.).

——**olivacea** Tul. (Sacc. **VII**:463 ; **319, I**:115). In ovaries of *Carex condensata*, North Western Himalaya (Gamble).

[**Ustilago**] **operta** Syd. & Butler (**452**:426 ; Sacc. **XXI**:502 ; **455**:249). In ovaries of *Panicum villosum*, Ootacamund (Barber) and of *P. prostratum*, Tukvar, Darjeeling (Hafiz Khan).

——**panici-frumentacei** Bref. (**82**:103, figs. ; Sacc. **XIV**:414 ; **452**:426 ; **111**:239, fig.). In ovaries of *Panicum frumentaceum*, Himalaya (Barclay) ; Nambur, Kistna District (Barber) ; Pusa (Butler).

——**panici-glauci** (Wallr.) Wint. (Sacc. **VII**:472 as *U. neglecta* Niessl ; **455**:249). In ovaries of *Setaria glauca*, Bellary (Barber) ; Amraoti District (Burkill).

——**panici-miliacei** (Pers.) Wint. (Sacc. **VII**:454 ; **455**:249 ; **111**:236, fig.). In inflorescences of *Panicum miliaceum*, Larkipur, Kashmir (Butler). This fungus was transferred to *Sorosporium* by Takahashi.

——**paradoxa** Syd. & Butler (**443, VI**:144, figs. ; Sacc. **XXIII**:611; **111**:239, figs. ; **282**). In ovaries of *Panicum frumentaceum*, Pusa (Butler) ; Sind (Kulkarni).

——**rabenhorstiana** Kuehn (Sacc. **VII**:471 ; **455**:249). In inflorescences of *Paspalum* (*Panicum*) *sanguinale*, Tukvar, Darjeeling (Hafiz Khan).

——**rottboelliae** Syd. & Butler (**453**:486 ; Sacc. **XXI**:497). In spikes of *Rottboellia compressa*, Pusa (Butler).

——**royleana** Syd. & Butler (**452**:426 ; Sacc. **XXI**:499). In spikes of *Paspalum royleanum*, Dehra Dun (Butler).

——**sacchari-ciliaris** Bref. (**82**:109, figs.; Sacc. **XIV**:418). In ovaries of *Saccharum ciliare* (*S. sara*), Calcutta (Cunningham) ; in inflorescences of the same host, Mozufferpore and Cawnpore (Butler).

——**schoenanthi** Syd. & Butler (**452**:425 ; Sacc. **XXI**:505). In ovaries of *Andropogon schoenanthus*, Palamcotta, Tinnevelli (Barber). Zundel (**526**:136) has transferred this to *Sphacelotheca*.

——**scitaminea** Syd. (**450**:281 ; *U. sacchari* Auct., non Rabenhorst ; **452**:424 ; **93**, figs. ; **111**:378, figs. ; **10**, figs. ; **206**, figs. ; **2**:23 ; **392**:4 ; **363**). In culms of *Saccharum officinarum* throughout India and Burma. See also *U. consimilis*.

——**shiraiana** P. Henn. (Sacc. **XVI**:369). On *Bambusa* sp., Dehra Dun (Hafiz Khan) ; Poona (Chibber).

——**spermophora** Berk. & Curt. (Sacc. **VII**:466 ; **453**:486). In inflorescences of *Eragrostis rhachitricha*, Pusa (Butler).

——**superflua** Syd. (**455**:249 ; Sacc. **XXIII**:607). In inflorescences of *Andropogon foveolatus*, Samalkota, Madras (Shaw). Zundel (**526**:138) transferred this to *Sphacelotheca*.

——**tenuis** Syd. (**452**:425 ; Sacc. **XXI**:506). In inflorescences of *Andropogon pertusus*, India (Wight) ; Hunsur, Mysore (Butler). Zundel (**526**:137) transferred this to *Sphacelotheca*.

——**tonglinensis** Tracy & Earle (Sacc. **XIV**:420 ; **455**:250). In inflorescences of *Ischaemum* sp., Samalkota, Madras (Shaw) ; Chatrapur, Ganjam (Fischer) ; Hoshangabad, Central Provinces (Butler).

[**Ustilago**] **trichophora** (Link) Kunze (Sacc. **VII**:462 ; **452**:426). In ovaries of *Panicum colonum*, Poona and Mysore (Butler) ; Nagpur (Pandit).

——**tritici** (Pers.) Jensen (Sacc. **IX**:283 ; **452**:424 ; **111**:164, figs. ; **467**:154 ; **2**:19 ; **392**:3 ; **363** ; **319** ; **I**:115 as *U. segetum* Wint.). In ovaries of *Triticum vulgare* throughout India.

—— —form **foliicola** Syd. & Butler (**452**:424). In the leaves of *Triticum vulgare*, Cawnpore (Butler).

——**tuberculiformis** Syd. (Sacc. **XVII**:473 ; **455**:248). In the leaves of *Polygonum chinense*, Darjeeling (McRae).

——**utriculosa** (Nees) Tul. (Sacc. **VII**:476 ; **453**:485, figs. ; **455**:248). In ovaries of *Polygonum tomentosum*, Pusa (Butler) ; Rangoon ; of *P. glabrum*, Pusa (Inayat). This fungus agrees fully with *U. koordersiana* Bref. (Sacc. **XI**:411) from Java, in the method of germination and characters of promycelia.

——**vittata** Berk. (**57**, No. 467 ; Sacc. **VII**:459). In ovaries of some Oplismenoid grass, near the summit of Parasnath, 4,000 ft., Bihar (Hooker f.) ; in ovaries of *Oplismenus compositus*, Vizagapatam (Gamble).

——**warneckeana** P. Henn. (Sacc. **XXI**:506 ; **445**:486). In inflorescences of *Andropogon contortus*, Taliparamba, North Malabar (McRae). Zundel (**526**:137) transferred this to *Sphacelotheca*.

——**zeae**(Beckm.) Unger (Sacc. **VII**:472 as *U. maydis* (DC.) Cda ; **455**:249 ; **111**:195, figs. ; **113** (21):65). In inflorescences, leaves, and culms of *Zea mays*, Harwan, Kashmir (Butler) ; Munsong, Darjeeling (McRae).

Uredinales.*

Aecidium adhatodae Syd. (**452**:440 ; Sacc. **XXI**:774 ; Syd. **IV**:93). On leaves of *Adhatoda vasica*, Dehra Dun (Butler).

——**aechmantherae** Syd. (**453**:504 ; Sacc. **XXI**:774 ; Syd. **IV**:93). On leaves of *Aechmanthera tomentosa*, Kumaon (Inayat) ; Mussoorie (Butler) ; of *A. gossypina* and *A.* sp., Kumaon (Inayat).

——**ajugae** Syd. (**453**:504 ; Sacc. **XXI**:772 ; Syd. **IV**:110). On leaves of *Ajuga* sp., Kumaon (Inayat).

——**amaryllidis** Syd. & Butler (**455**:274 ; Sacc. **XXIII**:867 ; Syd. **IV**:278). On leaves of *Amaryllis* sp., Dehra Dun (Butler) ; of *Crinum asiaticum*, Coimbatore (Fischer) ; of *C.* sp., Pusa (Dobbs) ; of *Pancratium* sp., Poona (Vaidaya).

——**argyreiae** Berk. & Broome (Sacc. **VII**:814 ; Syd. **IV**:129 ; **452**:441 ; **455**:274). On leaves of *Argyreia speciosa*, Sylhet (Butler) ; of *A. cymosa*, Chatrapur in Ganjam and Samalkota (Butler) ; of *A. argentea*, Wahjain, Assam (Butler) ; of *Lettsomia elliptica*, Mysore (Butler).

*The references to " Syd. I-IV " in this section of the list indicate Sydow's ' Monographia Uredinearum ' Vols. I-IV, indexed in the Bibliography under No. 442.

[Aecidium] blepharidis Har. & Pat. (Sacc. **XXI**:774 ; Syd. **IV**:94 ; **445**:488). On leaves of *Blepharis boerhaviaefolia*, Coimbatore (McRae).

——**brasiliense** Diet. (Sacc. **XIV**:383 ; Syd. **IV**:120). On leaves of *Cordia rothii*, Poona (Ajrekar).

——**breyniae** Syd. (**453**:506 ; Sacc. **XXI**:780 ; Syd. **IV**:185). On leaves of *Breynia rhamnoides*, Pusa (Butler) ; Samalkota (Shaw).

——**cassiae** Bres. (Sacc. **XI**:214 ; Syd. **IV**:212 ; **453**:505 as *A. torae* P. Henn). On leaves of *Cassia tora*, Mysore (Butler).

——**clerodendri** P. Henn. (Sacc. **XVI**:332 ; Syd. **IV**:116). On leaves of *Clerodendron* sp., India.

——**crataevae** Syd. (**452**:440 ; Sacc. **XXI**:755 ; Syd. **IV**:234). On leaves of *Crataeva religiosa*, Kanaighat, Sylhet (Butler).

——**crypticum** Kalchbr. & Cke (Sacc. **VII**:800 ; Syd. **IV**:41 ; **453**:504). On leaves of *Gerbera lanuginosa*, Mussoorie (Butler).

——**cunninghamianum** Barclay (**43**:224, fig. ; Sacc. **XI**:214 ; Syd. **IV**:224). On leaves of *Cotoneaster bacillaris*, near Simla (Barclay). Barclay (l. c.) thought that this might be a stage of *Gymnosporangium cunninghamianum*, but Sydow (l. c.) thinks it doubtful.

——**deutziae** Diet. (Sacc. **XIV**:376 ; Syd. **IV**:227 ; **453**:505). On leaves of *Deutzia staminea*, Kumaon (Inayat).

—— **dichrocephalae** P. Henn. (as *A.* "*dichrocephali*", the host being spelled "*Dichrocephalum*" ; Sacc. **XXI**:341 ; Syd. **IV**:35 ; **453**:504). On leaves of *Dichrocephala latifolia*, Khasi Hills (Butler).

——**esculentum** Barclay (**42**:161, figs. ; Sacc. **IX**:319 ; Syd. **IV**:209). In flowering twigs of *Acacia eburnea*, Poona (Wroughton). Prain (**42**) described the teratological effects on the fungus on the host.

——**flavescens** Barclay (**43**:226, fig. ; Sacc. **XI**:218 ; Syd. **IV**:60). On leaves of *Senecio rufinervis*, Mashobra, near Simla (Barclay).

——**flavidum** Berk. & Broome (Sacc. **VII**:795 ; Syd. **IV**:81 ; **262**:4 as *A. lasianthi* P. Henn.). Recorded on leaves of *Lasianthus* sp., North India (Warburg), but Sydow thinks that the host may be *Pavetta indica*, and the rust *A. flavidum*. (See note under *A. pavettae* below.)

——**girardiniae** Syd. (**452**:441 ; Sacc. **XXI**:782 ; Syd. **IV**:272). On leaves, petioles, and stems of *Girardinia heterophylla*, Mussoorie (Butler).

——**hedyotidis** Syd. (**445**:488 ; Sacc. **XXIII**:908 ; Syd. **IV**:77). On leaves of *Hedyotis nitida*, Chalisseri, South Malabar (McRae).

——**hemidesmi** Syd. (**455**:272 ; Sacc. **XXIII**:870 ; Syd. **IV**:133). On leaves of *Hemidesmus indicus*, Wahjain, Assam (Som).

——**infrequens** Barclay (**35, III**:105 ; Sacc. **IX**:321 ; Syd. **IV**:206 ; **455**:274). On leaves of *Geranium* (? *nepalense*), Simla (Barclay) ; of *G. nepalense*, Harwan, Kashmir (Butler).

[Aecidium], innatum Syd. & Butler (**455**:273 ; Sacc. **XXIII**:894 ; Syd. **IV**:191). On leaves of *Glochidion* sp., Matheran, Bombay (Ajrekar).

——**kaernbachii** P. Henn. (Sacc. **XVI**:343 ; Syd. **IV**:130 ; **452**:441 ; **455**:273). On leaves of *Ipomoea hederacea*, Kanaighat, Sylhet (Butler) ; of *I. biloba*, Godavari (S. N. Mitra) ; of *I. aquatica*, Sylhet (Butler) ; of *I.* sp., Dacca (Som).

——**leiocarpum** Syd. (**443, XV**:143 , Sacc. **XXIII**:898 ; Syd. **IV**:113).* On leaves of *Ocimum canum*, Koilpatti, Tinnevelli (McRae).

——**lepidagathis** Syd. (**453**:505 ; Sacc. **XXI**:773 ; Syd. **IV**:97). On leaves of *Lepidagathis hyalina*, Dehra Dun (Butler) ; Kumaon (Inayat) ; of *L.* sp., Nagpur (Pandit).

——?**leucospermum** DC. (Sacc. **VII**:669 ; **35, I**:361, figs. ; Syd. **I**:532 and **IV**:260). On leaves of *Anemone rivularis*, Simla (Barclay) ; of *A.* sp., Jaunsar (Gamble).

——**luculentum** Syd. (**453**:506 ; Sacc. **XXI**:781 ; Syd. **IV**:269). On leaves and rarely also on branches of *Loranthus longiflorus*, Yelwal, Mysore (Butler).

——**melaleucum** Syd. (**443, XV**:143 ; Sacc. **XXIII**:891 ; Syd. **IV**:148 ; **445**:327 as *A. bicolor* Sacc., which, however, differs microscopically). On leaves of *Maba buxifolia*, Tirupatti, Chittoor (McRae) ; of *M. buxifolia* var. *ebenus*, Tinakadu, Coimbatore (C. E. C. Fischer).

——**meliosmatis-myrianthi** P. Henn. & Shirai (Sacc. **XVI**:333 ; Syd. **IV**:178 ; **455**:273). On leaves of *Meliosma simplicifolia*, Vayitri, Wynaad (McRae). The species name of this rust was originally written " *meliosmae-myrianthae* ".

——**merenderae** Syd. (**452**:441 ; Sacc. **XXI**:784 ; Syd. **IV**:286). On leaves of *Merendera aitchisoni*, Salt Range, Punjab (Drummond).

——**microrhynchi** P. Henn. (**261**:154 ; Sacc. **XVI**:337 ; Syd. **IV**:48). On leaves of *Launaea pinnatifida* (*Microrhynchus sarmentosus*), near Patna (Schlagintweit).

——**montanum** Butler (**90**:(30), figs. ; Sacc. **XXI**:753 ; Syd. **IV**:248). On leaves of *Berberis lycium*, Mussoorie (Butler) ; Kangra (Burkill) ; on *B. coriaria* and *B. aristata*, Jaunsar (Hole) ; on *B.* sp., Verinag, Kashmir (Butler).

——**mori** Barclay (**43**:225; Sacc. **XI**:221 ; Syd. **IV**:275 ; **452**:441; **453**:507; **258** ;**392**: 4 ; *Caeoma mori* Barclay in **35, III**:97, fig.). On leaves of *Morus alba*, Simla (Barclay) ; Maymyo (Rhind) ; Bengal (Hector) ; Shillong ; on *M. indica*, Koppa, Mysore (Butler) ; on *M. serrata*, Maymyo (Schroff) ; on *M.* sp., Simla (Butler) ; Darjeeling.

Barclay (**43**:225) included the common rust of *Ficus* with this species, and was followed by Sydow & Butler (**452**:441), but the latter (**453**:507) subsequently pointed out that the species on *Ficus* differed from that on *Morus* and agreed with *Uredo fici* Cast. which Butler later transferred to *Kuehneola* as *K. fici* (Cast.) Butler. (See *Cerotelium fici*, below.)

——**myriacatidis** (Barclay) Syd. (Syd. **IV**:50 ; **35, I**:373, fig., as *A. compositarum* Mart. var. *myriacatidis* Barclay ; Sacc. **XXI**:765). On leaves of

Myriacatis nepalensis, Mashobra, near Simla (Barclay); Harwan, Kashmir (Butler).

[**Aecidium**] **nobile** Syd. (**452**:440; Sacc. **XXI**:770; Syd. **IV**:76). On leaves of *Coffea arabica*, Mysore (Butler).

——**ocimi** P. Henn. (Sacc. **XI**:218; Syd. **IV**:113; **445**:488). On leaves of *Ocimum canum*, Koilpatti, Madras (McRae).

——**orbiculare** Barclay (**43**:227, fig.; Sacc. **XI**:213; Syd. **IV**:256). On leaves, petioles, and stems of *Clematis grata*, Mattiain, near Simla (Barclay); of *C. orientalis* and *C. puberula*, India (Lace & Watt).

——**osmanthi** Syd. & Butler (**453**:505; Sacc. **XXI**:773; Syd. **IV**:142). On leaves of *Osmanthus fragrans*, Kumaon (Inayat).

——**paederiae** Diet. (Sacc. **XIV**:382; Syd. **IV**:81; **455**:273). On leaves of *Paederia foetida*, Chittagong (Sen). *Uredo paederiae* Syd. occurs on the same leaves and is perhaps genetically connected.

——**patulum** Syd. (**453**:505; Sacc. **XXI**:759; Syd. **IV**:225). On leaves, petioles, and fruits of *Crataegus coccinea*, Khasi Hills (Butler).

[——**pavettae** Berk. (Sacc. **VII**:795; Syd. **IV**:82; **467**:154; **455**:272). On leaves of *Pavetta* sp., Matheran, Bombay (Blatter); of *P. indica*, Belgaum (Ajrekar). Sydow (**455**:272) suggests that *A. flavidum* Berk. & Br. may be the same fungus. Petch (**367, IV**:166) also discusses this fungus, the type specimens of which seem to be lost.]

——**peristrophes** Syd. (**455**:272; Sacc. **XXIII**:867; Syd. **IV**:98). On leaves of *Peristrophe* sp., Kangra, Punjab (Mitter).

——**petchii** Sacc. & Trott. (Sacc. **XXI**:758; Syd. **IV**:200; **455**:273 as *A. paramignyae* Petch, non Racib.). On leaves of *Paramignya monophylla*, Wynaad (McRae).

——**phyllanthi** P. Henn. (Sacc. **XVI**:345; Syd.**IV**:192; **453**:505 as *A. phylanthinum* Syd.). On leaves of *Phyllanthus reticulatus*, Kanaighat, Sylhet (Butler).

——**plectranthi** Barclay (**35, III**:104; Sacc. **IX**:321; Syd. **IV**:113; **453**:504). On leaves of *Plectranthus coetsa*, Simla (Barclay); of *P. scrophularioides*, Mussoorie (Butler).

—— **polygoni-cuspidati** Diet. (Sacc. **XVII**:434; Syd. **IV**:267; **452**:441). On leaves of *Polygonum glabrum*, Kanaighat, Sylhet, and of *P. hydropiper*, Sylhet (Butler).

——**ponderosum** Syd. (**452**:440; Sacc. **XXI**:778; Syd. **IV**:139; **455**:273). On branches of *Vallaris heynii*, Dehra Dun (Butler); of *Cryptolepis buchanani* and of *Pergularia pallida*, Kumaon (Butler).

——**pygei** Syd. (**455**:272; Sacc. **XXIII**:907; Syd. **IV**:226). On leaves of *Pygeum* sp., Darjeeling (McRae).

——“ **ranunculacearum** DC.” (Syd. I; etc.). On *Ranunculus hirtellus*, Mundali (Gamble). The identity of this *Aecidium* needs confirmation.

[Aecidium] rhytismoideum Berk. (Sacc. **VII**:807 ; Syd. **IV**:145 ; **452**:441). On leaves of *Diospyros tomentosa*, Yelwal, Mysore (Butler) ; South India (Barber) ; Tinnevelli Hills (Ramaswami).

——**scutellariae** Syd. (**453**:504 ; Sacc. **XXI**:772 ; Syd. **IV**:114). On leaves of *Scutellaria angulosa* and *S. repens*, Kumaon (Inayat).

——**solani** Mont. (Syd. **IV**:109 ; **142**:61). On leaves of *Solanum* sp., Sutlej Valley (Gamble). Saccardo (Syll. **VII**:674) includes this *Aecidium* with *Puccinia physalidis* Peck, but Arthur (**20**:562) treats *P. physalidis* as a short cycle. rust.

——**spissum** Syd. (**455**:273 ; Sacc. **XXIII**:910 ; Syd. **IV**:202). On leaves of *Zanthoxylum* sp., Maymyo (Butler). Sydow (l.c.) remarks that the fungus suggests a *Caeoma*, but seems to be an *Aecidium* with feebly developed peridium.

——**stranvaesiae** Syd. (**453**:505 ; Sacc. **XXI**:758 ; Syd. **IV**:226). On leaves of *Stranvaesia glaucescens*, Kumaon (Inayat).

——**tweedianum** Speg. (Sacc. **VII**:817 ; Syd. **IV**:95 ; **455**:272). On leaves of *Dicliptera* sp., Mussoorie (Hafiz Khan) ; Kasauli (Butler).

——**urceolatum** Cke (**142**:61 ; Sacc. **VII**:775 ; Syd. **IV**:260). On leaves and stems of *Thalictrum* sp., Kanawar, 8000 ft. (Gamble). Sydow (l.c.) considers this near the aecidial stage of *Puccinia persistens* Plowr. Barclay (**35, I**:362, figs.) described as *A.* (?) *thalictri-flavi* DC. on *Thalictrum javanicum*, Simla, a species which is probably also *A. urceolatum*. He also mentions (**35, I**: 363, figs.) an *Aecidium* on *Thalictrum minus* found at Urni (about 126 miles from Simla on the Tibet road), which he thought might be quite possibly the same, although the effect on the host was more marked, and the spores somewhat smaller.

——**vangueriae** Cke (Sacc. **VII**:795 ; Syd. **IV**:89 ; **455**:273). On leaves of *Vangueria* sp., Pedang, Darjeeling (Burkill).

——**withaniae** Thuem. (Sacc. **VII**:811 ; Syd. **IV**:110 ; **455**:270 as *A. dietelianum* P. Henn.). On leaves of *Withania coagulans*, Nitr Sutlej Valley, Punjab (Watt).

Blastospora butleri Syd. (**455**:266, figs. ; Sacc. **XXIII**:665 ; Syd. **III**:164). On leaves of *Jasminum malabaricum*, Matheran, Bombay (Ajrekar). Ajrekar (**11**) states that this fungus is a *Uromyces*, but the examination of his material indicates that there are two rusts on *J. malabaricum* at Matheran, and that the form studied by him was not *B. butleri*, but possibly *Uromyces hobsoni* Vize.

——**hygrophilae** Syd. & Butler (**455**:265 ; Sacc. **XXIII**:665 ; Syd. **III**:164, fig.). On leaves of *Hygrophila salicifolia*, Chittagong (R. Sen).

Cerotelium alienum (Syd. & Butler) Arth. (**20**:698 ; Syd. **III**:322 as *Kuehneola aliena* Syd. & Butler ; Sacc. **XXIII**:789 ; **455**:267 as *Chrysomyxa aliena* Syd. & Butler ; **108**:79 ; **369**: 248 as *Uredo spondiadis* Petch ; **21**:510 as

Cerotelium spondiadis (Petch) Arth.). On leaves of *Spondias mangifera*, Chittagong (Sen).

[**Cerotelium**] **desmium** (Berk. &Broome) Arth. (**20**: 698 ; **111**:363, fig., as *Kuehneola desmium* (B. & Br.) Butler ; **21**: 510 as *Cerotelium gossypii* (Lagerh.) Arth. ; Syd. **III**:318, fig., as *Kuehneola gossypii* Arth. ; **453**:508 as *Uredo gossypii* Lagerh.; Sacc. **XI**:224). On leaves of *Gossypium* sp. cult., Pusa (Butler); Poona (Gammie); Dacca (Som); Assam (Taslim); Madras (Subramaniam).

——**fici** (Cast.) Arth. (**21**:509 ; **108**:76, fig., as *Kuehneola fici* (Cast.) Butler ; Sacc. **XXIII**:790 ; Syd. **III**:323; **453**:508 and **455**:271 as *Uredo fici* Cast.; **133**:14, figs., as *Trichobasis hobsoni* Vize ; Sacc. **IX**:334 as *Uredo hobsoni* (Vize) Sacc.). On leaves of *Ficus* sp., India (Hobson ; see Syd. **IV**:592, and *Marchalia ustulata* supra) ; of *F. glomerata*, Pusa (Butler); uredo only on leaves of *F. palmata*, *F. carica*, *F. religiosa*, and *F.* sp., in many localities in India and Burma. (See *Aecidium mori* above.)

——**lanneae** (v. Hoehn.) Arth.* (**21**:510 ; Sacc. **XXIIII**:789 as *Kuehneola butleri* Syd.; Syd. **III**:322 ; **108**:79 ; **455**:267 as *Chrysomyxa butleri* Syd.; **224**:57 as *Cerotelium butleri* (Syd.) Diet.). On leaves of *Odina wodier* (syn. *Lannea grandis*), Noakhali (Butler).

——**peregrina** (Syd. & Butler) Arth. (**21**:510 ; Sacc. **XXIII**:790 as *Kuehneola peregrina* (Syd. & Butler) Syd.; Syd. **III**:322 ; **455**:267 as *Chrysomyxa peregrina* Syd. & Butler ; **108**:79). On leaves of *Clerodendron* sp., Wahjain, Assam (Butler).

——**vitis** (Butler) Arth. (**21**:509 ; Sacc. **XXIII**:790 as *Kuehneola vitis* (Butl.) Syd.; Syd. **III**:321 ; **104**:158, fig., as *Chrysomyxa vitis* Butl.; **108**:79). On leaves of *Vitis latifolia*, Dacca (Som); Rangpur (Mitra); of *V. adnata*, Comilla (Inayat).

Chnoopsora butleri Diet. & Syd. (**220**:423, fig.; Sacc. **XXI**:600 ; Syd. **III**:398, fig.; **452**: 439). On leaves of *Adhatoda vasica*, Dehra Dun and Kumaon (Butler).

——**sancti-johannis** (Barclay) Diet. (**220**:421 ; Sacc. **XXI**: 600 ; Syd. **III**:397 ; **452**: 439 ; **35**, **III**:84 as *Melampsora sancti-johannis* Barclay). On leaves of *Hypericum cernuum*, Simla (Barclay) ; Mussoorie (Butler); Darjeeling (McRae) ; Jaunsar (Hole); Kumaon (Inayat); of *H. patulum*, Mussoorie (Butler; Gollan); of *H. elodeoides*, Assam (Butler); of *H.* sp., Kumaon (Inayat).

Chrysomyxa deformans (Diet). Jaczew. (**269** ; **211**:266, figs., as *Barclayella deformans* Diet.; Sacc. **IX**:316 ; Syd. **III**: 522 ; **89**:48 ; **90**, fig.; **30**, figs.; **35**, **III**:104 as *Aecidium thomsoni* Berk.; **137**:91, in part, and **140**:117, in part, as *Peridermium acicolum* Link). On leaves of *Picea morinda* (*Abies smithiana*), Himalayas near Simla (Barclay ; Gamble) ; Dalhousie, 7000-

* Arthur's genus *Cerotelium* is accepted by Sydow (see **458**:422) for *C. alienum*, *C. fici*, and *C. gossypii* above, but it is not certain whether the remaining species mentioned are best placed in this genus or retained in *Kuehneola*.

7500 ft. (Baden Powell); Sikkim (Hooker f.); Jaunsar (Oliver); reported from Kurram Valley by Collett; on cones of the same host, Jaunsar (MacIntosh). Dietel (**224**:44) agrees that this rust apparently is a *Chrysomyxa.* See *Peridermium thomsoni* Berk., with which this fungus was confused by early collectors.

[**Chrysomyxa**] **dietelii** Syd. (**453**:502 ; Sacc. **XXI**:716 ; Syd. **III**:511). Uredo and teleuto stages on leaves of *Rhododendron arboreum,* Kumaon (Inayat). Barclay (**45**) described a uredo on *Rhododendron lepidotum* collected by Lace in the Himalayas north of Simla, which perhaps belongs to this or the next species ; and also an *Aecidium on R. campanulatum.* Watt (**521**) also found an *Aecidium* on *R. lepidotum* which he reports Barclay considered to solve the relationship of these fungi. This suggests that it was identical with the aecidium on *R. campanulatum.* The occurrence of aecidia on *Rhododendron* in the Himalaya led Barclay to discuss the possibility of autoecism in *Chrysomyxa himalensis,* though the aecidial stages of *Chrysomyxa,* so far as known, all occur on *Picea.*

——**himalensis** Barclay (**39**:79, figs.; Sacc. **IX**:318 ; Syd. **III**:512 ; **45**, figs.; **90**, fig.; **452**:439 ; **453**:503). On leaves of *Rhododendron arboreum,* Simla (Barclay) ; Jaunsar (Hole); Kumaon (Inayat); Ranikhet and Mussoorie (Butler) ; of *R. campanulatum,* Jaunsar (Hole) ; of *R. hodgsoni,* Sikkim, alt. 11,500-12,000 ft. (Waddell). Teleuto stage only found, unless the uredo on *R. lepidotum* (see last species) should belong to this rust.

——**piceae** Barclay (**35, III**:94, figs.; Sacc. **IX**:318 ; Syd. **III**:520). On leaves of *Picea morinda* (*Abies smithiana*), Narkanda and Mashobra, near Simla (Barclay). It seems doubtful, from Barclay's description of spore germination, that this rust belongs to *Chrysomyxa.*

Coleosporium campanulae (Pers.) Lév. (Sacc. **VII**:753 ; Syd. **III**: 628 ; **35, III**: 90, figs.; **452**:439 ; **455**:270 ; **253**). On leaves of *Campanula colorata,* Simla (Barclay) ; Kurseong, near Darjeeling (McRae) ; Harwan, Kashmir, and Kasauli (Butler) ; Kumaon (Inayat); of *C. canescens,* Dehra Dun and Pusa (Butler); of *C. sylvatica,* Kumaon (Inayat); of *C. cana,* Bhim Tal, Kumaon (Inayat); of *C.* sp., Mussoorie (Kar); of *Cephalostigma schimperi,* Wynaad (McRae).

——**clematidis** Barclay (**35, III**:89, figs.; Sacc. **IX**:317 ; Syd. **III**:653 ; **453**:502 ; **57,** No. 464, as *Uredo clematidis* Berk.; **130**:75 ; **236**:272, figs.; **449**:56). On leaves of *Clematis nutans,* Parasnath, Bihar (Hooker f.) ; of *C.* sp., Himalayas north of Dehra Dun (Flemming); Wynaad (McRae); of *C. montana,* Simla (Barclay); Harwan, Kashmir (Butler) ; of *C. buchaniana,* Simla (Barclay); Kumaon (Inayat); of *C. grata,* Ranikhet, Kumaon (Butler); of *C. triloba,* Kumaon (Inayat).

——**datiscae** Tranz. (Sacc. **XXI**:722; Syd. **III**:646 ; **455**:271). On leaves of *Datisca cannabina,* Harwan, Kashmir (Butler); Garhwal (Gamble).

[**Coleosporium deeringiae** Patouill. (**361**:123 ; Sacc. **IX**:317). On leaves of *Bosea* (*Deeringia*) *amherstiana*, India. Sydow (**III**:656) points out that this is a *Cystopus*, probably *C. bliti.*]

——**inulae** (Kunze) Rabenh. (Sacc. **XVII**:461 ; Syd. **III**:609 ; **455**:271 ; **253**). Uredo stage on leaves of *Inula cappa*, Shillong (Butler) ; Almora, Kumaon (Inayat).

——**leptodermidis** (Barclay) Syd. (Syd. **III**:635 ; **455**:269 as *Chnoopsora leptodermidis* (Barcl.) Butler ; **35, III**: 86, figs., as *Melampsora leptodermidis* Barclay ; Sacc. **IX**:297). On leaves of *Leptodermis lanceolata*, Simla (Barclay) ; Mussoorie and Kumaon (Butler).

——**oldenlandiae** (Massee) Butler (**108**:79, figs.; Sacc. **XXIII**:860 ; Syd. **III**:636; **319, I**: 116 as *Uredo oldenlandiae* Massee ; **452**: 440 as *Aecidium oldenlandiae* (Massee) Syd.). On leaves of *Oldenlandia aspera*, Tehri Garhwal, Himalaya, 4,000 ft. (Gamble) ; Mysore and Pusa (Butler); Coimbatore (Fischer).

——**perillae** Syd. (Sacc. **XVI**:317 ; Syd. **III**:641 ; **453**:502). On leaves of *Perilla ocymoides*, Kumaon (Inayat).

——**plectranthi** Barclay (**35, III**:89, fig.; Sacc. **IX**:317 ; Syd. **III**:641 ; **453**:502). On leaves of *Plectranthus gerardianus*, Simla (Barclay) ; of *Ocimum* sp., Kanaighat, Assam (Butler).

Cronartium ribicola Fischer (Sacc. **VII**:598 ; Syd. **III**:567 ; **254**, figs.; **286**:272 as *Cronartium asclepiadeum* Fr. var. *ribis* Lév.). On leaves of *Ribes* sp., Himalaya (Jacquemont) ; of *R. rubrum*, Punjab (Hafiz Khan). The specimen referred to by Léveillé is represented at Paris by one leaf, bearing teleuto columns which resemble those of *C. ribicola*, but a *Peridermium* on the stem of *Pinus excelsa*, the only five-needled pine in the Himalaya, was found by Colley and Taylor to be different from the *Peridermium* stage of *C. ribicola* and was named by them *P. indicum* (q.v.). It is doubtful, therefore, if this fungus on *Ribes* is really *C. ribicola*.

Crossopsora zizyphi (Syd. & Butler) Syd. (**446**:243 ; Sacc. **XXIII**:854 ; **455**:268, figs., as *Cronartium zizyphi* Syd. & Butler ; Syd. **III**:579, fig.). On leaves of *Zizyphus rugosa*, Dacca (Som); Bombay (Ajrekar) ; of *Z. oenoplia*, Samalkota, Madras (Butler) ; Dacca (Som).

Cystopsora oleae Butler (**100**:448, figs.; Sacc. **XXI**:607 ; Syd. **III**:592, figs.; **8**, figs.). On leaves of *Olea dioica*, Khandala (Chibber) ; near Belgaum (Ajrekar) ; on *O.* sp., Wynaad (McRae) ; Glen Coorg, South India (Watt). Sydow (**IV**:321) states that the Ceylon fungus *Aecidium chionanthi* Berk. & Broome is the aecidial stage of *C. oleae*, the host plant being *Olea* instead of *Chionanthus* as reported. Petch (**367, VIII**:166) has collected it on *Olea polygama* in Ceylon. Ajrekar (**8**) gives a full description of this stage, and Petch also gives morphological details.

Diorchidium levigatum Syd. & Butler (**453**:500 ; Sacc. **XXI**:716). On leaves of *Oplismenus compositus*, Dehra Dun (Butler).

[Diorchidium] orientale Syd. & Butler (**453**:500; Sacc. **XXI**: 715; **455**:263). On leaves of *Panicum prostratum* and *P. ramosum,* Chittagong (Sen).

Gambleola cornuta Massee (**319, I**:115; Sacc. **XVI**:314; Syd. **III**: 584, fig.; **90,** fig.; **452**:440; **453**:503). On leaves of *Berberis nepalensis,* Chakrata, 7,000 ft. (Gamble); Nilgiris (Fischer); Mussoorie and Kumaon (Butler).

Gymnosporangium cunninghamianum Barclay (**38**:78, figs.; Sacc. **IX**:314; Syd. **III**:33; **90,** figs.; **452**:437; **35, I**:370 as *G.* (?) *clavariaeforme* (Jacq.); **35, III**:92). Aecidia on leaves of *Pyrus pashia,* Simla (Barclay); Dehra Dun and Mussoorie (Inayat); Dharampur (Burkill); Kumaon (Butler); of *P. variolosa,* Simla (Barclay); teleuto stage on twigs of *Cupressus torulosa,* Simla (Barclay); Jaunsar, 8,000 ft. (Kanji Lal); Mussoorie (Butler); Ranikhet.

Hamaspora longissima (Thuem.) Koern. (Sacc. **VII**:750; Syd. **III**:79; **314**:4, fig.; **452**:437). On leaves of *Rubus* sp., Wynaad (Butler); Syndai, Assam (Subramaniam).

Haplophragmium ponderosum Syd. & Butler (**455**:265, fig.; Sacc. **XXIII**:778; Syd.**III**:183; **326**:110; **111**:88, fig.). On twigs of *Acacia leucophloea,* Poona (Chibber); Salem and Coimbatore (McRae); Allahabad (Mitter).

Hemileia canthii Berk. & Broome (Sacc. **VII**:586; Syd. **III**:212; **452**:438). On leaves of *Plectronia parviflora,* Yelwal, Mysore (Butler); of *P. rheedii,* Belgaum (Ajrekar).

——**holarrhenae** Syd. (**443, XII**:196; Sacc. **XXIII**:792). On leaves of *Holarrhena antidysenterica,* Bombay (Ajrekar); Chittagong (R. Sen); Assam (Subramaniam).

——**indica** Massee (**320**:40, fig.; Sacc. **XXI**:598; Syd. **III**:219). On leaves of *Macropanax* sp., Belgaum (Hobson).

——**vastatrix** Berk. & Broome (Sacc. **VII**: 585; Syd. **III**:209; **89**:47; **135,** figs.; **452**:438; **111**:469, figs.; **487**:698; **392**:5; **393**:4; **132**:116; **320**:38). On leaves of *Coffea arabica* in many places in India and Burma; of *C. liberica,* Mysore (Butler); of *C. robusta,* Mysore (Anstead); of C. spp., Burma (Rhind); perhaps this species on *Gardenia gummiflora,* Sidapur, N. Kanara (Kulkarni), but see next entry.

——?**woodii** Kalchbr. & Cke (Sacc. **VII**:586; Syd. **III**:214). Thomas (**487**:699) records a rust on *Vangueria spinosa* and *Randia uliginosa,* Coorg and Mysore, which he considers morphologically identical with *H. vastatrix,* but he did not test it on coffee. *H. woodii* is recorded on species of *Vangueria,* and Pole Evans found this rust would not attack coffee (see Syd. **III**:215). *H. woodii* has been recorded on *Gardenia,* but Sydow (l.c.) refers the rust on this host to the similar *H. vastatrix.*

Masseeëlla capparidis (Hobson) Diet. (**213**:332, fig.; Sacc. **XIV**:292; Syd. **III**: **546,** fig.; **162**:89 as *Cronartium capparidis* Hobson; **315**:34, fig.). On leaves of *Capparis* sp., Belgaum (Hobson).

[**Masseeëlla**] **fluggeae** Syd. (Ann. Myc., **XXVI**:424, 1928). A specimen on *Fluggea microcarpa* was sent in from Padu, Burma to the Imperial Bureau of Mycology by Rhind.

Melampsora ciliata Barclay (**43**:223, fig.; Sacc. **XI**:183 ; Syd. **III**:345 ; **455**:269). On leaves of *Populus ciliata,* Simla (Barclay) ; Dubgaon, Kashmir (Butler).

[——**eucalypti** Rabenh. (**388**:150 ; Sacc. **VII**:595). On leaves of *Eucalyptus globulus,* Royal Botanic Garden, Calcutta (Kurz). Sydow (Monog. **III**:393) reports that no fungus is present.]

——**euphorbiae-gerardianae** W. Mueller (Sacc. **XXIII**:832 ; Syd. **III**:376). Recorded on *Euphorbia* from India. Some of the collections listed under *M. helioscopiae* may possibly belong here.

——**helioscopiae** (Pers.) Wint. (Sacc. **VII**:586 ; Syd. **III**:377 ; **452**:439 ; **455**:269). On leaves of *Euphorbia helioscopia,* Wazirabad, Punjab (Mitter) ; Gurdaspur (Burkill) ; Peshawar (Shaw) ; Kangra ; of *E. pilosa,* Mussoorie and Kumaon (Butler) ; of *E. dracunculoides,* Pusa (Butler) ; Sabour (Dutt) ; Lyallpur ; of *E. thyrsoidea,* Harwan, Kashmir (Butler) ; of *E. rothiana,* Bombay (Ajrekar) ; of *E.* sp., Lahore (B. Das).

——**hypericorum** (DC.) Schroet. (Sacc. **VII**:591 ; Syd. **III**:384 ; **319: I**:116 ; **142**: 61 as *Uredo hypericorum* DC.). On leaves of *Hypericum* sp. (uredo only), Simla (Gamble) ; of *H. cernuum,* Jaunsar, 7,000 ft. (Gamble).

——? **larici-caprearum** Kleb. (Syd. **III**:353 ; **319, I**:116 as *M. epitea* (Kze & Schm.) Thuem.; Sacc. **VII**:588 ; **453**:502 ; **142**:61 as *Lecythea epitea* Lév.; **453**:502 as *M.?farinosa* (Pers.) Schroet.; **140**:117 as *L. salicina* Lév.; **35, III**:88, figs., as *M. salicis-capreae* Wint.). On leaves of *Salix daphnoides* Simla (Gamble) ; of *S. elegans,* Deoban, 9,000 ft. (Gamble) ; Dalhousie (Baden Powell) ; Simla (Gamble); of *S. tetrasperma,* Ranikhet and Mussoorie (Butler) ; Bhowali and Bageshwar, Kumaon (Inayat) ; Kohat (Fletcher) ; of *S.* sp., Simla (Barclay) ; Kashmir (Butler). There are probably two or more species of *Melampsora* on *Salix* included here, and perhaps none of them is *M. larici-caprearum* : but until the aecidial stages are worked out, they cannot be named definitely. Most of the recorded determinations have had to be based on the uredo stage only.

——**lini** (Pers.) Lév. (Sacc. **VII**:588 ; Syd. **III**:381 ; **41** ; **521**:17 ; **452**:439 ; **89**:48 ; **326**:110 ; **111**:325, fig.; **363**). On leaves and stems of *Linum usitatissimum,* Pusa (Butler) ; Coimbatore (McRae) ; Saharanpur (Inayat) ; Noakhali, Bengal (Som) ; Khwang, Assam (Watt) ; Kashmir (Butler) ; Jubbalpur.

——**ricini** (Biv.) Pass. (Syd. **III**:391 ; Sacc. **VII**:596 as *Melampsorella ricini* (Biv.) de Toni ; **89**:48 ; **111**:330, figs.; **9**:1092 ; **452**:439 ; **133**:15 ; **326**:110). On leaves of *Ricinus communis,* Poona (Butler) ; Deccan districts, Bombay (Ajrekar) ; Coimbatore (McRae ; Subramaniam) ; Nagpur (Pandit). Only the uredo stage of this fungus is known ; it may perhaps be a *Melampsorella.*

?[**Melampsora**] **rostrupii** G. Wagner (Syd. **III**:343 ; **43**:223 as *M. ?aecidioides* (DC.) Schroet. ; Sacc. **VII**:590). On leaves of *Populus alba*, near Simla (Lace). Barclay noted that no paraphyses occurred in the uredo stage, the only stage found. This rust must remain uncertain until the aecidial stage is discovered in India.

——**salicis-albae** Kleb. (Syd. **III**:372 ; Sacc. **XVII**:266 as *M. alii-salicis-albae* Kleb. ; **452**:438). On leaves of *Salix alba*, Suket, Himalaya (Burkill).

——**yoshinagai** P. Henn. (Sacc. **XVII**:264 ; Syd. **III**:391 ; **455**:269). On leaves of *Wikstroemia canescens*, Mussoorie (Butler) ; Verinag, Kashmir (Butler).

Monosporidium andrachnis Barclay (**35, I**:371, figs.; Sacc. **IX**:297 ; Syd. **IV**: 365 ; **455**:274). On leaves of *Andrachne cordifolia*, Simla (Barclay) ; Kasauli (Butler) ; Mussoorie (Kar).

——**euphorbiae** Barclay (**35, I**:367, figs.; Sacc. **IX**:297 ; Syd. **IV**:364). On leaves of *Euphorbia pilosa* (*E. cognata*), Simla (Barclay).

Peridermium brevius (Barclay) Sacc. (Sacc. **IX**:327 ; Syd. **IV**:12 ; **35, III**:102, figs., as *Aecidium brevius* Barclay ; **89**:48 ; **90**, fig.; **452**:442 ; **140**:117, pro parte, as *P. orientale* Cke ; **254**). On needles of *Pinus excelsa*, Simla (Barclay) ; Jaunsar (Oliver) ; Mussoorie (Butler).

——**cedri** (Barclay) Sacc. (Sacc. **IX**:327 ; Syd. **IV**:2 ; **32**:223, figs., as *Aecidium cedri* Barclay ; **35, III**:104 ; **90** ; **111**:87, fig.; **452**:442 ; **491** ; **254**). On leaves of *Cedrus libani* var. *deodara*, near Simla (Barclay) ; common in the Himalayas (Troup ; Hafiz Khan).

——**ephedrae** Cke (**137**:95 ; Sacc. **VII**:835 ; Syd. **IV**:3 ; **89**:48 ; **90**, fig.; **452**:442 ; **254**). On young shoots and leaves of *Ephedra vulgaris*, Jaunsar (Hole).

——**himalayense** Bagchee (**26**:19, figs.; **35, III**:101, as *Aecidium complanatum* Barcl. var. *corticola* Barcl.; Syd. **IV**:12 ; **90** : (22) as *P. complanatum* Barcl. var. *corticola* ; **118; 27**). On branches of *Pinus longifolia*, Simla (Barclay) ; Garhwal (Osmaston) ; Naini Tal (Carr) ; Ranikhet (Butler) ; Almora (Champion) ; Ranikhet, Naini Tal, Pauri, Garhwal, Chakrata, and Simla (Bagchee). The *Cronartium* stage has recently been reported by Bagchee (**27**) on *Swertia alata*, *S. angustifolia*, and *S. cordata*, and we find the uredo stage of this *Cronartium* on these hosts to be identical with the type of *Uredo opheliae* Syd. (**440**:21 ; Syd. **IV**:434) on *S. angustifolia*, India.

——**indicum** Colley & Taylor (**127**: 329, fig.; **254**:439 ; **26**:17). On twigs and branches of *Pinus excelsa*, Kulu (Troup). See *Cronartium ribicola*, above.

——**orientale** Cke (**137**:91 ; Sacc. **VII**:836 ; Syd. **IV**:11 ; **140**:117 ; **211**:263 ; **489**:317 ; **26; 35, III**:101, figs., as *Aecidium complanatum* Barclay ; Sacc. **IX**:327, as *P. complanatum* Barclay ; **90**, fig.; **452**:442 ; **253; 118**). On needles of *Pinus longifolia*, Simla (Barclay ; Gamble ; Wroughton) ; Mussoorie (Butler ; Inayat) ; Kangra and Kasauli (Burkill) ; Almora, Dehra Dun, and elsewhere (Bagchee). Cooke included the *Peridermium* on *Pinus excelsa* needles (*Perid. brevius*, see above) with this species, but as pointed out by Thuemen (**489**:317), this must have been due to error or to mixing the material. The synonymy of *P. orientale* is discussed in **26**:2, footnote.

[**Peridermium**] **piceae** (Barclay) Sacc. (Sacc. **IX**:327 ; Syd. **IV**:4 ; **31**:140, figs., as *Aecidium piceae* Barclay ; **35, III**:104 ; **89**:48; **90**, fig.; **452**:442 ; **254**). On leaves of *Picea morinda* (*Abies smithiana*), Simla (Barclay) ; Jaunsar (Oliver) ; Mussoorie (Butler) ; Kashmir (Gammie).

——**thomsoni** Berk. apud Cke (**137**:94 ; Sacc. **VII**:837 ; Syd. **IV**:4 ; **54**:627, figs., as *Aecidium thomsoni* Berk.; **57**, after No. 468 ; **140**:117 ; **489**:320 ; **30**:1, figs.; **254**). On leaves of *Picea morinda* (*Abies smithiana*), Mahasu, near Simla (Gamble) ; north-west Himalaya, 8,000 ft. (Thomson) ; Sikkim, 9,000 ft. (Hooker f. & Thomson) ; Kulu (Troup).

This is probably, in part, the rust referred by Cooke (**137**:91 ; **140**:117) to "*Peridermium acicolum* Link." Barclay assumes the identity of the specimens from Simla and other north-western Himalayan regions with the species collected by Thomson in Sikkim and named *A. thomsoni* by Berkeley and he took it to be the aecidial stage of the fungus subsequently named by Dietel *Barclayella deformans* (*Chrysomyxa deformans* : q.v.). It is, however, doubtful that the two forms have any genetic connection.

Phakopsora cronartiiformis (Barclay) Diet. (**223**:385 ; Sacc. **XXIII**:844 ; Syd. **III**:412 ; **35, III**:98 as *Uredo cronartiiformis* Barcl.; **104**:153 as *P. vitis* Syd.). On leaves of *Vitis himalayana*, Simla (Barclay) ; Mussoorie (Butler).

——**phyllanthi** Diet. (**222**:469 ; Sacc. **XXI**:608 ; Syd. **III**:414 ; **455**:269). On leaves of *Phyllanthus distichus*, Godavari (Mitra) ; Tanjore (Sundararaman) ; Dacca (Som) ; Palghat (Subramaniam).

——**punctiformis** (Barclay & Diet.) Diet. (**213**:333 ; Sacc. **XIV**:289 ; Syd. **III**:408, fig.; **211**:267, fig., as *Melampsora punctiformis* Barcl. & Diet.; **308**:130). On leaves of *Galium aparine*, Simla (Barclay).

——**zizyphi-vulgaris** (P. Henn.) Diet. (**222**:469 ; Sacc. **XXI**:608 ; Syd. **III**:413 ; **455**:269 ; **453**:508 as *Uredo zizyphi-vulgaris* P. Henn.). On leaves of *Zizyphus jujuba*, Pusa (Butler) ; Nagpur (Pandit) ; Peshawar (Shaw) ; Jalalpur, Punjab (Cheema) ; Birbhum, Bengal (Basu) ; of *Z. rotundifolia*, Pusa (Butler).

Phragmidium assamense Syd. (**455**:264 ; Sacc. **XXIII**:824 ; Syd. **III**:150). On leaves of *Rubus lasiocarpus*, Shillong (Som).

——**barclayi** Diet. (**211**:264 ; Sacc. **IX**:315 ; Syd. **III**:150, fig.; **453**:501 ; **35, III**:81, figs., as *P. ?rubi* Pers.). On leaves of *Rubus lasiocarpus*, Simla (Barclay) ; Kumaon (Inayat) ; Achibal and Harwan, Kashmir (Butler) ; Simla (Mitter) ; Mussoorie (Kar).

——**burmanicum** Syd. (**455**:264 ; Sacc. **XXIII**:824 ; Syd. **III**:136). On leaves of *Rubus lasiocarpus*, Maymyo (Butler). Sydow (**448**:167) transfers this to the new genus *Phragmotelium*.

——**butleri** Syd. (**453**:501 ; Sacc. **XXI**:725 ; Syd. **III**:124, fig.). On leaves of *Rosa macrophylla*, Kumaon (Inayat).

[Phragmidium] disciflorum (Tode) James (Syd. **III**:115 ; Sacc. **VII**:746, pro parte, as *P. subcorticium* (Schrank) Wint. ; **140**:117 as *P. mucronatum* Fr.). Reported on leaves of *Rosa* sp., Kalatop Forest, Chamba, 7,000 ft. (Baden Powell). Cooke (**142**:61) records *Coleosporium pingue* Lév. (=*P. disciflorum*) on *Astilbe*, Simla (Gamble), but the fungus could scarcely have been this species and may have been *Pucciniostele* (q.v.).

——**egenulum** Syd. & Butler (**455**:263 ; Sacc. **XXIII**:823 ; Syd. **III**:124). On leaves of *Rosa webbiana*, Verinag, Kashmir (Butler).

——**incompletum** Barclay (**35, III**:83 ; Sacc. **IX**:316 ; Syd. **III**:151 ; **453**:501). On leaves of *Rubus paniculatus*, near Simla (Barclay) ; Mussoorie (Butler) ; Kumaon (Inayat).

——**laceianum** Barclay (**43**:220, fig.; Sacc. **XI**:207 ; Syd. **III**:97 ; **217**:130). On leaves of *Potentilla argyrophylla*, Bashahr (Lace) ; Narkanda (Barclay) ; Kulu (Murray).

——**nepalense** Barclay (**43**:220, figs.; Sacc. **XI**:207 ; Syd. **III**: 100 ; **217**:130). On leaves of *Potentilla nepalensis*, Mattiana, near Simla (Barclay).

——**octoloculare** Barclay (**43**:221, fig.; Sacc. **XI**:207 ; Syd. **III**:151). On leaves of *Rubus rosaefolius*, near Simla (Barclay).

——**orientale** Syd. (**453**:501 ; Sacc. **XXI**:730 ; Syd. **III**:152). On leaves of *Rubus ellipticus*, Ranikhet (Butler) ; of *R.* sp., Rangpur (Som).

——**potentillae** (Pers.) Karst. (Sacc. **VII**:743 ; Syd. **III**:97 ; **455**:263). On leaves of *Potentilla fragarioides*, Achibal, Kashmir (Butler). The fungus listed by Barclay (**35, III**:98) as "*Uredo ?eupatoriae* DC.", on all parts of *Potentilla* (*?kleiniana*), Simla (Barclay), probably belongs to this species. Cooke's reference (**130**:75) to "*Uredo potentillarum* DC. on leaves of *Agrimonia*" is included with *Pucciniastrum agrimoniae.*

——**quinqueloculare** Barclay (**35, III** :82 ; Sacc. **IX**:315 ; Syd. **III**:148). On leaves of *Rubus biflorus*, Simla (Barclay).

——**rosae-moschatae** Diet. (**217**:132, figs.; Sacc. **XXI**:725 ; Syd. **III**:125 ; **452**:437 ; **455**:263 ; **35, III**:79, figs., as *P. subcorticium* Schrank). On leaves of *Rosa moschata*, Simla (Barclay) ; Mussoorie, Murree, Kasauli, and Harwan (Butler) ; Peshawar (Brown) ; Sabathu, 4,000 ft. (Burkill) ; of *R. webbiana*, Harwan, Kashmir (Butler) ; of *R. centifolia*, Peshawar (Shaw) ; of *R.* sp., Dharampur (Cheema).

——**rubi** (Pers.) Wint. (Sacc. **VII**:745, pro parte ; Syd. **III**:141 ; **142**:61 as *Lecythea ruborum* Lév.). On leaves of *Rubus ellipticus* (*R. flavus*), Sutlej Valley, 3,000 ft. (Gamble) ; on leaves of *R. fruticosus*, Harwan, Kashmir (Butler).

Puccinia absinthii DC. (Sacc. **VII**:637 ; Syd. **I**:11 ; **455**:256). On leaves of *Artemisia* sp., Harwan, Kashmir (Butler).

——**aggregata** Syd. (**445**:326 ; Sacc. **XXIII**:667). On leaves of *Strobilanthes barbatus*, Attapady Valley, Coimbatore, Madras (Fischer).

[Puccinia] agrostidis Plowr. (Sacc. **XI**:202 ; Syd. **I**:717 ; **43**:226). On leaves of *Aquilegia vulgaris*, near Simla, 10,000 ft. (Lace).

——**?andropogonis** Schw. (Sacc. **VII**:664 ; Syd. **I**:720 ; **35, II**:246, fig., as *P. andropogi* Schw.). On leaves of *Andropogon tristis*, Simla (Barclay). This rust is not otherwise recorded outside of North America, and the host is not mentioned for any *Puccinia* in Volume I of Sydow. Barclay himself had some doubt as to the name of the rust.

——**anomala** Rostrup (Sacc. **XVII**:377 as *P. simplex* (Koern.)Erikss. & Henn. ; Syd. **I**:756 ; **111**:186, fig.). Rarely seen in India on *Hordeum vulgare*.

——**anthistiriae** Barclay (**35, II**:246, fig. ; Sacc. **IX** : 304 ; Syd. **I** :726, fig. ; **455**:261). On leaves, culms, and glumes of *Anthistiria anathera*, Simla (Barclay) ; Harwan, Kashmir (Butler).

——**apii** Desm. (Sacc. **XVII**:339 ; Syd. **I**:359 ; **43**:215, fig., as *P. ?castagnei* Thuem.). On leaves and petioles of *Apium graveolens*, Simla (Barclay).

——**apludae** Syd. (**452**:436 ; Sacc. **XXI**:707). On leaves of *Apluda mutica* (*A. aristata*), Dehra Dun (Butler).

——**arenariae** (Schum.) Wint. (Sacc. **VII**:683 ; Syd. **I**:553 ; **43**:219, fig., as *P. caudata* Barclay). On leaves of *Stellaria paniculata*, Narkanda, near Simla (Watt).

——**aristidicola** P. Henn. (Sacc. **XIV**:355 ; Syd. **I**:728 ; **453**:499). On leaves of *Aristida depressa*, Orai, United Provinces (Inayat).

——**artemisiella** Syd. (Syd. **I**:14 ; Sacc. **XVII**:279 ; **455**:256). On leaves of *Artemisia vulgaris*, Harwan, Kashmir (Butler) ; Mussoorie (Kar).

——**arthraxonis** (P. Henn.) Syd. & Butler (**453**:499 ; Sacc. **XXI**:707 ; **455**:261). On leaves of *Arthraxon lanceolatus*, Dehra Dun, Kumaon, and Kasauli (Butler) ; of *A.* sp., Mussoorie (Kar).

——**arundinellae** Barclay (**35, II**:245, fig. ; Sacc. **IX**:303 ; Syd. **I**:732, fig. ; **453**: 498 ; **455**:261). On leaves of *Arundinella setosa*, Simla (Barclay) ; of *A. wallichii*, Simla (Barclay) ; Wahjain, Assam (Butler) ; of *A. brasiliensis*, Dehra Dun (Butler) ; of *A.* sp., Maymyo (Butler) ; Kumaon (Inayat).

——**barbeyi** (Roum.) P. Magn. (Sacc. **XVI**:305 ; Syd. **I**:618 ; **452**:433). On stems of *Asphodelus fistulosus*, Lyallpur (Butler).

——**belamcandae** (P. Henn.) Diet. (**219**:305 ; Sacc. **XXI**:672 ; **453**:508 as *Uredo belamcandae* P. Henn.). On leaves of *Belamcanda punctata* (*B. chinensis*), Kumaon (Inayat). Saccardo (**XXI**:804) lists "*Uredo pulchra* Syd." on "*Gladiolus* spec., Kumaon," etc. This mistake (the description being of *U. belamcandae*) is discussed by Sydow (**IV**:511).

——**bupleuri-falcati** (DC.) Wint. (Sacc. **XVII**:393 ; Syd. **I**:364 ; **35, III**:98 as *Uredo bupleuri* Barclay). On leaves of *Bupleurum falcatum*, Simla (Barclay) ; of *B. lanceolatum*, Raiengarh, Punjab (Gamble).

——**burmanica** Syd. & Butler (**455**:261 ; Sacc. **XXIII**:746). On leaves of *Anthistiria imberbis* (*Themeda triandra*), Maymyo (Butler).

[**Puccinia**] **butleri** Syd. (**452**:431 ; Sacc. **XXI**:651 ; **116**:15 ; **111**:90, fig. ; **192**:102). On leaves and stalks of *Launaea asplenifolia*, Bengal, Bihar, Central India, Rajputana, and United Provinces (Prain) ; Pusa, Cawnpore, and Burma (Butler).

——**cacao** McAlp. (Sacc. **XXI**:697). On leaves of *Rottboellia compressa*, Pusa (Butler) ; Ranchi (Mitra).

——**calospermae** Syd. & Butler (**452**:432 ; Sacc. **XXI**:670). On leaves, petioles, stems, and flowers of *Deeringia celosioides*, Dehra Dun (Butler) ; Dacca (Som).

——**caricis** (Schum.) Rebent. (Sacc. **VII**:626 ; Syd. **I**:648 ; **453**:497 ; **35, II**:244 ; **34**:29, figs., as *Aecidium urticae* Schum. var. *himalayense* Barclay and *P. urticae* Barclay ; **35, I**:368). Aecidia on leaves and stems of *Urtica parviflora*, Simla (Barclay) ; Kumaon and Mussoorie (Butler) ; teleuto stage on *Carex setigera*, Simla (Barclay).

——**caricis-asteris** Arth. (Sacc. **XVII**:371 ; Syd. **I**:665). On *Aster asperulus*, Mussoorie (Gamble).

——**caricis-filicinae** Barclay (**35, II**:250, figs. ; Sacc. **IX**:303 ; Syd. **I**:678, fig.). On leaves and sheaths of *Carex filicina*, Simla (Barclay).

——**carthami** (Hutzelm.) Corda (Sacc. **VII**:646 ; Syd. **I**:35 ; **452**:431 ; **455**:256). On leaves of *Carthamus oxyacantha*, Kangra, Punjab ; of *C. tinctorius*, Pusa (Butler) ; Punjab (Cheema).

——**centaureae** Mart. (Sacc. **XVII**:286 ; Syd. **I**:39 ; **455**:257). On leaves of *Centaurea calcitrapa*, Harwan, Kashmir (Butler) ; Peshawar (Shaw).

——**cephalandrae-indicae** Syd. (**452**:433 ; Sacc. **XXI**:618). On leaves of *Coccinia* (*Cephalandra*) *indica*, Pusa and Nadiad, Bombay (Butler).

——**chaerophylli** Purt. (Sacc. **XVI**:281 ; Syd. **I**:367). On *Anthriscus nemorosa*, Punjab (Gamble).

——**chrysanthemi** Roze (Sacc. **XVI**:296 ; Syd. **I**:46 ; **452**:431). On leaves of *Chrysanthemum indicum*, The Droog, Nilgiris (Butler).

——**chrysopogi** Barclay (**35, II**:247, figs. ; Sacc. **IX**:304 ; Syd. **I**:746 ; **453**:497 ; **35, I**:363, figs., as *Aecidium jasmini* Barclay ; **455**:273 ; **48**:241, figs., as *P. jasmini-chrysopogonis* Barclay ; **319, I**:116 as "*Aecidium clematidis* DC."). Aecidia on leaves of *Jasminum humile*, Simla (Barclay) ; Kumaon (Inayat) ; Jaunsar, 9,000 ft. (Rogers) ; of *J.* sp., Harwan, Kashmir (Butler) ; teleuto stage on leaves of *Chrysopogon gryllus*, Simla (Barclay) ; Harwan, Kashmir (Butler).

——**cichorii** (DC.) Bellynck (Sacc. **XVII**:311 ; Syd. **I**:49 ; **455** :257). On leaves of *Cichorium intybus*, Harwan, Kashmir (Butler).

——**cipurae** Syd. (**453**:496 ; Sacc. **XXI**:672). On leaves of *Cipura paludosa*, Royal Botanic Garden, Calcutta (Butler).

——**circaeae** Pers. (Sacc. **VII**:686 ; Syd. **I**:422 ; **35, II**:235, fig.). On leaves of *Circaea alpina*, Chor, near Simla, 9,000 ft. (Barclay).

[**Puccinia**] **citrulli** Syd. & Butler (**455**:259; Sacc. **XXIII**:714). On leaves of *Citrullus colocynthis,* Coimbatore (Subramaniam).

——**collettiana** Barclay (**40**:87, figs.; Sacc. **IX**:306; Syd. **I**:226; **453**:495; **35,** **II**:241, figs., as "*P. helvetica* L."). On leaves of *Rubia cordifolia,* Simla (Barclay); Ranikhet (Butler); Mussoorie (Kar); Kasauli (Butler). Léveillé (**285**:69) records "*Puccinia galiorum* Lk" [*P. punctata*] on leaves of "*Rubia munjista*" [*R. cordifolia*], India. It may have been *P. collettiana.*

——**coronata** Corda (Sacc. **VII**:623, pro parte; Syd. **I**:699; **455**:260; **319, I**:116). Aecidia on leaves of *Rhamnus purpurea,* Deoban, 9000 ft. (Gamble); on *R. procumbens,* Mussoorie (Butler); uredo and teleuto stages on leaves of *Stipa* sp., Mussoorie; of *Poa flexuosa* and *Agropyron* sp., Harwan, Kashmir (Butler).

——**cressae** (DC.) Lagerh. (Sacc. **IX**:307; Syd. **I**:320; **455**:272 as *Aecidium cressae* DC.). On leaves of *Cressa cretica,* Gokak Government Farm, Bombay (Kulkarni).

——**curculigonis** Racib. (Sacc. **XVI**:307; Syd. **I**:604; **452**:433). On leaves of *Curculigo orchioides,* Wynaad (Butler); Dehra Dun (Kar).

——**cynodontis** Desm. (Sacc. **VII**:661; Syd. **I**:748; **268**:217; **452**:436). On leaves of *Cynodon dactylon,* Saharanpur (Butler); Lyallpur (Hafiz Khan); Peshawar (Shaw); Ganeshkhind Botanic Garden, Poona (Gammie); Cawnpore, Surat, Kashmir (Butler). The aecidia on *Plantago* have not yet been recorded for India.

——**dactylidina** Bubak (Sacc. **XXI**:698; **455**:261). On leaves of *Dactylis glomerata,* Harwan, Kashmir (Butler).

——**dispersa** Erikss. & Henn. (Syd. **I**:709; **263**:323 as *P. rubigo-vera* (DC.) Wint.; Sacc. **VII**:624). "On leaves of an Asperifoliaceae" Siwalik Range (Gollan). A doubtful record.

——**dissiliens** Cke (**130**:75; Sacc. **VII**:723; Syd. **I**:582). On leaves of *Rumex* sp. (resembling *R. acetosa*), Himalaya north of Dehra Dun (Fleming).

——**dovrensis** Blytt (Sacc. **XIV**:311; Syd. **I**:80; **453**:495). On leaves of *Erigeron alpinus,* Kumaon (Inayat). Barclay (**43**:218) referred a rust on *E. alpinus* var. *multicaulis,* near the banks of the Sutlej, Bashahr (Lace), doubtfully to *P. doloris* Speg. As Sydow points out (loc. cit.), it was probably *P. dovrensis.*

——**droogensis** Butler (**90**:(30), fig.; Sacc. **XXI**:616; **452**:432; **369**:231). On leaves of *Berberis aristata,* The Droog, Nilgiris (Butler).

——**duthiae** Ellis & Tracy (**227**:283; Sacc. **XIV**:352; Syd. **I**:726; **452**:435; **89**:45). On leaves of *Andropogon pertusus,* Dehra Dun (Duthie); Poona (Chibber); Ranchi (Mitra); Phulgru (Watt); Dharwar and Kasauli (Butler); of *A. intermedius,* India (Duthie).

[**Puccinia**] ? **ellisii** de Toni (Sacc. **VII**:651 ; Syd. **I**:356 ; **43** :215, fig.). On leaves of *Angelica glauca*, Phagu, near Simla (Watt). Sydow (l.c.) states that this is probably a distinct species.

——**engleriana** P. Henn. (Sacc. **XVII**:336 ; **455**:258). On leaves of *Tabernaemontana heyneana*, Matheran, Bombay (Ajrekar).

——**epilobii-tetragoni** (DC.) Wint. (Sacc. **VII**:608 ; Syd. **I**:424 ; **455**:257). On leaves of *Epilobium* sp., Harwan, Kashmir (Butler).

——**eulaliae** Barclay (**43**:216 ; Sacc. **XI**:199 ; Syd. **I**:797 ; **268**:220). On leaves of " *Pollinia japonica* ", Simla (Barclay). It would seem possible that Barclay had the grass formerly known as *Eulalia japonica*, now called *Miscanthus sinensis*. *P. eulaliae* is recorded on the latter host from Japan.

——**excelsa** Barclay (**43**:216, fig. ; Sacc. **XI**:194 ; Syd. **I**:286). On leaves of *Phlomis bracteosa* (*P. lamiifolia*), Mahasu and Huttoo Peak, near Simla (Barclay).

——**expallens** Syd. (**453**:496 ; Sacc. **XXI**:673). On leaves of *Hypoxis aurea*. Dehra Dun (Butler).

——**fagopyri** Barclay (**41**:261 ; **35, III**:107, fig. ; Sacc. **IX**:306 ; Syd. **I**:566, fig.). On leaves of *Fagopyrum esculentum*, Simla (Barclay) ; Mussoorie (Kar).

——**ferruginosa** Syd. (Sacc. **XVII**:280 ; Syd, **I**:13, fig. ; **455**:256). On leaves of *Artemisia* sp. (*A. ?vulgaris*), Shillong (Butler).

——**flavipes** Syd. (**453**:497 ; Sacc. **XXI**:684). On leaves of *Fimbristylis miliacea*, Yelwal, Mysore (Butler).

——**fragariae** Barclay (**35, I**:359, figs. ; Sacc. **IX**:309 ; Syd. **I**:483 ; **35, II**:244). On leaves of *Fragaria vesca*, Simla (Barclay).

——**fusca** (Pers.) Wint. (Sacc. **VII**:669 ; Syd. **I**:530 ; **319, I**:116). On *Anemone rivularis*, Himalaya (Gamble).

——**gentianae** (Strauss) Link (Sacc. **VII** : 604 ; Syd. **I**:340 ; **35, III**:108). On leaves of *Gentiana kurroo*, near Simla (Barclay).

——**geranii-silvatici** Karst. (Sacc. **VII**:682 ; Syd. **I**:465, fig. ; **455**:257 ; **37**:27, figs., as *P. geranii-silvatici* var. *himalensis* Barclay ; **35,II**:236 ; **211**:269). On leaves and stems of *Geranium nepalense*, Simla (Barclay) ; Verinag, Kashmir (Butler). Sydow (**I**:466-8) is unable to maintain Barclay's variety. The name is written in error " var. *nepalensis* Barclay " in Sydow's Monograph (l.c.).

——**glumarum** (Schm.) Erikss. & Henn. (Sacc. **XVII**:380 ; Syd. **I**:706 ; **116**, figs. ; **452**:434 ; **111**:152, figs. ; **393**:3 ; **363** ; **230**). On leaves, culms, and glumes of *Triticum* spp. cult. and *Hordeum vulgare* throughout northern India ; on leaves of *Phalaris minor*, Lyallpur, Hissar, and Dehra Dun (Butler) ; of *Brachypodium sylvaticum*, Simla (Butler).

——**gracilenta** Syd. & Butler (**455**:263 ; Sacc. **XXIII**:729). On leaves of *Bambusa* sp., Darjeeling (McRae).

——**graminis** Pers. (Sacc. **VII**:622 ; Syd. **I**:692 ; **35, I**:367, figs. ; **35, II**:249, figs. ; **319, I**:116 ; **116**, figs. ; **452**:434 ; **455**:260 ; **111**:152, figs. ; **392**:3 ; **393**:3 ; **49** ;

140:117 as *Aecidium berberidis* Pers.). Aecidia on leaves of *Berberis vulgaris*, *B. aristata*, *B. lycium*, and *B. umbellata* in the Himalayas ; uredo and teleuto stages on leaves, culms, and glumes of *Triticum* spp. cult. and *Hordeum vulgare* throughout northern and central India and Burma ; of *Festuca gigantea*, Simla (Barclay) ; Harwan, Kashmir (Butler) ; of *F. kashmiriana* and *Brachypodium sylvaticum*, Harwan, Kashmir (Butler).

[**Puccinia**] **heterospora** Berk. & Curt. (Sacc. **VII**:695 ; Syd. **I**:472, fig. ; **452**:432). On leaves of *Sida humilis*, Dehra Dun (Butler) ; Bombay (Dastur) ; Benares (Muckerji) ; of *S. spinosa*, Poona, Dharwar, and Hunsur, Mysore (Butler) ; Manigaon, Berars (Burkill) ; of *S. mysorensis*, Poona, and Mudigere, Mysore (Butler) ; of *S. cordifolia*, Dacca (Som) ; of *S.* sp., Nagpur (Shrivastan) ; Fraserpet, Coorg (Butler) ; Ranchi (Mitra).

——**hieracii** (Schum.) Martius (Sacc. **VII**:633, pro parte ; Syd. **I**:95). On leaves of *Hieracium crocatum*, Harwan, Kashmir (Butler).

——**himalensis** (Barclay) Diet. (**215**:63 ; Sacc. **XVII**:386 ; Syd. **I**:738 fig. ; **47**:227, figs., as *P. coronata* Cda var. *himalensis* Barclay ; **453**:498 ; **268**:189 ; **35, I**:358, figs., as *P. ?coronata* Cda ; **35, II**:248, figs.). Aecidia on leaves and stems of *Rhamnus dahurica*, Simla (Barclay) ; Harwan, Kashmir, and Kumaon (Butler) ; of *R. virgata*, Simla (Tucker) ; uredo and teleuto states on leaves and sheaths of *Brachypodium sylvaticum*, *Oryzopsis* (*Piptatherum*) *holciforme*, and *Festuca gigantea*, Simla (Barclay). In one paper (**35, II**:247, fig.) Barclay describes a teleuto stage also on *Agrostis hookeriana* as being almost identical with that on *Piptatherum* and *Festuca*, but in his final paper (**47**) it is not included.

Dietz (**225**) considers that *P. himalensis* cannot be maintained distinct from *P. lolii* as it occurs in America on the ground of host reactions, since the American oat rust was able to infect *Rhamnus dahurica ;* but Dietel (**215**) and Sydow (**I**:738) consider that there are morphological differences between the two.

——**hookeri** Syd. (Syd. **I**:723, fig. ; Sacc. **XVII**:379). On leaves of *Andropogon echinulatus*, Himalaya (Hooker f. & Thomson).

——**hydrocotyles** (Link) Cke (Sacc. **VII**:641 ; Syd. **I**:388 ; **452**:432). On leaves of *Hydrocotyle polycephala*, Wahjain, Assam (Butler).

——**inayati** Syd. (**453**:494 ; Sacc. **XXI**:651). On leaves of *Launaea nudicaulis*, Kumaon (Inayat) ; Burma (Butler) ; Wynaad and Darjeeling (McRae).

——**incompleta** Syd. (**455**:261 ; Sacc. **XXIII**:737). On leaves of *Ischaemum ciliare* var. *wallichii*, Panora, Wynaad (McRae) ; Chittagong (Sen).

——[**insidiosa** Berk. (**57**, No. 470 ; Sacc. **VII**:690 ; Syd. **I**:545). On leaves of *Clematis nutans*, Parasnath (Hooker f.), associated with *Uredo clematidis* (stage of *Coleosporium clematidis* Barclay). Sydow (**449**:56) found that no distinct species of *Puccinia* is represented in the original material, and it must be deleted.]

[**Puccinia**] **invenusta** Syd. (**453**:498 ; Sacc. **XXI**:686). On leaves of *Phragmites karka*, Pusa (Butler) ; Kamrup District, Assam (Taslim).

——**investita** Schw. (Sacc. **VII**:707 ; Syd. **I**:88, fig. ; **453**:494). On leaves of *Anaphalis* sp., The Droog, Nilgiris (Butler).

——**iridis** (DC.) Wallr. (Sacc. **VII**:657 ; Syd. **I**:598 ; **35**, **III**:105, fig. ; **455**:260). On leaves of *Iris florentina* and *I. pallida*, Simla (Barclay) ; of *I. kashmiriana*, Harwan, Kashmir (Butler) ; of *I.* sp., Srinagar, Kashmir (Butler) ; Mussoorie (Kar).

——**komarovi** Tranz. (Syd. **I**:451 ; Sacc. **XVII**:350 ; **35**, **III**:106, fig., as "*P. ?argentata* Schultz" ; **211**:261 as *P. argentata* (Schultz). On leaves of *Impatiens amphorata*, Simla (Barclay) ; of *I.* sp., Mussoorie (Butler).

——**kozukensis** Diet. (Sacc. **XVII**:379 ; Syd. **I**:721 ; **455**:260). On leaves of *Andropogon micranthus*, Maymyo (Butler).

——**kuehnii** (Krueg.) Butler (**108**:81, figs. ; Sacc. **XXIII**:744 ; Syd. **IV**:608 ; **111**:380, figs. ; **452**:445 as *Uredo kuehnii* (Krueg.) Wakker & Went). On leaves of *Saccharum spontaneum*, Bassein, Burma (with teleuto stage ; Inayat) ; Lyallpur (A. Khan) ; Pusa (Butler) ; of *S. fuscum*, Cawnpore and Saharanpur (Butler) ; of *S. arundinaceum*, Dehra Dun, Poona, Mozufferpore, and Pusa (Butler) ; Lahore (B. Das) ; Dehra Dun and Lyallpur (A. Khan) ; of *S. narenga*, Mussoorie (Butler) ; of *S. sara*, Lahore (B. Das) ; Wazirabad, Punjab (Cheema) ; of *S.* sp., Bombay (Chibber). This rust occurs ordinarily only in the uredo stage.

——**lateripes** Berk. & Rav. (Sacc. **VII**:649, pro parte ; Syd. **I**:234 ; **455**:257). On leaves of *Ruellia* sp., Pusa (Butler).

——**lateritia** Berk. & Curt. (Sacc. **XIV**:321 ; Syd. **I**:211, fig. ; **452**:431 ; **453**:495). On leaves of *Hedyotis vestita* and *H. auricularia*, Wahjain, Assam (Butler) ; of *Spermococe stricta*, Coimbatore, Dehra Dun, and Yelwal, Mysore (Butler) ; of *S.* sp., Hoshangabad (Butler).

——**leonotidicola** P. Henn. (Sacc. **XVII**:328 ; Syd. **I**:280 ; **455**:271 as *Uredo leonotidis* P. Henn. ; Syd. **IV** :566). On leaves of *Leonotis nepetaefolia*, Islampur, Bombay (Chibber) ; Belgaum, Bombay (Kulkarni).

——**leptodermidis** Barclay (**35**, **III**:87, 109, figs. ; Sacc. **IX**:303 ; Syd. **I**:222 ; **455**:257). On leaves and stems of *Leptodermis lanceolata*, Simla (Barclay) ; Kasauli (Butler).

——**leucadis** Syd. (Sacc. **XVII** : 329 ; Syd. **I**:281, fig. ; **445**:487). On leaves of *Leucas urticaefolia*, Coimbatore (McRae) ; of *L.* sp., Mahableshwar, Bombay (Chibber).

——**leucophaea** Syd. & Butler (**455**:258 ; Sacc. **XXIII**:749). On leaves of *Colquhounia coccinea*, Mussoorie (Butler).

——**lolii** Niels. (Sacc. **XI**:203 as *P. coronifera* Kleb. ; Syd. **I**:704 ; **453**:498 ; **111**:179, fig.). On leaves of *Avena sativa*, Pusa (Butler). The crown rust of oats has been reported only once in India, in 1907, when there was a moderate epidemic at Pusa. *P. himalensis* (q.v.) is a similar or possibly identical rust occurring in the Himalayas.

[**Puccinia**] **lycoctoni** Fckl (Sacc. **XVII**:358 ; Syd. **I**:527 ; **453**:496). On leaves of *Aconitum lycoctonum*, Simla (Watt).

——**macrorrhynchi** Rabenh. (**387**:89 ; Sacc. **VII**:673 ; Syd. **I**:120, fig.). On leaves of "*Macrorrhynchus asplenifolius*", Royal Botanic Garden, Calcutta (Kurz). The host may perhaps have been *Microrrhynchus asplenifolius*, now placed in *Launaea*.

——**maydis** Béreng. (Sacc. **VII**:659 as *P. sorghi* Schw. ; Syd. **I**:830 ; **452**:434 ; **111**:191, figs. ; **43**:214 as *P. sorghi* Schw.). On leaves of *Zea mays*, Mashobra, near Simla (Barclay) ; Manjri Farm near Poona, and Kashmir (Butler).

——**melanocephala** Syd. (**453**:500 ; Sacc. **XXI**:685). On leaves of *Arundinaria* sp., Wahjain, Assam (Butler).

——**menthae** Pers. (Sacc. **VII**:617 ; Syd. **I**:282 ; **35, II**:242 ; **452**:432 ; **453**:495 ; **455**:258). On leaves of *Origanum vulgare*, Simla (Barclay) ; Kumaon (Inayat) ; Mussoorie (Kar) ; Harwan and Beri Nag in Kashmir, Ranikhet, and Kasauli (Butler) ; of *Calamintha umbrosa*, Mussoorie (Butler) ; of *C. clinopodium*, Harwan, Kashmir (Butler) ; of *Mentha sylvestris*, Bilaspur (Burkill) ; Peshawar (Shaw) ; Shadipur, Kashmir (Butler).

——**mysorensis** Syd. & Butler (**452**:434 ; Sacc. **XXI**:683 ; **369**:234). On leaves and culms of *Kyllinga triceps*, Hunsur, Mysore (Butler).

——**nakanishikii** Diet. (Sacc. **XXI**:691 ; **452**:435). On leaves of *Andropogon nardus*, Hunsur, Mysore (Butler) ; Wynaad (McRae) ; of *A. laniger*, Punjab (Cheema).

——**nepalensis** Barclay & Diet. (**211**:265 ; Sacc. **IX**:309 ; Syd. **I**:578, fig. ; **35, II**:240, fig., as *P. acetosae* Schum. ; **455**:259). On leaves of *Rumex nepalensis*, Simla (Barclay) ; of *R. orientalis*, Harwan, Kashmir (Butler) ; of *R.* sp., Kasauli (Butler) ; Mussoorie (Mitra).

——**neyraudiae** Syd. (**455**:260 ; Sacc. **XXIII**:741). On leaves of *Neyraudia madagascarensis*, Mungpoo, Darjeeling (Kawakami).

——**nitida** Barclay (**35, III**:107, fig. ; Sacc. **IX**:307 ; Syd. **I**:574 ; **455**:259). On leaves of *Polygonum amplexicaule*, Mashobra, near Simla (Barclay) ; of *P. chinensis*, Tukwar, Darjeeling (McRae) ; of *P.* sp., Mussoorie, and Verinag, Kashmir (Butler).

——**obscura** Schroet. (Sacc. **VII**:629 ; Syd. **I**:645 ; **453**:497). On leaves of *Luzula campestris*, Cherrapunji, Assam (Butler).

——**oligocarpa** Syd. & Butler (**455**:262 ; Sacc. **XXIII**:746). On leaves of *Stipa* sp., Harwan, Kashmir (Butler).

——**oplismeni** Syd. (**452**:436 ; Sacc. **XXI**:702). On leaves of *Oplismenus compositus*, Mussoorie (Butler) ; of *O.* sp., Kumaon (Butler).

——**oryzopsidis** Syd. & Butler (**453**:498 ; Sacc. **XXI**:702). On leaves of *Oryzopsis molinioides*, Ranikhet, Kumaon (Butler).

[**Puccinia**] **pachypes** Syd. (**455**:262 ; Sacc. **XXIII**:745). On leaves of *Spodiopogon albidus*, Vayitri, Wynaad (McRae).

——**paspali** Tracy & Earle (Sacc. **XIV**:348 ; Syd. **I**:774 ; **452**:436 ; **455**:262 ; **89**:45). On leaves of *Paspalum* (*Panicum*) *sanguinale*, Dehra Dun, Hoshangabad, and Bassein, Bombay (Butler) ; Lyallpur (Hafiz Khan) ; of *P. sanguinale* var. *ciliare*, Pusa (Butler) ; of *P. sanguinale* var. *debile*, Royal Botanic Garden, Calcutta (Butler).

——**penniseti** Zimmerm. (Sacc. **XVII**:390 ; **452**:435 ; **111**:224, figs.). On leaves of *Pennisetum typhoideum*, Nagpur, Pusa, Poona, Cawnpore, Trichinopoli, and Dohad Farm, Bombay (Butler) ; Dhulia, Bombay. Barclay (**41**:257 ; **43**:215 ; see also Watt, **521**:35, 124, 127) gave the name " *P. penniseti* Barclay " in 1890, without formal description, to a rust on *Pennisetum typhoideum* which, however, he confused with *Puccinia purpurea* (q.v.), since he states that the rust on *Zea mays* (*P. maydis*) is distinct from the rust on *Sorghum* and *Pennisetum* (*P. penniseti*). (See also Syd. **I**:805).

——**peraffinis** Syd. (**445**:326 ; Sacc. **XXIII**:666). On leaves of *Justicia diffusa*, Nandi Droog, Mysore (Anstead).

——**persistens** Plowr. (Sacc. **IX**:312 ; Syd. **I**:825 ; **35**, **I**:362, figs., as *Aecidium ?thalictri-flavi* DC.). On leaves of *Thalictrum javanicum*, Simla (Barclay).

——**phlogacanthi** Syd. (**443**, **VI**:143 ; Sacc. **XXIII**:667). On leaves of *Phlogacanthus guttatus*, Sylhet (Griffith and Wallich).

——**phragmitis** (Schum.) Koern. (Sacc. **VII**:630 ; Syd. **I**:787 ; **130**:75 as *Aecidium rubellum* Pers. ; **236**:273). On leaves of *Rumex* sp. (" the common dock "), Himalaya north of Dehra Dun (Fleming).

——**pimpinellae** (Strauss) Martius (Sacc. **VII**:616, pro parte ; Syd. **I**:408 ; **35**, **I**:356, figs. ; **35**, **II**:244, fig. ; **453**:495). On leaves of *Pimpinella diversifolia*, Simla (Barclay) ; Kumaon (Inayat) ; Harwan, Kashmir (Butler) ; of *P.* sp., Mussoorie (Butler).

——**polliniae** Barclay (**35**, **II**:243 ; Sacc. **IX**:313 ; Syd. **I**:798 ; **33**:21, figs., as *Aecidium strobilanthis* Barclay ; **35**, **I**:369 ; **452**: 436). Aecidia on leaves of *Strobilanthes dalhousianus*, Simla (Barclay) ; of *S.* sp., Dehra Dun (Butler) ; Mussoorie (Kar) ; Amboli, Belgaum ; uredo and teleuto stages on leaves of *Pollinia nuda*, Simla (Barclay).

——**polygoni-amphibii** Pers. (Sacc. **XVII**:394 ; Syd. **I**:569 ; **455**:259 ; **286**:271 as *P. polygonorum* Link ; **403**:301 as *P. ?polygoni* Pers.). " India orient. Jacquemont " (this specimen was not found at Paris) ; on leaves of *Polygonum* sp., Mussoorie (Kar) ; of *P. persicaria*, Harwan, Kashmir (Butler).

——?**porri** (Sowerby) Wint. (Sacc. **VII**:605 ; Syd. **I**:610 ; **113**(19):54 as " *P. porri* Saw."). Aecidiospores and teleutospores of a rust on leaves on onion (*Allium cepa*), Lyallpur, were reported by McRae to resemble *P. porri* very closely.

——**prainiana** Barclay (**44**:67, figs. ; Sacc. **XI**:197 ; Syd. **I**:635, figs. ; **36**:37, figs., as *Caeoma smilacinis* Barclay ; **35**, III:95 as *C. smilacis* Barclay ; **211**:269).

On leaves of *Smilax aspera*, Simla (Barclay); Mussoorie, Kumaon, and The Droog, Nilgiris (Butler); of *S. elegans*, Ranikhet, Kumaon (Butler); of *S. macrophylla*, Mahableshwar, Bombay (Chibber); of *S.* sp., Noakhali, and Bilin, Burma (Butler); Wynaad (McRae); Khasi Hills (Subramaniam).

[**Puccinia**] **prenanthis-purpureae** (DC.) Lindr., var. **himalensis** Barclay (**44**:65, figs.; Sacc. **XI**:189; Syd. **I**:136). On leaves of *Prenanthes brunoniana* and *Lactuca macrorhiza*, Simla (Barclay). Sydow (l.c.) thinks this variety may deserve specific rank.

——**princeps** Syd. (**452**:432; Sacc. **XXI**:659). On branches of *Pogostemon* sp., Dehra Dun (Butler).

——**propinqua** Syd. & Butler (**453**:499; Sacc. **XXI**:692). On leaves of *Andropogon* sp., Ranikhet, Kumaon (Butler).

——**prunicolor** Syd. & Butler (**452**:435; Sacc. **XXI**:693). On leaves of *Andropogon serratus*, Dehra Dun (Butler).

——**pruni-persicae** Hori (Sacc. **XXIII**:770; **445**:487). On leaves of *Prunus persica*, Coimbatore (McRae). This specimen has not been seen by us, and was determined by Sydow from the uredo stage, which alone was present. A specimen similarly determined by McRae from Pusa has been examined and found to consist of typical *P. pruni-spinosae* uredosori. Specimens of Hori's fungus distributed as Sydow, Fungi exotici exsiccati No. 471 "in foliis *Pruni persicae*, Prov. Totomi, Yaizu, Japonia, K. Hara" have been examined and found to bear uredospores intermixed with the teleuto and of a quite different type from those of *P. pruni-spinosae*. They are also quite different from the uredo stage of *P. pruni-persicae* as described and figured by Hori in Phytopath., II: p. 144, 1912, nothing resembling which was found in Hara's specimen. The uredo described by Hori agrees with that of *P. pruni-spinosae* and it would appear that he was mistaken in thinking that the two stages he describes belonged to the same fungus. As the Indian record is based on the uredo, it should probably be deleted.

——**pruni-spinosae** Pers. (Sacc. **VII**:648; Syd. **I**:484; **144**:96, as *Uromyces amygdali* Cke; **452**:432; **455**:258). On an unnamed host, Belgaum (Hobson); on leaves of *Prunus persica*, Pusa (Butler); and several other collections in India and Burma (some of which may be the preceding species); of *P. armeniaca*, Shillong (Som); Maymyo (Butler); of *P. communis*, Verinag, Kashmir (Butler); of *P. puddum*, Bageswar, Kumaon (Inayat). Barclay (**35**, I:363) mentions an *Aecidium* on *Thalictrum minus* which he collected in the interior of the Himalaya towards Tibet, and which was characterized by the production of considerable distortion and hypertrophy of the host. This may have been the aecidial stage of *P. pruni-spinosae*. (See Syd. **IV**:350.)

——**pulvinata** Rabenh. (Sacc. **VII**:711; Syd. **I**:76, fig.; **43**:219; **455**:257). On leaves of *Echinops niveus*, Simla (Watt); of *E. echinatus*, Lahore (B. Das); of *E. cornigerus*, Verinag, Kashmir (Butler).

[Puccinia] punctata Link (Sacc. **XVII**:393 ; Syd. **I**:213; **35, II**:239, figs., as *P. galii* Pers.). On leaves of *Galium aparine*, Simla (Barclay). Sydow (**I**:217) includes this record under *P. ambigua* (A. & S.) Lagerh., presumably because the host is *Galium aparine* : but Barclay describes and figures uredospores, which are unknown in *P. ambigua*. He states that there is no aecidial stage, but both *P. ambigua* and *P. punctata* have aecidia. As for the host, *P. punctata* is described on *Galium aparine* in North America. See *P. collettiana*.

——**purpurea** Cke (**133**:15, with figure referred to " *Trichobasis purpureum* Cke " ; Sacc. **VII**:657 ; Syd. **I**:803 ; **452**:434 ; **111**:206, fig. ; **392**:2). On leaves of *Andropogon sorghum* (*Sorghum vulgare*), India (Hobson) ; Coimbatore, Poona, Ranchi, Surat, and Maymyo, Burma (Butler) ; Arvu and Phulgaon (Watt) ; of *A. halepensis*, Mandalay (Shaw) ; Lahore (Cheema) ; Pusa and Dehra Dun (Butler). Barclay first (**41**) referred the rust on *A. sorghum* to *Puccinia sorghi*, but later (**43**:214 ; see Watt, **521**) to *P. penniseti* Barclay (non Zimm.) ; see above.

——**pusilla** Syd. (**452**:435 ; Sacc. **XXI**:693). On leaves of *Andropogon assimilis*, Dehra Dun (Butler).

——**romagnoliana** Maire & Sacc. (Sacc. **XVII**:374 ; Syd. **I**:682 ; **453**:497 ; **452**:443 as *Uredo ? cypericola* P. Henn.). On leaves of *Cyperus rotundus*, Calcutta, Surat, Pusa, Dehra Dun, Saharanpur, Chittagong, Hoshangabad, and Samalkota Farm (Butler) ; Lyallpur (Hafiz Khan) ; Rangpur (Mitra) ; Lahore (B. Das) ; of *C. tegetum*, Pusa (Butler) ; of *C. compressus* and *C. arenarius*, Chatrapur, Ganjam (Butler) ; of *C. capitatus*, Surat, and *C. tuberosus*, Samalkota (Butler) ; of *C.* sp., Nagpur (Pandit) ; Ranchi (Mitra) ; Hmawbi, Burma (Butler).

——**rosae** Barclay (**35, II**:233, figs. ; Sacc. **IX**:299 ; Syd. **I**:487). On leaves and branches of *Rosa macrophylla*, Narkanda, near Simla (Barclay). Sydow **448**:168) transfers this to the new genus *Teloconia*.

——**roscoeae** Barclay (**35, II**:237, fig. ; Sacc. **IX**:307 ; Syd. **I**:596 ; **455**:260). On leaves of *Roscoea alpina*, Simla (Barclay) ; of *R.* sp., Mussoorie (Butler).

——**rostrata** Cke (**132**:116 ; Sacc. **VII**:693 ; Syd. **I**:516 ; **130**:75 as *P. cruciferarum* Cke ; **236**:273, figs. ; **449**:58). On leaves of some member of the Cruciferae, Himalaya north of Dehra Dun (Fleming) ; redescribed by Sydow (**449**).

——**ruelliae** (Berk. & Broome) Lagerh. (Sacc. **XIV**:324 ; Syd. **I**:235 ; **452**:432). On leaves of *Ruellia longifolia*, Cawnpore (Butler) ; of *R. prostrata*, Pusa, and Hunsur, Mysore (Butler) ; Nagpur (Pandit) ; Calcutta ; of *R.* sp., Benares (Muckerji) ; Godavari (Mitra) ; Samalkota (Subramaniam) ; Belgaum (Kulkarni).

——**rufipes** Diet. (Sacc. **XVII**:377 ; Syd. **I**:757 ; **452**:436). On leaves of *Imperata arundinacea*, Dehra Dun, Kumaon, South Sylhet, Pusa, and Moulmein. (Butler) ; Saharanpur (B. Das) ; Samalkota (Subramaniam) ; of *I.* sp., Wynaad.

[**Puccinia**] **saniculae** Grev. (Sacc. **VII**:618 ; Syd. **I**:413 ; **453**:495 ; **35**, **I**:352, figs., as *Aecidium saniculae* Barclay). On leaves of *Sanicula europaea,* Simla (Barclay) ; Kumaon (Inayat).

——? **saussureae** Thuem. (Sacc. **VIII**:708 ; Syd. **I**:140 ; **455**:257). On leaves of *Saussurea* sp., Nagpur (Pandit). Uredo only, so the identification is doubtful.

——**saxifragae-ciliatae** Barclay (**35, II**:234, figs. ; Sacc. **IX**:299 ; Syd. **I**:506, fig.). On leaves of *Saxifraga ligulata* var. *ciliata,* Simla (Barclay) ; on *S.* sp., Mussoorie (Kar).

——**saxifragae-micranthae** Barclay (**43**:218 ; Sacc. **XI**:185 ; Syd. **I**:506). On leaves of *Saxifraga micrantha,* Bashahr, near Simla, 9,500 ft. (Lace).

——**schirajewskii** Tranz. (Sacc. **XXIII**:698 ; **455**:257). On leaves of *Serratula pallida,* Harwan, Kashmir (Butler).

——**scirpi** DC. (Sacc. **VII**:659 ; Syd. **I**:688 ; **455**:260 ; **455**:274 as *Aecidium nymphoidis* DC.). Aecidia on leaves of *Limnanthemum* sp., Comilla (Subramaniam) ; Wular Lake, Kashmir (Butler) ; uredo and teleuto stages on leaves of *Scirpus articulatus,* Godavari (Sundararaman) ; Bassein, Bombay (Chibber) ; of *S. barbatus,* Ganjam (Butler) ; of *S.* sp., Lahore (B. Das). Sydow (**455**:274) thinks it very doubtful that the *Aecidium* really belongs to *P. scirpi.*

——**solanacearum** Sacc. & Syd. (Sacc. **XIV**:358 ; Syd. **I**:274 ; **142**:61 as *P. solani* Cke (non Schw.) ; **449**:58, with revised diagnosis). On leaves and stems of *Solanum* sp., Sutlej Valley, Himalaya (Gamble). Cooke mentioned that the teleuto stage was associated with "*Aecidium solani,*" but Sydow (**449**) considers that the two stages probably belong to the same rust.

——**solmsii** P. Henn. (Sacc. **XIV**:357 ; Syd. **I**:568, fig. ; **453**: 496). On leaves of *Polygonum chinense,* Koppa in Mysore, Wynaad, and Wahjain, Assam (Butler) ; Darjeeling (McRae) ; Mahableshwar, Bombay (Chibber) ; of *P. alatum,* Kumaon (Inayat) ; of *P.* sp., Nangpo, Assam, and Maymyo (Butler) ; Darjeeling (McRae).

——**sonchi** Rob. (Sacc. **VII**:638 ; Syd. **I**: 154, fig. ; **453**:494). On leaves of *Launaea nudicaulis,* Koppa, Mysore and Srinagar, Kashmir (Butler) ; Kumaon (Inayat).

——**spongiosa** Berk. & Broome (Sacc. **VII**:703 ; Syd. **I**: 228, fig. ; **452**:431 ; **326**: 110). On leaves of *Webera corymbosa,* Hunsur, Mysore (Butler) ; Coimbatore District (McRae).

——**suaveolens** (Link) Rostrup (Sacc. **VII**:633, pro parte ; Syd. **I**:53, 856 ; **455**:257). On leaves of *Cnicus argyracanthus,* Shadipore, Kashmir (Butler), of *Cirsium arvense,* Pusa (Kar) ; of *C.* sp., Kamrup, Assam (Taslim) ; Lahore (B. Das).

——**taraxaci** (Rebent.) Plowr. (Sacc. **IX**:305 ; Syd. **I**:164 ; **452**:430 ; **455**:257 ; **35, II**:238, figs., as *P. flosculosorum* Alb. & Schw.). On leaves of *Taraxacum officinale,* Simla (Barclay) ; Chindi, Himalaya (Burkill) ; Gulchina in Kumaon, Kasauli, and Harwan, Kashmir (Butler) ; of *T. wattii,* Ranikhet, Kumaon (Butler).

[Puccinia] thwaitesii Berk. (Sacc. **VII**:720; Syd. **I**:233, fig.; **452**:431). On leaves of *Justicia gendarussa*, Koppa in Mysore, Tellicherry in Malabar, Comilla, and Insein, Burma (Butler); Dacca (Som); Wynaad (McRae); Calcutta; Rangoon.

——**triticina** Erikss. (Sacc. **XVII**:376; Syd. **I**:716; **116**, figs.; **452**:434; **111**: 154, figs.; **392**:3; **363**; **49**; **230**). On leaves of *Triticum* spp. cult., common in Bengal, Bihar, United Provinces, Central Provinces, Punjab, Kashmir, and Burma.

——**urticae** Barclay (**34**:38, figs.; Sacc. **IX**:299; Syd. **I**:590, fig.; **35, II**:234, fig.; **455**:259). On leaves of *Urtica parviflora*, Simla (Barclay); Kasauli (Butler); of *U.* sp., Mussoorie (Kar). The *Aecidium* which Barclay found on the same host is now believed to belong to *P. caricis* (q. v.).

——**ustalis** Berk. (**57**, No. 469; Sacc. **VII**:691; Syd. **I**:546; **43**:217, fig.). On leaves of *Ranunculus pulchellus*, Momay Samdong, Sikkim, 15,500 ft. (Hooker f.); of *R. hirtellus*, Mattiana, near Simla (Barclay). Berkeley states "some species of *Ranunculus pulchellus*," so he may have had uncertainty as to the host.

——**versicolor** Diet. & Holw. (Sacc. **XIV**:352; Syd. **I**:724, fig.; **452**:436). On leaves of *Andropogon contortus*, Belgaum, Dharwar, Poona, near Pusa, and Maymyo (Butler); Kumaon (Inayat); Wynaad (McRae); of *A.* sp., Maymyo (Butler).

——**violae** (Schum.) DC. (Sacc. **VII**:609; Syd. **I**:439; **35, II**:234; **453**:495; **521**: 84; **142**:61 as *Aecidium violae* Schum.; **35, I**:354, figs., as *A. ? violae* Schum.). On leaves of *Viola* sp., Simla (Gamble); Mussoorie (Kar); on leaves, petioles, and stalks of *V. serpens*, Simla (Barclay); Ranikhet in Kumaon, and Harwan, Kashmir (Butler); on *V. suaveolens* (Watt). Watt states (**521**:84, footnote) that Barclay thought the form on *Viola suaveolens* might possibly prove to be a new species. Where it was collected is not mentioned.

——**wattiana** Barclay (**35, III**:109; Sacc. **IX**:298; Syd. **I**:544, fig.). On leaves and petioles of *Clematis puberula*, Sutlej Valley, near Suni (Watt); of *C. connata*, Harwan, Kashmir (Butler); of *C. buchaniana*, Kumaon (Inayat); of *C. gouriana*, Dehra Dun and Almora (Inayat); of *C.* sp., Mussoorie (Butler).

——**xanthopoda** Syd. (**453**:496; Sacc. **XXI**:683). On leaves of *Scleria* sp., Ranikhet, Kumaon (Butler).

——**xanthosperma** Syd. (**452**:437; Sacc. **XXI**:686). On leaves of *Bambusa* sp., Mussoorie (Butler).

Pucciniastrum agrimoniae (Schw.) Tranzsch. (Syd. **III**:446; **452**:439; **455**:270; **43**:229 as *Uredo agrimoniae* Schroet.; Sacc. **VII**:839; **130**:75 as "*U. potentillarum* DC."). On leaves of *Agrimonia* sp., Himalaya (Fleming); of *A. eupatorium*, Simla (Barclay); Kumaon (Inayat); Mussoorie (Butler); of *A. pilosa*, Simla; of *A.* sp., Cherrapunji, Assam (Subramaniam).

[**Pucciniastrum**] **castaneae** Diet. (Sacc. **XVII**:401 ; Syd. **III**:453 ; **455**:270). On leaves of *Castanopsis javanica*, Maymyo (Butler).

——**celastri** Syd. (**453**:503 ; Sacc. **XXI**:732 ; Syd. **III**:454). On leaves of *Celastrus paniculatus*, Kumaon (Inayat).

——**coriariae** Diet. (Sacc. **XVI**:320 ; Syd. **III**:452 ; **453**:503). On leaves of *Coriaria nepalensis*, Almora (Inayat) ; Mussoorie (Kar).

Pucciniostele clarkiana (Barclay) Diet. (**214**:564, fig. ; Syd. **III**:327, fig. ; **224**:95, fig. ; **455**:267, fig. ; **216**:20 ; **221, III**:312 ; **43**:222, fig., as *Xenodochus clarkiana* Barclay ; Sacc. **XI**:208). On leaves of *Astilbe rivularis*, Cheog forest, Simla (Barclay) ; Darjeeling (Burkill) ; of *A. japonica*, Jaunsar (Gamble) ; of *A.* sp., Dumpep, Khasi Hills, Assam (Burkill).

Komarov and Tranzschel (see **276**) confused this species with one found in Manchuria : that two species occur was pointed out by Dietel (**216**). *P. clarkiana* Kom. & Tranz. refers to the species now called *P. mandschurica* Diet. Saccardo, Syll. **XVI**:321, refers in part to the latter species, which has not been found in India.

Cooke (**142**:61) perhaps saw *P. clarkiana*, as noted above under *Phragmidium disciflorum*.

Ravenelia breyniae Syd. (**453**:501 ; Sacc. **XXI**:774 ; Syd. **III**:291, fig. ; **369**:239). On leaves of *Breynia rhamnoides*, Pusa (Butler) ; Wynaad (McRae) ; Allahabad (Mitter : Sydow, Fungi exot. exs., No. 163).

——**emblicae** Syd. (**452**:438 ; Sacc. **XXI**:744 ; Syd. **III**:293). On leaves of *Phyllanthus emblica* (*Emblica officinalis*), Dehra Dun and Maymyo (Butler) ; Nagpur (Pandit). Cunningham (**186**:230, pl. XI, fig. 13) described and figured what was evidently this fungus from Tanjore, Madras. The host was subsequently stated (**189**) to be *P. emblica*. The rust referred by Cooke (**144**:96; **146**:386, fig.) to *R. aculeifera* Berk., on leaves of an unknown tree, Belgaum (Hobson), appears to be *R. emblicae*, and according to Mr. C. E. C. Fischer, of Kew, the host is really *P. emblica*.

——**hobsoni** Cke (**146**:386, figs. ; Sacc. **VII**:772 ; Syd. **III**:287, fig. ; **218**:400 ; **452**:437 as *R. stictica* Berk. & Broome ; **133**:15 ; **189,** figs. ; **369**:238). On leaves of *Pongamia glabra*, India (Hobson) ; Calcutta (Cunningham) ; Madras and Nagina (Butler) ; Bandra, Bombay (Dastur). Dietel (**218**) and Sydow restrict the name *R. stictica* to the rust occurring on *Mundulea*, which has not been recorded in India.

——**indica** Berk. (**55**:132, fig. ; Sacc. **VII**:772 ; Syd. **III**:257, fig. ; **57,** No. 465 ; **146**:385, fig. ; **218**:391 ; **212**:51). This name has become much confused. Berkeley originally described it as " on pods of *Acacia* sp., Dunway Pass, Behar " (Hooker f.). Mr. C. E. C. Fischer, of Kew, kindly examined the two pods constituting the type, and he finds it to be not *Acacia*, but *Albizzia*, and evidently *Albizzia procera*. The rust appears to be that now known as *R. sessilis*. Subsequently Berkeley (**57**, No. 465) recorded *R. indica* " on pods and stems of *Abrus*, Paras Nath " (Hooker f.). Mr. Fischer examined these specimens, and found them to be not *Abrus* sp., but *Cassia absus*. The rust, which has teleutospores rather similar in aspect to those of the preceding collection, is still known as *R. indica*. Berkeley and Broome (**67**:93) later recorded *R. indica* from Ceylon on " *Bauhinia tomentosa* and *Cassia abrus*". The latter specimen is at Kew, labelled " *Cassia absus* ", the " *abrus* " being a misprint. The rust here also is what is known now as *R. indica*. The *Bauhinia* reference was evidently an error (**218**; **369**:238). Arthur (**20**:196), Dietel, and Sydow record *R. indica* as occurring only in Ceylon and Mexico. India should be included, and the name *R. indica* may perhaps best be conserved for the rust on *Cassia absus*, despite the fact that the type was on another host, and was evidently really *R. sessilis*.

[**Ravenelia**] **mitis** Syd. (**443, XIV**:257 ; Sacc. **XXIII**:805). On leaves of *Tephrosia purpurea*, N. W. Himalayas (Hooker f. & Thomson). See *Uredo tephrosiae* Rabenh., below.

——**ornata** Syd. (**452**:437 ; Sacc. **XXI**:738 ; Syd. **III**:234, fig. ; **369**:238). On leaves of *Abrus pulchellus*, Dehra Dun ; Wahjain, Assam (Butler).

——**sessilis** Berk. (Sacc. **VII**:773 ; Syd. **III**:248 ; **189**, figs. ; **218**:384, fig. ; **146**:386, fig. ; **212**, fig. ; **452**:437). On leaves and fruit of *Albizzia lebbek*, Calcutta (Cunningham) ; Dehra Dun, Raipur, Pusa, Cawnpore, Noakhali (Butler) ; Nagpur (Pandit) ; Ramhati, Bengal (Bannerji) ; Ramgarh, Kumaon (Mitra) ; Chittagong (Sen) ; Arrah (Bhattachariya). Sydow (**448**:165) transfers this to the genus *Haploravenelia*. See also *Sphaerophragmium acaciae* below, and *R. indica* above.

Schroeteriaster cingens Syd. (**455**:270 ; Sacc. **XXII**:841 ; Syd. **III**:404). On leaves of *Bridelia tomentosa* var. *chinensis*, Rangpur, Bengal (Som) ; Assam (Subramaniam) ; on *B.* sp., Mussoorie (Kar).

——**ehretiae** (Hirats.) Syd. & Butler (**455**:270 ; Sacc. **XXIII**:842 ; Syd. **III**:405). On leaves of *Ehretia acuminata*, Chittagong (Sen).

Sphaerophragmium acaciae (Cke) P. Magn. (**307**:121, figs. ; Sacc. **XI**:209 ; Syd. **III**:185, fig. ; **144**:94 as *Triphragmium acaciae* Cke). On leaves of *Albizzia lebbek*, Belgaum (Hobson) ; Calcutta (Cunningham) ; Dacca (Som) ; Myitnge, Burma (Shaw). The teleutospores of this fungus were thought by Cunningham (**189**:21, figs.) to be a second teleuto form of *Ravenelia sessilis*.

Thekopsora gaultheriae Syd. (Syd. **III**:466 ; **453**:503 as *Pucciniastrum gaultheriae* Syd. ; Sacc. **XXI**:733). On leaves of *Gaultheria nummularioides*, Kumaon (Inayat).

Triphragmium thwaitesii Berk. & Broome (Sacc. **VII**:770 and **XVI**:322 ; Syd. **III**:180, fig. ; **455**:264). On leaves of Araliaceae, Pussimbing, Darjeeling (McRae) ; of *Heptapleurum* sp., Nilgiris (Barber). Petch (**367**, IV:163) deals fully with the confusion between this species and *T. clavellosum*. Sydow (**448**:170) transferred *T. thwaitesii* to the genus *Nyssopsora* Arth.

Uredo acori Racib. (Sacc. **XVI**:357 ; Syd. **IV**:521 ; **452**:443). On leaves of *Acorns calamus*, Gauhati, Assam (Butler) ; Bombay (Chibber).

——**apludae** Barclay (**35, III**:99 ; Sacc. **IX**:333 ; Syd. **IV**:531). On leaves of *Apluda aristata*, Simla (Barclay). This stage may belong to one of the species of *Uromyces* or *Puccinia* found on *Apluda*.

——**cajani** Syd. (**452**:442 ; Sacc. **XXI**:792 ; Syd. **IV**:473). On leaves of *Cajanus indicus*, Pusa (Butler).

——**colebrookiana** Barclay (**43**:227 ; Sacc. **XI**:225 ; Syd. **IV**:417). On leaves of *Colebrookea oppositifolia*, Suni, near Simla (Watt).

——**courtoisiae** Syd. (**440**:22 ; Sacc. **XVII**:456 ; Syd. **IV**:522). On leaves of *Courtoisia cyperoides*, Khasi Hills (Hooker f. & Thomson).

—— ? **cypericola** P. Henn. (Sacc. **XIV**:405 ; Syd. **IV**:523 ; **452**:443). On leaves of *Cyperus rotundus*, Dehra Dun, Calcutta, Cawnpore, Pusa, and Hunsur, Mysore (Butler) ; Saharanpur (Gollan) ; of *C. capitatus*, Surat, and *C. tuberosus*, Samalkota (Butler).

——**deutziae** Barclay (**35, III**:100 ; Sacc. **IX**:328 ; Syd. **IV**:489). On leaves of *Deutzia corymbosa*, Simla (Barclay).

——**dioscoreae** P. Henn. (Sacc. **XIV**:402 ; Syd. **IV**:512 ; **455**:276). On leaves of *Dioscorea* sp., Darjeeling (McRae).

——**dioscoreae-sativae** Syd. (**455**:271 ; Sacc. **XXIII**:925 ; Syd. **IV**:514). On leaves of *Dioscorea sativa*, Chaumuhani, Noakhali District (Butler).

——**ehretiae** Barclay (**43**:228 ; Sacc. **XI**:225 ; Syd. **IV**:423). On leaves and petioles of *Ehretia acuminata* (*E. serrata*), Tons River near Chakrata (Gamble). An anomalous form with pycnidia and cup-like sori ; perhaps an *Aecidium*.

——**erythrinae** P. Henn. (Sacc. **XXI**:790 ; Syd. **IV**:480). On leaves of *Erythrina* sp., Noakhali, Bengal (Butler).

——**exasperata** (Cke) de Toni (Sacc. **VII**:846 ; Syd. **IV**:555 ; **143**:95, as *Trichobasis exasperata* Cke). On leaves of an unknown tree, Belgaum (Hobson).

——**fuirenae** P. Henn. (Sacc. **XVI**: 359 ; Syd. **IV**:525 ; **452**:443). On leaves of *Fuirena umbellata*, Wynaad, of *F. glomerata*, Bassein, Burma, and *F.* sp., Tellicherry, Malabar (Butler).

——**gomphrenae** Barclay (**35, III**:99, fig., written *U.* "*gomphrenatis*" ; Sacc. **IX**: 327 ; Syd. **IV**:494 ; **453**:491 as *?Uromyces cyathulae* P. Henn.). On leaves and stems of *Gomphrena globosa*, Simla (Barclay) ; on leaves of *Cyathula capitata*, Kumaon (Inayat) ; Mussoorie (Kar).

——**hemidesmi** Syd. (**455**:271 ; Sacc. **XXIII**:918 ; Syd. **IV**:428). On leaves of *Hemidesmus indicus*, Chittagong (Sen).

——**ichnocarpi** Barclay (**43**:228 ; Sacc. **XI**:226 ; Syd. **IV**:432 ; **452**:442). On leaves of *Ichnocarpus frutescens*, Suni, near Simla (Barclay) ; Kanaighat in Sylhet, and Yelwal, Mysore (Butler) ; Dacca and Dehra Dun (Kar). Sydow and Petrak (**458**:426) suggest that this form may be the uredo stage of *Achrotelium ichnocarpi* Syd., nov. gen., nov. sp., found in the Philippines.

[**Uredo**] **ignobilis** Syd. (**452**:444 ; Sacc. **XXI**:807 ; Syd. **IV**:548). On leaves of *Sporobolus diander*, Pusa (Butler) ; of *S.* sp., Comilla (Inayat).

——**ischaemi** Syd. & Butler (**453**:509 ; Sacc. **XXI**:807 ; Syd. **IV**:540). On leaves of *Ischaemum timorense*, Chittagong (Sen).

——**lipocarphae** Syd. (**453**:509 ; Sacc. **XXI**:806 ; Syd. **IV**:526). On leaves of *Lipocarpha sphacelata*, Yelwal, Mysore (Butler).

——**microspora** (Vize) Sacc. (Sacc. **IX**:330 ; Syd. **IV**:401 ; **133**:14, fig., as *Trichobasis microspora* Vize). On leaves " of Compositae". Locality and collector not stated, but the packet is marked India and the collector is given on it as Hobson who collected mainly around Belgaum. Sydow (l. c.) considers that the host may be one of the Leguminosae, and the rust perhaps a *Ravenelia*.

——**operta** Syd. & Butler (**453**:509 ; Sacc. **XXI**:808 ; Syd. **IV**:536 ; **111**:242, fig.). On leaves of *Coix lachryma-jobi*, Wynaad (Butler) ; of *C.* sp., Wynaad (McRae)

——**ophiuri** Syd. & Butler (**452**:445 ; Sacc. **XXI**:812 ; Syd. **IV**:542). On leaves of *Ophiurus corymbosus*, Dehra Dun.

——**paederiae** Syd. (Sacc. **XVII**:439 ; Syd. **IV**:407 ; **455**:271). On leaves of *Paederia foetida*, Chittagong (Sen). *Aecidium paederiae* Diet. occurs on the same leaves and is perhaps genetically connected.

——**panacis** Syd. (**440**:22 ; Sacc. **XVII**:443 ; Syd. **IV**:442). On leaves of *Aralia* (*Panax*) *pseudo-ginseng*, Sikkim (Hooker f. & Thomson).

——**paspali-scrobiculati** Syd. (**452**:444 ; Sacc. **XXI**:808 ; Syd. **IV**:544 ; **453**:509 ; **111**:240, fig.). On leaves of *Paspalum scrobiculatum*, Kanaighat in Sylhet, and Kumaon (Butler).

——**pileae** Barclay (**43**:228 ; Sacc. **XI**:227 ; Syd. **IV**:500). On leaves of *Pilea trinervia*, Mashobra, near Simla (Barclay).

——**pouzolziae** Syd. (**452**:443 ; Sacc. **XXI**:803 ; Syd. **IV**: 500). On leaves of *Pouzolzia pentandra*, Nilgiris (Butler).

——**punctoidea** Cke (**130**:75 ; Sacc. **VII**:858 ; Syd. **IV**:487 ; **236**:270, figs.). On pinnate leaves of a tree belonging to the Leguminosae, Himalaya north of Dehra Dun (Fleming).

——?**rottboelliae** Diet. (Sacc. **XVII**:457 ; Syd. **IV**:546). On *Rottboellia compressa*, Pusa (Butler). This rust may be *U. ophiuri*, which occurs on a related grass.

——**sesbaniae** P. Henn. (Sacc. **XXI**:791 ; Syd. **IV**:484). On leaves of *Sesbania aegyptica*, Manjri, near Poona (Ajrekar).

——**sissoo** Syd. & Butler (**452**:442 ; Sacc. **XXI**:791 ; Syd. **IV**:477 ; **453**:507). On leaves of *Dalbergia sissoo*, Dehra Dun and Pusa (Butler) ; Wynaad (McRae) ; Poona (Kulkarni).

——**socotrae** Syd. (Sacc. **XVII**:448 ; Syd. **IV**:474 ; **452**:442). On leaves of *Cassia sophora*, Chittagong (Sen) ; Ramhati, Bengal (Bannerji). Sydow at first (**441**:332) considered the Indian rust somewhat different from the type, but later (**452**) placed it in this species.

[**Uredo**] **spinulosa** (Cke) Sacc. (Sacc. **IX**:333 ; Syd. **IV**:553 ; **133**:15, fig., as *Trichobasis spinulosa* Cke). On undetermined leaves, Belgaum. Sydow (l. c.) has redescribed the fungus, but the host remains unknown.

——**tectonae** Racib. (Sacc. **XVI**:362 ; Syd. **IV**:422 ; **452** : 443 ; **89**:48 ; **388**:150 as *Aecidium effusum* Niessl). On leaves of *Tectona grandis*, Royal Botanic Garden, Calcutta (Kurz) ; Mysore, Pusa, Calcutta (Butler) ; Rangpur (Som ; Mitra) ; Comilla (Inayat) ; Birbhum (Basu).

——**tephrosiae** Rabenh. (Syd. **IV**:485). On *Tephrosia purpurea*, India. The name only was given by Rabenhorst in Fungi Europaei No. 2375, and in Sacc. **VII**:861. Sydow (**443, XIV**:257) thinks it possible that it may belong to *Ravenelia mitis* Syd.

——**tephrosiicola** P. Henn. (Sacc. **XVII**:446 ; Syd. **IV**:485 ; **113**(20):69). Recorded by McRae on *Tephrosia candida*, India.

——**valerianae-wallichii** Diet. (**219**:303 ; Sacc. **XXI**:794 ; Syd. **IV**:404 ; **452**:443 ; **453**:508 ; **35, I**:352, as "*Uromyces ?valerianae* Schum." ; **35, III**:77 ; **211**:264). On leaves of *Valeriana wallichii*, Simla (Barclay) ; Kumaon (Inayat) ; of *V. leschenaultii*, Kumaon (Inayat) ; Shillong (Butler) ; of *V.* sp., Mussoorie (Butler) ; Shillong (Subramaniam). An *Aecidium* on the specimen collected by Barclay, and on some of the later collections, was thought by Sydow (**452**:443) to be perhaps genetically connected, although Barclay (**35, I**:354) was quite convinced that there was no relationship between the two forms.

——**vicatiae** Syd. (**453**:508, as *U.* "*viaticae*" owing to a typographical error ; Sacc. **XXI**:789 ; Syd. **IV**:441). On leaves of *Vicatia coniifolia*, Kumaon (Inayat).

——**wedeliae-biflorae** Syd. (Sacc. **XXI**:796 ; Syd. **IV**:400). On leaves of *Wedelia urticaefolia*, Palghar, Bombay (Ajrekar).

Uromyces achrous Syd. (**453**:491 ; Sacc. **XXI**:549 ; Syd. **II**:91, fig.). On leaves and occasionally on branches of *Dalbergia sissoo*, Pusa (Butler).

——**agropyri** Barclay (**43**:212 ; Sacc. **XI**:181 ; Syd. **II**:317). On *Agropyron* sp., Bashahr, near Simla, 7,000 ft. (Lace).

——**aloës** (Cke) P. Magn. (Syd. **II**:265, fig. ; **445**:487 ; **17**, figs. ; Sacc. **XI**:227 as *Uredo*). On leaves of *Aloe spicata*, Coimbatore (McRae) ; of *A. vera*, Talegaon, Poona District (Ajrekar & Tonapy).

——**ambiens** Cke (**130**:75 ; Sacc. **VII**:584 ; Syd. **II**:152 ; **236**:272, figs. ; **43**:213 ; **449**:54, revised diagnosis). On leaves of *Buxus* probably *sempervirens*, near Dunooltie above Dehra Dun (Fleming) ; on leaves of *B. sempervirens*, Bashahr, near Simla, 6,000 ft. (Lace).

——**andropogonis-annulati** Syd. & Butler (**453**:492 ; Sacc. **XXI**:592 ; Syd. **II**:320, fig.). On leaves of *Andropogon annulatus*, Pusa, Orai, Cawnpore, Kumaon, Dehra Dun, and Samalkota (Butler) ; Bassein, Bombay (Chibber) ; Saharanpur (B. Das) ; Poona (Mitra) ; Lyallpur (A. Khan).

[**Uromyces**] **anthyllidis** (Grev.) Schroet. (Sacc. **VII**:551 ; Syd. **II**:64 ; **453**:490 as *U. trigonellae* Pass. ; **111**:362). On leaves of *Trigonella foenum-graecum*, Poona (Butler).

——**apludae** Syd. & Butler (**453**:493 ; Sacc. **XXI**:591 ; Syd. **II**:321, fig.). On leaves of *Apluda aristata*, Bassein, Bombay (Butler) ; Amritsar (Inayat).

——**appendiculatus** (Pers.) Link (Sacc. **VII**:535 ; Syd. **II**:120 ; **452**:428 ; **455**:256 ; **111**:260, fig.). On leaves of *Vigna catjang*, Coimbatore (Subramaniam) ; Myingyan, Burma (Butler) ; Dacca (Som) ; Mandalay (Shaw) ; Pusa (Mitra) ; Nagpur (Pandit) ; Poona (Ajrekar) ; of *V. vexillata*, Dharmsala, Punjab (Mitter) ; of *Dolichos lablab*, Dauracherra, Sylhet (Butler) ; Nagpur (Pandit) ; of *Phaseolus sp.*, Auda Tode, Wynaad, (McRae) ; of *P. mungo*, Samalkota, Madras (Barber) ; of *P. mungo* var. *radiatus*, Palghat, Madras (Subramaniam); of *P. vulgaris*, Godavari (Sundararaman) ; Anand, Bombay (Chibber).

——**arisaemae** Cke (Syd. **II**:295). On *Arisaema* sp., Chittagong (Sen).

——**behenis** (DC.) Unger (Sacc. **VII**:559 ; Syd. **II**:218 ; **455**:256). On leaves of *Lychnis indica*, Harwan, Kashmir (Butler).

——**bidentis** Lagerh. (Sacc. **XIV**:278 ; Syd. **II**:3 ; **452**:428). On leaves of *Bidens pilosa*, Nilgiris (Butler) ; Wynaad (McRae).

——**blainvilleae** Berk. (Sacc. **VII**:576 ; Syd. **II**:4 ; **455**:255). On leaves of *Blainvillea rhomboidea* (*B. latifolia*), Samalkota (Subramaniam).

——**chenopodii** (Duby) Schroet. (Sacc. **VII**:548 ; Syd. **II**:233). On *Suaeda fruticosa*, Lahore (Kashyap).

——**ciceris-arietini** (Grogn.) Jacz. & Boyer (Sacc. **XI**:175 ; Syd. **II**:84 ; **111**:271, fig.). On leaves of *Cicer arietinum*, Pusa (Butler) ; Jessore (Annett) ; Gokak Farm, Bombay (Kulkarni). Barclay (**41**) referred a rust of *C. arietinum* doubtfully to *U. pisi* ; it was probably *U. ciceris-arietini*.

——**comedens** Syd. (Syd. **II**:37 ; Sacc. **XXI**:571). On leaves of *Jasminum pubescens*, Manipur (Watt).

——**commelinae** Cke (Sacc. **VII**:573 ; Syd. **II**:292 ; **452**:429 ; **452**:443 as *Uredo ochracea* Diet., and *Uredo davaoensis* Syd. ; Syd. **IV**:595-596). On leaves of *Commelina bengalensis*, Tellicherry in Malabar, and Bangalore (Butler) ; of *C. obliqua*, Nilgiris (Butler) ; of *C.* sp., Darjeeling and Wynaad (McRae) ; Kanaighat in Sylhet, and Mandalay (Butler) ; Chittagong (Sen) ; Chikodi, Bombay (Kulkarni) ; of *Cyanotis tuberosa*, Yelwal in Mysore, and The Droog, Nilgiris (Butler). Sydow at first considered the uredo stage on *Cyanotis* to be distinct, but later (Monog. **IV**:596) thought it no more than biologically different.

——**decoratus** Syd. (**453**:491 ; Sacc. **XXI**:549 ; Syd. **II**:88, fig. ; **111**:374, figs.). On leaves of *Crotalaria juncea*, Samalkota (Butler) ; Dacca (Som) ; Godavari (Sundararaman) ; Dehra Dun (Kar) ; Sialkot (Cheema).

——**echinulatus** Niessl (**388**:149 ; Sacc. **VII**:577 ; Syd. **II**:348 ; **111**:69, fig., as *Uredo*.) On leaves of *Bassia latifolia*, Royal Botanic Garden, Calcutta

(Kurz); Palghar, Bombay (Chibber); Burdwan (Mitra); Pusa (Kar). Sydow (**444**:139) finds that only the uredo stage is known for this fungus.

[**Uromyces**] **eragrostidis** Tracy (Sacc. **XI**:182 ; Syd. **II**:326, fig. ; **453**:494). On leaves of *Eragrostis cynosuroides*, near Pusa (Butler); Chenab bank, Punjab (Cheema).

—**eriochloae** (Syd.) Syd. & Butler (**453**:492 ; Sacc. **XXI**:590 ; Syd. **II**:327, fig. ; **452**:444 as *Uredo eriochloae* Syd.) On leaves and culms of *Eriochloa polystachya*, Pusa, and Burhogah, Saran district, Bihar (Butler).

——**fabae** (Pers.) de Bary (Sacc. **VII**:531, pro parte ; Syd. **II**:103 ; **452**:428 ; **455**:255 ; **392**:6 ; **111**:29, 255, figs.). On leaves of *Vicia faba*, Pusa (Butler) ; Saharanpur and Lyallpur (Mitter) ; of *Pisum sativum*, Pusa (Butler) ; Maymyo, Burma (Rhind) ; of *P. arvense*, Pusa and Maymyo (Butler) ; of *Lathyrus sphaericus*, Pusa and Kumaon (Butler) ; on leaves and stems of *Lens esculenta*, Cawnpore (Hayman) ; Pusa (Butler). It may have been *U. fabae* which Barclay (**41**) saw, and referred to as " *U. pisi* on *Lathyrus sativus* ".

——**floscopae** (P. Henn.) Syd. (Sacc. **XXIII**:637 ; Syd. **IV**:596 ; **453**:509 as *Uredo assamensis* Syd.). On leaves of *Floscopa scandens*, Kanaighat, Sylhet (Butler).

——**geranii** (DC.) Otth & Wartm. (Sacc. **VII**:535 ; Syd. **II**:190 ; **455**:256). On leaves of *Geranium wallichianum*, Harwan, Kashmir (Butler).

——**heterogeneus** Cke (**132**:115 as *U.* " *heterogenis* " Cke, with fig. as *U.* " *heterogenum* " ; Sacc. **VII**:583 ; Syd. **II**:59). On leaves of *Hibiscus* sp., Kolapore, Bombay (Hobson).

——**hobsoni** Vize (**132**:115, fig. ; Sacc. **VII**:583 ; Syd. **II**:38 ; **319**, **I**:115 ; **452**:429 ; **111**:103, fig. ; **35**, **III**:76, figs., and **46**:150, figs. , as *U. cunninghamianus* Barclay). On stems of *Jasminum* sp., Kolapore, Bombay (Hobson) ; Kashmir (Butler) ; of *J. grandiflorum*, Sairi, near Simla (Barclay) ; Kumaon (Inayat) ; Mysore (Butler) ; Poona ; Dehra Dun ; of *J. arborescens*, Pusa (Ghosh) ; of *J. officinale*, Kangra District, Punjab (Mitter) ; of *J. malabaricum*, Mysore (Butler). Ajrekar (**11**) perhaps had this rust from Matheran (see *Blastospora butleri* Syd.).

——**inayati** Syd. (**453**:493 ; Sacc. **XXI**:590 ; Syd. **II**:321, fig.). On leaves of *Apluda aristata*, Kumaon (Inayat) ; Mussoorie (Butler) ; Wynaad (McRae) ; of *A. varia*, Kumaon (Inayat).

——**indicus** Patouill. (**360**:81, fig. ; Sacc. **VII**:558 ; Syd. **II**:307, fig.). On leaves of *Scirpus affinis*, India (Jacquemont). There is a good specimen of this rust at Paris. It was presumably found by Patouillard amongst the collections made by Jacquemont at an early date.

——**leptodermus** Syd. (**452**:430 ; Sacc. **XXI**:593 ; Syd. **II**:334 ; **453**:493 ; **452**:444 as *Uredo isachnes* Syd. ; **452**:444 as *Uredo panici-prostrati* Syd.). On leaves of *Panicum javanicum*, Dehra Dun, Pusa, Dohad Farm in Bombay, and Hoshangabad (Butler) ; Wynaad (McRae) ; of *P. isachnes*, Phulgaon, Deccan (Watt) ; Samalkota (Hafiz Khan) ; Nagpur (Pandit) ; Ganeshkhind Botanic

Garden, Poona (Gammie); Poona (Butler); of *P. prostratum*, Nilgiris and Pusa (Butler); Darjeeling (McRae).

[Uromyces] **lespedezae-procumbentis** (Schw.) Lagerh. (Syd. **II**:108; **455**:255; Sacc. **VII**: 549). On leaves of *Lespedeza bicolor*, Harwan, Kashmir (Butler).

——**linearis** Berk. & Broome (Sacc. **VII**:575; Syd. **II**:336, fig.; **453**:493; **455**:256; **111**:233, fig.). On leaves of *Panicum miliare*, Poona (Gammie); of *P. repens*, Poona, Pusa, Mandalay, Coimbatore, and Mysore (Butler); Ranchi (Mitra).

——**mac-intirianus** Barclay (**35**, **III**:79; Sacc. **IX**:293; Syd. **II**:25, fig.; **452**:429). On leaves of *Hemigraphis latebrosa*, Simla (Barclay); Parasnath, Bihar (Watt).

——**mimusops** Cke (Sacc. **VII**:579; Syd. **II**:39; **445**:487). On leaves of *Mimusops elengi*, South Kanara (McRae); Bombay (Ajrekar).

——**mucunae** Rabenh. (**387**:62; Sacc. **VII**:568; Syd. **II**:117; **455**:255; **111**:268, figs.). On leaves of *Mucuna ? pruriens*, Royal Botanic Garden, Calcutta (Kurz); on *M.* (*Stizolobium*) *deeringiana*, Pusa (Butler); Nagpur (Shrivastan); of *M.* sp., Maymyo and Dehra Dun (Butler); Wynaad (McRae); Poona.

——**mussooriensis** Syd. (**452**:430; Sacc. **XXI**:589; Syd. **II**:342, fig.). On leaves of *Stipa sibirica*, Mussoorie (Butler).

——**orientalis** Syd. (**453**:490; Sacc. **XXI**:547; Syd. **II**:102, fig.; **455**:255). On leaves and stems of *Indigofera linifolia*, Pusa and Hoshangabad (Butler); Gujrat, Punjab (Cheema); of *I. cordifolia*, Poona (Butler); on leaves of *I. glandulosa*, Poona (Kulkarni); Hoshangabad (Butler).

——**polygoni** (Pers.) Fckl (Sacc. **VII**:533; Syd. **II**:236; **455**:256). On leaves of *Polygonum aviculare*, Harwan, Kashmir (Inayat).

——?**proëminens** (DC.) Lév. (Sacc. **VII**:553, proparte; Syd. **II**:158; **452**:429 as *U. euphorbiae* Cke & Peck). On *Euphorbia* sp., Dharwar (Burkill); Nagpur (Pandit). The identity of this rust remains uncertain. Sydow does not include India for any of the twenty-seven species of *Uromyces* listed upon *Euphorbia*. Barclay's record (**43**:213) of *U. ? pulvinatus* Kalch. & Cke, on leaves of *Euphorbia hypericifolia* var. *indica*, Bashahr, near Simla, 6,000 ft. (Lace), was considered doubtful by Barclay himself. *U. pulvinatus* is now considered a synonym of *U. proëminens*, and *Euphorbia hypericifolia* L. is listed by Arthur (**20**:259) as a host of *U. proëminens* in North America.

——**rottboelliae** Arth. (**19**:229; Sacc. **XVII**:262; Syd. **II**:338; **452**:430). On leaves of *Rottboellia speciosa*, Jaunsar, 7,000 ft., (Duthie); Simla and Kashmir (Butler).

—— ?**scirpi** (Cast.) Burr. (Sacc. **VII**:558; Syd. **II**:302). A doubtful specimen on *Scirpus* sp., Lahore (Das).

——**schoenanthi** Syd. (**452**:429; Sacc. **XXI**:588; Syd. **II**:319). On leaves of *Andropogon schoenanthus*, Poona and Dharwar (Butler); Wynaad (McRae).

——**setariae-italicae** (Diet.) Yoshino (Syd. **II**:339, fig.; **452**:444 as *Uredo setariae-italicae* Diet.; Sacc. **XVII**:448; **111**:223, fig.). On leaves of *Setaria italica*, Poona and Pusa (Butler); Godavari (Mitra); Wynaad (McRae); of *S. inter*-

media, Poona (Butler); of *S. verticillata*, Poona (Butler); Jalu, Darbhanga District (Taslim); of *S. glauca*, Poona, Dehra Dun, Mysore, Hoshangabad, Saharanpur, and Pusa (Butler); Darjeeling (McRae).

[**Uromyces sojae** (P. Henn.) Syd. (Sacc. **XXI**:539; Syd. **II**:128; **452**:429). The rust reported to be on *Glycine soja*, Poona, from which the combination *Uromyces sojae* was made, proves to be upon *Mucuna*, and the fungus is *U. mucunae* Rabenh. No rust on *Glycine soja* has in reality been found in India.]

——**solidaginis** (Sommerf.) Niessl (Sacc. **VII**:566; Syd. **II**:10; **35**, **III**:77; **211**:269). On leaves of *Solidago virgaurea*, Simla (Barclay).

——**sphaerophleus** Cke (**132**:115, fig.; Sacc. **VII**:582; Syd. **II**:359; **210**:23, fig.). On leaves, apparently of *Ononis*, Kolapore, Bombay (Hobson).

——**striatus** Schroet. (Sacc. **VII**:542; Syd. **II**:115; **452**:428; **89**:45). On leaves of *Medicago sativa*, Poona, and Hissar, Punjab (Butler); Pusa (Shaw); Jullundur (Sahai); of *M. denticulata*, Hoshiarpur, Punjab (Hafiz Khan).

——**strobilanthis** Barclay (**35**, **III**:78; Sacc. **XXI**:564; Syd. **II**:26). On leaves of *Strobilanthes dalhousianus*, Simla (Barclay).

——**superfluus** Syd. (Syd. **II**:337, fig.; Sacc. **XXI**:593; **452**:430 as " *U. linearis* B. & Br. "). On leaves of *Panicum antidotale*, Dehra Dun (Butler).

——**trifolii** (Hedw. f.) Lév. (Sacc. **VII**:534, pro parte; Syd. **II**:132; **455**:256). On leaves of *Trifolium pratense*, Harwan, Kashmir (Butler); of *T. resupinatum*, Peshawar (Shaw).

——**vestergreni** Syd. (Syd. **II**:74; Sacc. **XXI**:354; **452**:429 as *U. verruculosus* B. & Br. (non Schroet.). On leaves of *Bauhinia tomentosa*, Yelwal, Mysore (Butler); of *B. acuminata*, Palghat, Madras (Subramaniam).

——**vignae** Barclay (**43**:211, fig.; Sacc. **XXI**:540; Syd. **II**:124). On leaves of *Vigna vexillata*, Tara Devi, near Simla (Barclay).

——**vossiae** Barclay (**35**, **III**:76, fig.; Sacc. **IX**:295; Syd. **II**:344). On leaves of *Vossia speciosa*, Simla (Barclay).

HYMENOMYCETES.

Agaricus burkillii (Massee) Sacc. & Trott. (Sacc. **XXIII**:302; **319**, **VI**:122 as *Psalliota burkillii* Mass.). On the ground under a wall, Calcutta (Burkill).

——**campester** L. (Sacc. **V**:997; **260**:153 as *Psalliota campestris* Fr.; **520**; **74**; **324**:197, figs.). On the ground, Pengi, North-western Himalaya (Marten); Pusa (McRae).

——**comosus** (P. Henn.) Sacc. & D. Sacc. (Sacc. **XVII**:84; **263**:331 as *Psalliota comosa* P. Henn.). On the ground, Botanic Garden, Saharanpur (Gollan).

——?**cretaceus** Fr. (Sacc. **V**:995; **263**:330 as *Psalliota* cfr. *cretacea* Fr.). On the ground, Botanic Garden, Saharanpur (Gollan).

——**exaltatus** Berk. (**57**, No. 291; Sacc. **V**:993). On clay and earthy banks, Darjeeling, 7,000 ft. (Hooker f.).

[Agaricus] ?elvensis (Berk. & Broome) Sacc. (Sacc. **V**:993 ; **263**:330 as *Psalliota* cfr. *elvensis* Berk. & Br.). On the ground, Botanic Garden, Saharanpur (Gollan).

——**fulviceps** Berk. (**57**, No. 403 ; Sacc. **V**:1010 as *A.* "*fulvipes* Berk." misprint for *A. fulviceps*). On the ground, Sikkim (Hooker f.).

——**latipes** Berk. (**57**, No. 383 ; Sacc. **V**:1000). On the ground, Nunklow, Khasi Hills, 4,000 ft. (Hooker f.).

——**rimosus** (P. Henn.) Sacc. & D. Sacc. (Sacc. **XVII**:83 ; **263**:331 as *Psalliota rimosa* P. Henn.). On the ground, Botanic Garden, Saharanpur (Gollan).

——**squalidus** Massee (**319**, **XIV**:255 ; Sacc. **XXIII**:300). Calcutta (Burkill).

——**silvaticus** Schaeff. (Sacc. **V**:1000 ; **57**, after No. 291). On earth, Darjeeling, 7,500 ft. (Hooker f.).

——**woodrowii** Massee (**319**, **III**:151 ; Sacc. **XVII**:83 ; **524**:363, fig.). On the ground, Poona (Woodrow).

Aleurodiscus acerinus (Pers.) v. Hoehn. & Lits. (Sacc. **VI**:587 as *Stereum acerinum* Pers.). Reported on dead wood, India.

Amanita caesarea Scopoli (Sacc. **V**:8 ; **57**, after No. 360, as *Agaricus caesarius* Scop.). Khasi Hills (Hooker f.).

Amanitopsis berkeleyi (Hooker f.) Sacc. (Sacc. **V**:24 ; **57**, No. 243 as *Agaricus* (*Amanita*) *berkeleii* Hooker f.). On the ground, Darjeeling, 7,500 ft. (Hooker f.).

——**eriophora** (Berk.) Sacc. (Sacc. **V**:26 ; **57**, No. 242 as *Agaricus* (*Amanita*) *eriophorus* Berk.). On the ground, Darjeeling, 7,500 ft. (Hooker f.). Berkeley remarks that the stem, though bulbous, exhibits no trace of a volva.

——**fritillaria** (Berk.) Sacc. (Sacc. **V**:26 ; **57**, No. 261 as *Agaricus* (*Amanita*) *fritillarius* Berk.). Khasi Hills (Hooker f.). Saccardo placed this in *Amanitopsis*, although Berkeley describes a broad ring, and mentions no volva, but states that the stem is bulbous at the base.

——**regalis** (Berk.) Sacc. (Sacc. **V**:25 ; **57**, No. 241 as *Agaricus* (*Amanita*) *regalis* Berk.). On the ground, Jalapahar, Darjeeling, 7,500 ft. (Hooker f.).

——**vaginata** (Bull.) Roze (Sacc. **V**:21 ; **57**, after No. 360, as *Agaricus vaginatus* Bull.). Below Nunklow, Khasi Hills, 4,000 ft. (Hooker f.).

Annularia burkillae Massee (**319**, **XIV**:255 ; Sacc. **XXIII**:181 as *A. burkilli* Mass.). At the roots of trees, Maidan, Calcutta (Burkill).

Anthracophyllum nigrita (Lév.) Kalchbr. (Sacc. **V**:1139 ; **57**, No. 408 as *Xerotus lobatus* Berk. ; **84**;222). On dead wood, Khasi Hills (Hooker f.). Petch (**367**, IV :153) states that *A. nigrita* is a *Panus* identical with *P. melanophyllus* Fr. See also *X. lateritius*.

Armillaria adelpha Berk. (**57**, No. 251 ; Sacc. **V**:84). On dead wood, Darjeeling, 7,000-8,000 ft. (Hooker f.).

——**dichupella** Berk. (**57**, No. 247 ; Sacc. **V**:83). On dead wood, Darjeeling, 7,500 ft. (Hooker f.).

[Armillaria] duplicata Berk. (**57**, No. 248 ; Sacc. **V**:83). On dead wood, Darjeeling, 7,500 ft. (Hooker f.).

——**horrens** Berk. (**57**, No. 245 ; Sacc. **V**:82). On bark of old trees, etc., Darjeeling, 7,500 ft. (Hooker f.).

——**mellea** (Vahl) Quél. (Sacc. **V**:80 ; **266**:437). On *Picea morinda*, Deoban, Jaunsar, United Provinces (Hole). See also p. xiv above.

——**multicolor** Berk. (**57**, No. 249 ; Sacc. **V**:84). On dead wood, Jalapahar, Darjeeling, 7,500 ft. (Hooker f.).

——**omnituens** Berk. (**57**, No. 250 ; Sacc. **V**:84). On dead wood, Darjeeling, 8,500 ft. (Hooker f.).

——**rhizopoda** Cke (**162**:89 ; Sacc. **V**:79). On clay banks, mostly attached to roots of grass, sedges, etc., Afghanistan boundary (Aitcheson).

——**vara** Berk. (**57**, No. 246 ; Sacc. **V**:83). On rotten timbers, Sinchul, Himalaya, 8,600 ft. (Hooker f.).

Auricularia epitricha Berk. in Herb. (**179**:15 ; Sacc. **XI**:143 ; *A. carteri* Berk. [MS?]). On bark, Bombay ; Nilgiris.

——**mesenterica** Fr. (Sacc. **VI**:762 ; **467**:154 ; **288**, No. **60**:6). On dead wood, Bandra, Khandala, Bombay (Blatter) ; India (P. D. Master).

——**rugosissima** (Lév.) Bres. (**84**:231 ; **57**, No. 345, as *Phlebia reflexa* Berk. ; Sacc. **VI**:500 ; **288**, No. **46**:7 ; **287**:1263 as " *Auricula reflexa* " ; **319**, **IV**:94 as *Auricularia butleri* Massee). On wood, Great Runjeet River, 7,000 ft., and Tonglo, Sikkim, 10,000 ft. (Hooker f.) ; Dehra Dun (Butler) ; India (Bose).

——**vespertilis** Fr. subspecies **venulosa** Fr. (**240**:113 ; Sacc. **VI**:764). On trunks, Sikkim.

Bolbitius grandiusculus Cke & Massee (Sacc. **IX**:143 ; **319**, **III**:151). On the ground, Poona (Woodrow).

Boletus areolatus Berk. (**57**, No. 396 ; Sacc. **VI**:44 ; **57**, after No. 414). In open pastures, Kala Pani, Khasi Hills, 5,500 ft. (Hooker f.).

——**delphinus** Berk. (**57**, No. 331 ; Sacc. **VI**:28). On exposed ground, Darjeeling, 7,500 ft. (Hooker f.). A specimen was also sent to Montagne, and is now in Paris.

——**emodensis** Berk. (**57**, No. 329 ; Sacc. **VI**:20). On the ground, Darjeeling, 7,500 ft. (Hooker f.). A specimen was sent to Montagne. Berkeley gives a figure reference, but as he explains after No. 360, this was not published " but will appear in Sir W. J. Hooker's ' Icones ' ". Saccardo (l. c.) lists " Hook. tab. DCCCLXX", and occasionally refers to figures with other species collected by Hooker and described by Berkeley. These drawings were not published, but are filed at Kew.

——**flavipes** Berk. (**57**, No. 412 ; Sacc. **VI** : 28). On the ground, Myrong, Khasi Hills (Hooker f.).

——**fragicolor** Berk. (**57**, No. 394 ; Sacc. **VI**:19). Khasi Hills (Hooker f.).

[**Boletus**] **furfurasceus** Berk. (**57**, No. 392, and after No. 411 ; Sacc. **VI**:28). On clay banks, Moflong, Khasi Hills, 5,500 ft. (Hooker f.).

——**gigas** Berk. (**57**, No. 395 ; Sacc. **VI**:33). In copses of *Andromeda* and birch, Lachen River, Sikkim, 1,200 [?12,000] ft. (Hooker f.).

——**pusillus** Berk. (**57**, No. 413 ; Sacc. **VI**:46). On the ground, Moflong, Khasi Hills (Hooker f.).

——**scrobiculatus** Berk. (**57**, No. 397, and after No. 414 ; Sacc. **VI**:37). On soil in open places, Moflong, Khasi Hills (Hooker f.).

——**squamatus** Berk. (**57**, No. 393, and after No. 414 ; Sacc. **VI**:18). In woods, Myrong, Khasi Hills, 6,000 ft. (Hooker f.).

——**ustalis** Berk. (**57**, No. 330 ; Sacc. **VI**:20). On rotten tree trunks, Darjeeling, 7,500 ft. (Hooker f.).

——**verrucarius** Berk. (**57**, No. 414 ; Sacc. **VI**:33). On the ground, Sikkim (Hooker f.).

Calocera dilatata Mont. (**351**, No. 602 ; Sacc. **VI**:733 ; **350**:152 as *Clavaria dilatata* Mont.). On the ground, edge of hill forest, Madura (Bélanger). Montagne notes : " An *C. hostmanni* Lév. eadem ?" Specimen not found in Herb. Montagne.

—**sphaerobasis** Berk. (**57**, No. 347 ; Sacc. **VI**:737). On the ground, apparently springing from a twig, Darjeeling, 7,500 ft. (Hooker f.).

——**viscosa** (Pers.) Fr. (Sacc. **VI**:732 ; **196**:127 ; **144**:96). Belgaum (Hobson) ; Sibpur (Kurz).

Cantharellus cibarius Fr. (Sacc. **V**:482 ; **263**:328 ; **288**, No. **65**:4). India (Kashyap) ; on the ground under *Pinus longifolia*, Arnigadh, Mussoorie (Gollan).

——? **congregatus** Mont. (not *C. congregatus* (Pat.) Sacc. & Syd., Sacc. **XIV**:100) (**349**:21 ; **351**, No. 421). On dead trunks and wood, Ootacamund (Perrottet?). A specimen, in poor condition, is in Herb. Montagne in Paris. Montagne himself was uncertain about the species. The name *C. congregatus* Mont. was not published in Saccardo's Sylloge.

——**infundibuliformis** Fr. (Sacc. **V**:490 ; **57**, after No. 389). On the ground, Myrong, Khasi Hills (Hooker f.).

Cladoderris dendritica Pers. (Sacc. **VI**:549 ; **57**, after No. 450, as *Thelephora dendritica* Pers.). On rotten wood, Nunklow, Khasi Hills (Hooker f.).

——**mussooriensis** (P. Henn.) Sacc. (Sacc. **XVII**:163 ; **263**:324 as *Lachnocladium mussooriense* P. Henn.). On the ground. Arnigadh, Mussoorie (Gollan). Bresadola (**84**:60) states that this is a *Stereum* near *S. junghuhnii* Fr. Lloyd (**298**:11) also gives notes on the hymenium and spores, and finds it to be a *Stereum*.

Clavaria botrytis Pers. var. **concolor** Berk. (**57**, after No. 399). Khasi Hills (Hooker f.).

——**corniculata** Schaeff. (Sacc. **VI**:694 ; **352** as " *Calocera corniculata* "). Sonamarg, Kashmir (R. R. Stewart).

[Clavaria] formosa Pers. (Sacc. **VI**:699 ; **57**, after No. 399). Khasi Hills (Hooker f.).

——**fusiformis** Sowerby (Sacc. **VI**:718 ; **287**:955). India (Cave).

——**gollani** P. Henn. (**260**:151 ; Sacc. **XVI**:208). On the ground, Saharanpur (Gollan).

——**jacquemontii** Lév. (**284**:179, fig. ; **285**:214 ; Sacc. **VI**:698). On the ground, Kashmir (Jacquemont). The type specimen at Paris is a finely-branched fungus, still in good condition.

——**miltina** Berk. (**57**, No. 400 ; Sacc. **VI**:727). On rotten timber in wet woods, Kala Pani, Khasi Hills, 5,000 ft. (Hooker f.).

——**pyxidata** Pers. (Sacc. **VI**:698 ; **263**:324 ; **288**, No. **44**:4). On wood, Arnigadh, Mussoorie (Gollan) ; India (Legere).

——**stricta** Pers. (Sacc. **VI**:705 ; **57**, after No. 399). Khasi Hills (Hooker f.).

Clitocybe incongrua Berk. (**57**, No. 253 ; Sacc. **V**:195). On the ground, Jallapahar, Darjeeling, 7,500 ft. (Hooker f.).

——**laccata** (Scop.) Sacc. (Sacc. **V**:197 ; **263**:335 ; **57**, after No. 363, as *Agaricus laccatus* Scop.). On the ground, Arnigadh, Mussoorie, 5,500 ft. (Gollan) ; in pine woods, Sikkim, 11,000 ft. (Hooker f.).

——**pumila** Massee (**319**, **XIV**:254 ; Sacc. **XXIII**:62). About ants' nests under a wall, Calcutta (Burkill).

Collybia albuminosa (Berk.) Petch (**367**, **III**:268, with synonymy ; **248**:15 as *Armillaria eurhiza* Berk. ; Sacc. **V**:85 ; **71**:349, fig. ; **74**:643 as *Lepiota albuminosa* Berk.). This species grows from termites' nests, and is discussed by Bose (**78**). Edible, and known in Bengal as " Patal Kour " (Bose). Occurs also in Central Provinces and Berar, with several vernacular names (Graham).

——**ambusta** Fr. (Sacc. **V**:247 ; **70**, **I**:112). On burnt ground, Calcutta (Bose).

——**antitypa** Berk. (**57**, No. 263 ; Sacc. **V**:230). On mossy trunks, Darjeeling, 8,500 ft. (Hooker f.).

——**blandula** Berk. (**57**, No. 364 ; Sacc. **V**:219). In pine woods, Sikkim, 11,000 ft. (Hooker f.).

——**camptopoda** Berk. (**57**, No. 264 ; Sacc. **V**:231). On wood, Darjeeling, 7,500 ft. (Hooker f.).

——**dryophila** (Bull.) Fr. var. **caespitis** Berk. (**57**, No. 365 ; Sacc. **V**:234). Amongst grass and moss, Lachen, Himalaya, 14-16,000 ft. (Hooker f.).

——**lutea** Massee (**319**, **VI**:122 ; Sacc. **XXIII**:77). On a wall, Calcutta (Burkill).

——**longipes** (Bull.) Berk. (Sacc. **V**:202 ; **57**, after No. 401). Khasi Hills (Hooker f.).

——**macra** Berk. (**57**, No. 366 ; Sacc. **V**:236). On the ground in pine woods, Sikkim, 11,000 ft. (Hooker f.).

——**maculata** (Alb. & Schw.) Fr. (Sacc. **V**:207 ; **57**, after No. 363). In pine woods, on *Picea morinda* (*Abies smithiana*), Lachen, Himalaya, 9,000 ft. (Hooker f.).

——**mimica** W. G. Sm. (Sacc. **V**:214 ; **70**, **I**:112). In grassy fields, Calcutta (Bose).

——**napipes** Hook. f. in Berk. (**57**, No. 254 ; Sacc. **V**:201 as *C. napipes* Berk.). On the ground, Darjeeling, 7,500 ft. (Hooker f.).

[Collybia] papaveracea Berk. (**57**, No. 259 ; Sacc. **V**:225). On dead sticks in moss, Darjeeling, 7,500 ft. (Hooker f.).

——**podagrosa** Berk. (**57**, No. 260 ; Sacc. **V**:211). On clay banks, Sinchul, Himalaya, 8,000 ft. (Hooker f.).

——**radicata** (Rehl.) Berk. var. **superbiens** Berk. in litt. (Sacc. **V**:201). Recorded by Saccardo as a common form in the Khasi Hills.

——**raphanipes** Berk. (**57**, No. 255 ; Sacc. **V** : 202). On the ground, Jallapahar, Darjeeling, 7,000 ft. (Hooker f.).

——**rhodella** Berk. (**57**, No. 262 ; Sacc. **V**: 236). On wood, Darjeeling, 7,500 ft. (Hooker f.). Not *C. rhodella* Pat., Sacc. **V**: 233, **XIV** : 81.

——**rupicola** Massee (**319**, **I** :114 ; Sacc. **XVI**:25). Amongst rocks, Tehri Garhwal, Himalaya, 7,500 ft. (Gamble).

——**stillaticia** Berk. (**57**, No. 256 ; Sacc. **V** :231). On dead and living tree trunks, Jallapahar, Darjeeling, 8,000 ft. (Hooker f.).

——**stipitaria** Fr. (Sacc. **V**:216 ; **260**:153 ; **263** :335 ; **467** :158). On grass stems and roots, Botanic Garden, Saharanpur (Gollan) ; on wood, Simla (Blatter).

——**triplicata** Berk. (**57**, No. 258 ; Sacc. **V** :221). Habitat and exact locality not given, Sikkim (Hooker f.).

——**undabunda** Berk. (**57**, No. 257 ; Sacc. **V**:201). On old timber in woods, Darjeeling, 7,500 ft. (Hooker f.).

——**ustipes** Berk. (**57**, No. 261 ; Sacc. **V**:234). On the ground, Darjeeling, 8,000 ft. (Hooker f.).

——**velutipes** (Curt.) Fr. (Sacc. **V** :212 ; **57**, after Nos. 260 and 363, as *Agaricus velutipes* Curt.). On dead wood, Darjeeling, 7-8,000 ft., and in pine woods, Sikkim, 11,000 ft. (Hooker f.).

Coniophora indica Massee (**313** :134 ; Sacc. **IX** :241). On wood, Bombay.

——**membranacea** DC. (Sacc. **VI** :649 ; **144** :93). On walls of gaol, Simla (Prof. Balfour).

Coprinus comatus Fr. (Sacc. **V**:1079 ; **57**, after No. 311). Bombay (J. D. Campbell) ; on grassy earth, Darjeeling (Hooker f.).

——**fimbriatus** Berk. & Broome (Sacc. **V**:1105 ; **71**:352, fig.). Usually on dung, Howrah and Hooghly Districts, Bengal (Bose).

——**hookeri** Berk. (**57**, No. 312 : not in Sacc.). In grassy places, Jalapahar, 7,500 ft. (Hooker f.).

——**niveus** Fr. (Sacc. **V**:1088 ; **70**, I and II, figs.). On dung and heaps of rotten straw, common in Bengal (Bose).

——? **spraguei** Berk. & Curt. (Sacc. **V** :1101 ; **263** :329). On the ground, Botanic Garden, Saharanpur (Gollan).

——**vellereus** Berk. (**57**, No. 313 ; not in Sacc.). On dead wood and earth, Darjeeling (Hooker f.).

Corticium coeruleum Fr. (Sacc. **VI** :614 ; **319**, **I** :114 ; **467** :154 ; **379** :287). Khandala, Bombay (Blatter) ; on old dry wood (phosphorescent), Dehra Dun (Gamble).

——**dealbans** Tunstall (**504** :51, without formal description). India, " found in all districts " on bark of *Thea sinensis* (Tunstall).

——**incarnatum** (Pers.) Fr. (Sacc. **VI** :625 ; **263** :323). On dead twigs, Arnigadh, Mussoorie (Gollan).

——**invisum** Petch (**373**, **III** :316 ; **500**:257 ; **498**:53, figs.; **504**:48 ; **494**:121 as *Hypochnus theae* Bernard). The cause of the black rot of *Thea sinensis* in India, at first identified as *Hypochnus theae* Bernard, is now recognised to be *C. invisum.* Tunstall (**498** :55 ; see also **500** and **504**) notes, however, that there is another *Corticium* on tea which may be Bernard's species.

——**koleroga** (Cke) v. Hoehn. (**264**, No. 468 ; **111**:477, figs.; **487**:702 ; **126**, figs.; **517**, figs.; **134**:2, figs., as *Pellicularia koleroga* Cke ; Sacc. **IV**:149 ; **136** ; **135**, figs.). On twigs and leaves of *Coffea* spp., South India ; on *Gardenia gummifera, Plectronia* (*Canthium*) *parviflora, Lawsonia alba, Dendrocalamus* sp., *Jasminum* sp., *Pleopeltis linearis*, and *Niphobolus fissus*, Mysore (Venkatarayan).

——**leve** Pers (Sacc. **VI** :611 ; **57**, after No. 453). On decayed wood, Nangki, E. Nepal, 10,000 ft. (Hooker f.).

——**levigatum** Fr. (Sacc. **VI** :628 ; **196**:127). Yomah, Burma (Kurz).

——**repens** Berk. (**63** :811 ; not in Sacc.; **377**:1). The common fungus on tea and other plants in north-eastern India, known as " thread blight", was thus described by Berkeley in 1873 : " *Corticium repens* B. Hypothallo filiformi repente albo, hymenio pallide rufo. Spreading widely over living shrubs, on which it forms white linear creeping threads, which run off from the bark to the leaves ; hymenium of a very pale rufous colour. At present it has not been observed apparently in its most perfect state......" He reported that it occurred in India on tea and chestnuts, collected by Mr. Grote. In a later paper (**64**) Berkeley did not mention the name *C. repens*, and no one since has succeeded in assigning the fungus to a definite position, although it is recognised to be like a *Corticium* (**111**:456, figs. ; discussed by Petch, **377**)."

——**salmonicolor** Berk. & Broome (Sacc. **VI** :620 ; **367**, **III**:278 ; **379** :281 ; **111** : 102, 500, figs.; **498**:57 ; **392**:5 ; **394**:9 ; **344**; **500** ; **113**(21):59 ; **347**). On living stems of *Hevea, Thea, Coffea*, and other plants in Assam, South India, Burma, and the Andaman Islands ; on *Cinchona ledgeriana*, Mungpoo and Munsong (McRae) ; on *Citrus aurantium*, Assam (S. K. Mitra).

——**solani** Bourdot & Galzin (**113**(19);**459** ; **111**:21, 263, figs., as *Hypochnus solani* Prill. & Del.; Sacc. **XXI**:414 ; **409**:139, figs., as *C. vagum* Berk. & Curt.; **413** ; **392**:2 as *Rhizoctonia solani* Kuehn ; **393** :2 ; **113**(17):55). On living *Arachis hypogaea, Solanum tuberosum, Lycopersicum esculentum, Vigna catjang,*

Trichosanthes cucumerina, *Piper betle*, *Trifolium alexandrinum*, *Sesamum indicum*, and other plants in northern and western India, and in Burma.

[Corticium] violaceo-lividum (Sommerf.) Fr. (Sacc. **VI** :627 ; **196** :127). Yomah, Burma (Kurz).

Cortinarius emodensis Berk. (**57,** No. 384 ; Sacc. **IX** :121 and **XI** :65). In pine woods, on *Abies webbiana*, Lachen, Himalaya, 10,000 ft. (Hooker f.).

——**flammeus** Berk. (**57,** No. 386 ; Sacc. **IX** :127 and **XI** :66). In pine woods, Sikkim, 11,000 ft. (Hooker f.).

——**saniosus** Fr. (Sacc. **V**:980 ; **57,** after No. 386). In pine woods, Sikkim, 11,000 ft. (Hooker f.).

——**vinosulus** Sacc. (Sacc. **IX** :121 as " *C. vinosulus* Berk." ; **57,** No. 385 as *C. vinosus* Berk.; not *C. vinosum* Cke). In pine woods, Sikkim, 11,000 ft. (Hooker f.).

——**violaceus** Fr. (Sacc. **V** :924 ; **57,** after No. 385). In woods, Myrong, Khasi Hills (Hooker f.).

Crepidotus alveolus (Lasch) Fr. (Sacc. **V** :877 ; **263** :331). On the ground, Botanic Garden, Saharanpur (Gollan).

——**applanatus** (Pers.) Fr. (Sacc. **V**:878 ; **263**:331). On dead branches, Arnigadh, Mussoorie (Gollan).

Cyclomyces turbinatus Berk. (**57,** No. 445 ; Sacc. **VI**:390). On decayed wood, Nunklow, Khasi Hills (Hooker f.).

Daedalea andamani Berk. in Herb. (**174** :93 ; Sacc. **XI** :100). On trunks, Andaman Islands.

——**boseii** Lloyd (**287**:1069, 1109, fig.; **70, VII** :32). On dead branches of *Mangifera indica*, Bengal (Bose).

——**discolor** Fr. (Sacc. **VI** :383 ; **196**:125). Yomah, Burma (Kurz).

——**emodensis** Berk. (**57,** No. 444 ; Sacc. **VI**:374). On dead wood, Lebong, Darjeeling (Hooker f.).

——**flavida** Lév. (Sacc. **VI**:381 ; **70, V**:24, figs.; **287**:1010 ; **70, VII**:33 as *D. microzona* Lév.=form of *D. flavida* Lév.; **288,** No. **60**:6 as " *Trametes flavida* "; **287**:1069). India (P. D. Master) ; on logs and bamboos, Calcutta (Bose) ; the form *microzona*, which is stated to have more regular pores and a thinner substance, Sunkna, Darjeeling (Bose). Lloyd (**288,** No. **36**:3) suggested that *D. flavida* is the same as *Lenzites ochroleuca* Lév. but he continued to use the former name also. See " *Lenzites flavida.*"

——**gollanii** Massee (**319, VIII** :217 ; Sacc. **XXIII** :450). On dead wood, Mussoorie (Gollan).

——**hobsoni** Berk. (Sacc. **VI**:379 ; **70, II**:143, figs.). Bengal (Bose). Perennial.

——**latissima** Fr. (Sacc. **VI** :383 ; **274** :482 ; **51** :382). Recorded from India (Wight), possibly in error for *D. sinulosa*.

——**pruinosa** Lév. (Sacc. **VI** :379 ; **84** :231 ; **286** :145 as *Sistotrema ochroleucum* Lév.; **288,** No. **28** :2 ; **286**:143 as *Hexagonia glabra* Lév.; **57,** No. 409 as

Lenzites ochrophyllus Berk.; **174** :93 as *Daedalea flabellum* Berk. in Herb.; **57,** No. 340, as *Trametes lobata* Berk.; **196**:124 ; **467**: 157 as *Lenzites ochroleuca* Lév.; **287**:503, fig.; **288,** No. **31**:3, and No. **49**:3 ;**293**:31, figs., as *Hexagonia ochroleuca*). On trunks, Bombay (Polydore Roux) ; on dead charred wood, Darjeeling, 7,500 ft., and on dead wood, Mungdurbi, Darjeeling, 4,000 ft. (Hooker f.) ; on old logs, Sikkim Terai (Kurz) ; Khandala, Bombay (Blatter) ; Bengal (Hutchings) ; India (Kirtikar ; Cave). Lloyd (**293**) discusses the synonymy of this polymorphic fungus. In its most usual form it is a *Lenzites*. Lloyd considers that *Daedalea flavida* Lév. (see above) may be the same species. Bresadola (**84**:230-231) considers *Daedalea pruinosa* the name to apply to this species. The type specimens of *Hexagonia glabra* and *Sistotrema ochroleucum* are in good condition at Paris, with notes by Bresadola.

[**Daedalea**] **quercina** (L.) Pers. (Sacc. **VI**:370 ; **70, I** :113). On dead wood, Calcutta (Bose).

——**sinulosa** Klotzsch (**239** :495 ; Sacc. **VI** :384). On wood, India. See *D. latissima*.

——**suberosa** Massee (**319, IV**:94 ; Sacc. **XXI**:351). On wood, Mysore (Butler).

——**subsulcata** Berk. & Broome (Sacc. **VI** :372). On dead wood, Narcondam, Andaman Islands (C. G. Rogers).

——**tenuis** Berk. (Sacc. **VI** :376 ; **57,** after No. 443 ; **196** :125). On dead wood Parasnath, and Khasi Hills (Hooker f.); Yomah, Burma, and South Andaman Island (Kurz).

——**unicolor** (Bull.) Fr. (Sacc. **VI** :377 ; **70, IX** :41). On prostrate trunks and stumps, Pashok, Darjeeling and Kawnghka, Burma (Bose).

——**zonata** Schwein. (Sacc. **VI** :382 ; **196** :125). Pellowa, Toukyeghat, Burma (Kurz).

Deconia atro-rufa Schaeff. (Sacc. **V**:1059 ; **263**:330 as *Psilocybe atrorufa* (Schaeff.) Fr. form *minor*). On the ground, Botanic Garden, Saharanpur (Gollan).

Eccilia blandfordii P. Henn. (**260** :153 ; Sacc. **XVI** :82). On the ground, Kalsia (J. H. Blandford) ; Botanic Garden, Saharanpur (Gollan).

——**griseo-rubella** (Lasch) Sacc. (Sacc. **V** :730 ; **263**:333). A form, on the ground, Botanic Garden, Saharanpur (Gollan).

Elmerina vespacea (Pers.) Bres. (**84** :231 ; **240** :101 as *Hexagonia macrotrema* Jungh.; Sacc. **VI**:369). Recorded by Fries from Java and India, but Lloyd (**293**:30) found no Indian specimen ; yet he (**288,** No. **28**:2) records *Hexagonia albida* Berk., which Bresadola (**84** :230) considers to be *E. vespacea*, from Bengal (Hutchings) ; see also *Lenzites alutacea*.

Entoloma cystopodum Berk. (**57,** No. 285 ; Sacc. **V**:698). On dead leaves, twigs, moss, etc., Darjeeling, 8,000 ft. (Hooker f.).

——**euthelum** Berk. (**57,** No. 378 ; Sacc. **V**:694). In pine woods, Sikkim, 11,000 ft. (Hooker f.).

[Entoloma] goliath Hook. fil. (**57,** No. 284 ; Sacc. **V** :680 as *E. goliath* Berk.). In woods, Darjeeling, 7,500 ft. (Hooker f.).

——**microcarpum** Berk. & Broome (Sacc. **V** :687 ; **78** :256). Growing from old termite nests or from the soil, Hooghly District and elsewhere in Bengal (Bose). Commonly eaten by the villagers.

Exidia bursaeformis Berk. (**57,** No. 348 ; Sacc. **VI**:773). On moss and tree trunks, Darjeeling, 7,000-10,000 ft. (Hooker f.).

——**glandulosa** (Bull.) Fr. (Sacc. **VI** :773 ; **349** :23). Nilgiris (Perrottet). A small specimen is in Montagne's Herbarium.

———— var. **fuliginosa** Mont. (**350** :152). On old trunks, near Kaschaou (Bélanger).

Exobasidium assamense Syd. & Butler (**455** :275 ; Sacc. **XXIII** :556). On leaves of *Camellia drupifera*, Dumpep, Khasi Hills (Burkill & Banerjee).

——**butleri** Syd. (**455** :279 ; Sacc. **XXIII** :554). On leaves of *Rhododendron arboreum*, Kumaon Himalaya (Inayat).

——**cinnamomi** Petch (Sacc. **XXI** :419 ; **367, III** :279). It is perhaps a fungus identical with this Ceylon species to which Gamble (**242** :138) referred as *E. cinnamomi* Massee to be published (a nomen nudum, as the description was never published by Massee), on leaves of *Cinnamomum tamala*, Himalaya. (A mention of this fungus is also given in **357** :133). No Indian specimen was found at Kew.

——**euryae** Syd. & Butler (**455** :275, figs.; Sacc. **XXIII** :555 ; **111** :85, fig.). On inflorescence of *Eurya acuminata*, Khatamandu, Nepal (Manners Smith).

——**indicum** Syd. & Butler (**455** :279, figs.; Sacc. **XXIII** :555 ; **511** :36, fig.). On leaves of *Symplocos theaefolia*, Darjeeling (McRae).

——**pieridis** P. Henn. (**455**:277 ; **511**:37). On leaves of *Pieris ovalifolia*, Ranikhet (Butler) ; Darjeeling (Hafiz Khan) ; jungles in the neighbourhood of Kalimpong, near Darjeeling (Tunstall).

——**vexans** Massee (**318** :109 ; Sacc. **XVI** :198 ; **111** :422, figs.; **310,** figs.; **322,** figs.; **323,** figs.; **455** :274 ; **511** ; **499** ;**501** ; **503**). On leaves and twigs of *Thea sinensis*, Assam and Darjeeling (Mann ; McRae ; Tunstall ; Bose ; Butler, et al.).

Favolus bengala Bose in Lloyd (**287**:1147, fig., 952 and 1010 ; **79** :138). India (Bose).

——**brasiliensis** Fr. (Sacc. **VI** :394). Recorded by Saccardo as from Sikkim.

——**boucheanus** Klotzsch (Sacc. **VI**:392). Narcondam, Andaman Islands (C. G. Rogers).

——**jacobaeus** Sacc. & Berl. (Sacc. **IX**:203 ; **287** :1126). South India (D. Maruda Rajan).

——**scaber** Berk. & Broome (Sacc. **VI** :393 ; **70, I**:113). On rotten wood, Calcutta (Bose). Petch (**372**:28, 57) states that this is a *Hexagonia*, and renames it *H. scabra* (B. & Br.) Petch.

[**Favolus**] **septiporus** Berk. (**57,** No. 446; Sacc. **VI** :402). On dead wood, Nunklow, Khasi Hills (Hooker f.).

——**spathulatus** (Jungh.) Bres. (**84** :230; **57,** after No. 445, as *F. multiplex* Lév.; Sacc. **VI** :398; **287** :955 as *Polystictus vibecinus* Fr.). Churra (Hooker f.); India (Cave).

——**tenerrimus** Berk. (**57,** No. 341; Sacc. **VI** :396). Darjeeling (Hooker f.).

——**tessellatus** Mont. (Sacc. **VI** :393; **263** :327). On *Barringtonia acutangula*, Botanic Garden, Saharanpur (Gollan).

Fistulina hepatica Fr. (Sacc. **VI** :54; **57,** after No. 343). Darjeeling, 4,000 ft. (Hooker f.).

Flammula chrysomyces Berk. (**57,** No. 289; Sacc. **V** :825). On dead wood, Darjeeling, 7-8,000 ft. (Hooker f.).

——**dilepis** Berk. & Broome (Sacc. **V** :812; **71** :351, figs.). Very common in stumps and holes in palms and large trees in Bengal (Bose).

——**flavida** (Schaeff.) Fr. (Sacc. **V**:820; **57,** after No. 379, as *Agaricus flavidus* Schaeff.). In pine woods, Sikkim, 11,000 ft. (Hooker f.).

——**macrophala** (Berk.) Sacc. (Sacc. **V**:817; **57,** No. 293, as *Agaricus* (*Hypholoma*) *macrophalus* Berk.). On tree trunks, Darjeeling, 7-8,000 ft. (Hooker f.).

——**phlegmatica** Berk. (**57,** No. 379; Sacc. **V** :815). In pine woods, Sikkim, 11,000 ft. (Hooker f.).

——**sapinea** Fr. (Sacc. **V**:824; **57,** after No. 401; **263**:333; **319, III**:151). Simla (Thomson); on wood ?, Botanic Garden, Saharanpur (Gollan); Mirga Forest, Chitral Relief Expedition, 9,000 ft. (Duthie).

Fomes adamantinus (Berk.) Sacc. (**57,** No. 426 as *Polyporus* (*Placodermei*) *adamantinus* Berk. (Sacc. **VI**:204; **70, VII**:30; **299**:235, fig.). On dead wood, Khasi Hills and Darjeeling .(Hooker f.); at the base of a palm tree, Royal Botanic Garden, Calcutta (Bose).

——**annosus** Fr. (Sacc. **VI**:197; **265**: 191; **266**:435; **88,** figs.; **89**:48; **70, IX**:39). On roots and collar of *Cedrus libani* var. *deodara*, Himalaya, and on *Abies pindrow* and *Picea morinda*, Deoban, Jaunsar (Hole); at the base of stumps of pine trees, and on pine wood paling, Shillong (Bose).

——**annularis** Lloyd (**70, II**:141, figs.). On tree trunks, Darjeeling (Bose).

——**badius** Berk. (Sacc. **VI**:175; **89**:48; **299**:249; **516,** figs., as *F. pappianus* Bres.). Parasitic on *Acacia arabica*, Berar (Vahid) and Sind. Lloyd (**299** and **288,** No. **60,** note 383) considers this species to be very close to *F. rimosus*.

——**brunneo-pictus** Berk. (Sacc. **VI**:155; **196**:122, figs., as *Polyporus* (*Pleuropus*) *brunneo-pictus* Berk. var.). On old wood, Arracan, Kolodyne Valley (Kurz).

——**durissimus** Lloyd (**73**:130, figs.; **287**:1069). On dead stem of *Artocarpus* and on fern stem, Calcutta (Bose).

——**elegans** Wakefield (**519, XXIV**:207; Sacc. **XXIII**:399). On living *Shorea robusta*, Singhbhum, Bengal (Hole). Not recorded as especially injurious.

[Fomes] endophaeus Berk. (**57,** No. 427 as *Polyporus* (*Placodermei*)*endophaeus* Berk.; Sacc. **VI**:178). On dead wood, Khasi Hills (Hooker f.). Lloyd (**299**:280) states that no type exists, but that it was probably *F. melanoporus* (q. v.).

——**fastuosus** Lév. (Sacc. **VI**:172 ; **70, VII**:31 ; **287**:1069). On trunks, Jalpaiguri, Bengal (Bose). Bose considers this practically equivalent to *F. senex*, and notes that Lloyd (**299**:250, 277) considers *F. fastuosus* very similar to *F. pseudosenex*.

——**fomentarius** (L.) Fr. (Sacc. **VI**:179 ; **299**:281, 235, fig.; **288,** No. **49**:3 ; **287**: 1295 ; **57,** after No. 425, as *Polyporus fomentarius* ; **352** as *Elfingiella fomentaria* ; **157**:2 as *Polyporus introstuppeus* Berk. & Cke). On dead wood, Khasi Hills and Darjeeling, growing to an enormous size on poplars in the more northern Himalaya (Hooker f.) ; on branches of *Juglans*, north-west India ; India (G. H. Cave ; B. Sahni ; P. L. Dey) ; on *Betula*, Sonamarg, Kashmir (R. R. Stewart). Bresadola considers "*P. introstuppeus* Berk." to be "*Fomes inzengae* Fr." (**84**:225).

——**geotropus** Cke (Sacc. **VI**:166 ; **287**:1186). India (Bose).

——**holosclerus** Berk. (Sacc. **VI**:193 ; **196**:123 as *Polyporus* (*Placodermei*) *holosclerus* Berk.). Yomah, Myodwine, Burma (Kurz). Lloyd (**300**:380) states that this species is *Polyporus gilvus*.

——**hypoplastus** Berk. (Sacc. **VI**:151 ; **196**:121 as *Polyporus* (*Mesopus*) *hypoplastus* Berk.). On bamboo, Howrah District, Bengal (Kurz).

——**igniarius** (L.) Fr. (Sacc. **VI**:180; **57,** after No. 426, as *Polyporus igniarius* Fr.). On dead trees, Darjeeling (Hooker f.).

——**inzengae** de Not. (Sacc. **VI**:175). See *F. fomentarius* above.

——**lamaoensis** (Murr.) Sacc. & Trott. (Sacc. **XXI**:287 ; **287**:1069, 1186, 1266 ; **70, VII**:29 ; **506**:28 ; **493**:6, figs., as *Hymenochaete noxia* Berk.; **111**:429, figs.; **494**:115 ; **393**:4 ; **394**:4 ; **433**). On roots of many trees and shrubs, including tea, coffee, rubber, and cinchona, Assam and Bengal (Tunstall) ; Sundribunds, Bengal (Bose); Burma (Rhind) ; India (D. Maruda Rajan).

——**lignosus** Klotzsch (Sacc. **VI**:146 as *Polyporus* ; **70, VII**:28 as *F. lignosus* form *kamphoeveneri* Fr. ; **392**:5 ; **506**:35). On stumps of various trees and other plants, Sukna, Darjeeling (Bose) ; almost unknown on rubber in Burma (Rhind) ; rare on tea, northeast India (Tunstall). Discussed by Lloyd (**299**:230), who considers *F. kamphoeveneri* Fr. to be a synonym.

——**marginatus** Fr. (Sacc. **VI**:168 ; **57,** after No.427, as *Polyporus marginatus* Fr.; **196**:123). On dead trees, Khasi Hills and Sikkim (Hooker f.) ; in pine forests, Bookee, Karen Country, Burma, 4-6,000 ft. (Kurz). Lloyd (**299**:219) states that this is only a form of *F. pinicola* Fr. occurring on frondose trees. He states that he has seen specimens of *F. pinicola* from India, but does not indicate the type of host of the specimens.

——**melanoporus** Mont. (Sacc. **VI**:196 ; **70, VII**:30 ; **299**:240, 279 ; **287**:1069 ; **157**:2 as *F. cornu-bovis* Cke, *Polyporus phaenus* Berk. in Herb.). In Dooars

Forests, Jalpaiguri, Bengal (Bose); on rotten branches, Khasi Hills. As mentioned, Lloyd suggests that F. *endophaeus* Berk. is probably this species.

[Fomes] nigro-laccatus Cke (Sacc. **VI**:177 ; **287**:1165 ; **288**, No. **38**:8 ; **299**:265). India (H. Val Ryan). Lloyd considers this species to be a *Ganoderma* similar to *G. applanatum*.

——**pachyphlaeus** Patouill. (Sacc. **IX**:174 ; **70, V**:22, figs.; **299**:260, 279 ; **70, VII**:32 as *F. elmeri* (Murr.) Sacc. & Trott.). On outer bark of *Ficus bengalensis*, Royal Botanic Garden, Calcutta, and on dead wood, in grassland forest, Cox's Bazaar, Bengal (Bose). Bose keeps the two species separate, although he recognizes that Lloyd considered them identical.

——**pallidus** Petch (Sacc. **XXIII**:399 ; **70, III**:2, figs.). On wood, Tipperah, Bengal (Bose).

——**pectinatus** Klotzsch (Sacc. **VI**:193 ; **70, V**:23, figs.; **299**:253, figs.; **274**:485 as *Polyporus pectinatus* Klotz.; **51**:388). India (Wight); on bark of *Glycosmis pentaphylla*, Bengal (Bose) ; " Used by the villagers as a curative against eczema of the car " (Bose).

——**peguanus** Mont. (**350**:148 ; Sacc. **VI**:179). On trunks of *Nauclea*, Burma (Bélanger). Lloyd (**299**:284) states that the type is in Berlin, and is the same as *F. caliginosus* Berk.

——**pinicola** Fr. (Sacc. **VI**:167 ; **299**:219). Lloyd's mention of India is noted under *F. marginatus*. See also *F. ungulatus*.

——**pseudoferreus** Wakefield (Sacc. **XXIII**:399 ; **394**:3 ; **386**:647). On roots of *Hevea brasiliensis*, Burma (Rhind ; Pinching).

——**pseudosenex** (Murr.) Sacc. & Trott. (Sacc. **XXI**:292 ; **288**, No. **60**:7). North-west Himalaya, 6,500 ft. (W. T. Saxton.)

——**pudens** Berk. (**57**, No. 418 as *Polyporus* (*Pleuropus*) *pudens* Berk.; Sacc. **VI**:162). In woods, Myrong, Khasi Hills, 6,000 ft. (Hooker f.). Lloyd (**296**:126) states that this species is known from a single, young, half specimen.

——**rimosus** Berk. (Sacc. **VI**:181 ; **260**:151 ; **70, V**:23, fig.; **435**:69 ; **287**:1125) On dead trunks of *Dalbergia sissoo*, Kalsia (J. H. Blandford) ; on dead stem of *Heritiera minor*, Sundribuns, Bengal (Bose) ; on a living *Acacia* sp., Coimbatore (**435**) ; India (D. Maruda Rajan).

——**rufolaccatus** Bose (**73**:129, fig.; **287**:1069, 1147, fig.). On a dead tree, Simla District (Bose).

——**semitostus** Berk. (**57**, No. 430 as *Polyporus* (*Placodermei*) *semitostus* Berk. Sacc. **VI**:200 ; **391**:140 ; **299**:221 ; **287**:1126 as *Trametes semitosta*). On dead wood, Khasi Hills (Hooker f.) ; South India (D. Maruda Rajan) ; Nicobar Islands. Bresadola (**84**:235) considers *Polyporus plebejus* Berk. from India to be *F. semitostus*, but Lloyd (**299**:220, 227 ; **288**, No. **68**:10) does not agree, and refers both species to *Trametes* (see *T. plebeja* below). *Fomes semitostus* has also been confused with *F. lignosus*.

[**Fomes**] **senex** Nees & Mont. (Sacc. **VI**:164 ; **57,** after No. 426, as *Polyporus senex* Nees & Mont.; **472**:151 ; **467**:157 as *Polyporus ? senex* Mont.; **70, IV**:2, figs.; **288,** No. **38**:8, No. **60**:6, No. **69**:5 ; **299**:259, fig.). On tree trunks, Sikkim ; on dead trees, Khasi Hills and Darjeeling (Hooker f.) ; on old stems, St. Xavier's College, Bombay (Blatter) ; on dead stumps, Khulna and Calcutta (Bose) ; India (H. V. Ryan ; P. D. Master ; Cave).

——**spadiceus** Berk. (**51**: 388 as *Polyporus spadiceus* Berk.; Sacc. **VI**:193 ; **300**:358). On trunks, India (Wight). The fungus referred by Troup (**490**:19) to *Fomes fulvus* (Scop.) Bres. on *Xylia dolabriformis*, Burma, is considered by Weir (in litt.) to be probably *F. spadiceus.*

——**subresinosus** Murrill (Sacc. **XXI**:285). Stated by Lloyd (**299**:215) to occur in India.

——**thomsoni** Berk. (**57,** No. 428 as *Polyporus* (*Placodermei*) *thomsoni* Berk.; Sacc. **VI**:170). On trunks, probably of pines, Simla (Thomson). Lloyd (**299**:286) states that the type is old and poor, and represents a *Trametes*, possibly *T. persoonii.*

——**ungulatus** (Schaeff.) Sacc. (Sacc. **VI**:167 ; **352**). Sonamarg, Kashmir (R. R. Stewart). Lloyd (**299**:286) says *F. ungulatus* is *F. pinicola.*

——**velutinosus** Lloyd (**299**:260, fig.). Bengal (Hutchings) ; India (Kirtikar). Lloyd states that this has the main characters of *F. senex*, but is thin, and possibly a *Polyporus.*

——**zeylandicus** Cke (Sacc. **IX**:168 as *Polyporus* ; **288,** No. **45**:2). India (Irani), " compared with type at Kew " (Lloyd).

Galera burkillii Massee (**319, X**:2 ; not in Sacc.). On the ground, Sureil, near Darjeeling (Burkill).

——**delicatula** Massee (**319, XIII**:189 ; Sacc. **XXIII**:279). In grass, Government Experimental Farm, Dacca (Burkill).

——**lateritia** Fr. (Sacc. **V**:860 ; **263**:331). On the ground, Botanic Garden, Saharanpur (Gollan).

——**tenera** (Schaeff.) Fr. (Sacc. **V**:860 ; **57,** after No. 290, as *Agaricus tener* Schaeff.). On the ground, Jalapahar and Darjeeling, 7,500 ft. (Hooker f.).

——**vinolenta** Berk. (**57,** No. 382 ; Sacc. **V**:861). In moss and on decayed wood in pine forest, Sikkim, 11,000 ft. (Hooker f.).

——**zeylanica** Petch (Sacc. **XXIII**:281 ; **71**:351, figs.). On the ground in grass, Hooghly, Bengal (Bose).

Ganoderma amboinensis (Lam.) Pat. (Sacc. **VI**:156 as *Fomes amboinensis* (Lam.) Fr.; **196**:122 as *Polyporus amboinensis* Fr.). Karen Country, Burma, 4,000 ft. (Kurz). Currey states that it is probably not distinct from *G. lucidum.*

——**applanatum** (Pers.) Pat. (Sacc. **VI**:176 as *Fomes applanatus* (Pers.) Wallr.; **263**:325 ; **467**:155 ; **70, III**:4, figs.; **506**:32 ; **70, V**:22, figs., as *Fomes leucophaeus* Mont.; **299**:264 ; **287**:1010, 1069, 1263 ; **196**:123 as *Polyporus* (*Placodermei*) *applanatus* Fr.). Toukyeghat, Toungoo, Burma(Kurz); on tree trunks,

Botanic Garden, Saharanpur (Gollan); Khandala, Bombay (Blatter); Bengal (Bose); India (H. V. Ryan); on tea, Darrang (Tunstall). Lloyd considers *F. leucophaeus* to be only a form of *G. applanatum*. He (**299**:265) interprets *F. nigro-laccatus* Cke as a form of *G. applanatum*, and records it for India (**288,** No. **38**: 8), although the " types " are different and Bresadola (**84**:226) refers Cooke's species to *Polyporus galegensis* Mont.

[Ganoderma] australe (Fr.) Pat. (Sacc. **VI**:176 as *Fomes australis* Fr.; **263**:325; **299**:265; **288,** No. **31**:2, No. **47**:2; **2**:55; **57,** after No. 425, as *Polyporus australis* Fr.). On wood, East Nepal and Khasi Hills (Hooker f.); Arnigadh, Mussoorie (Gollan); on *Grevillea* and *Coffea* roots, South India (McRae); India (Cave; Bashambar). Lloyd (**287**:1165) states that many of the older determinations are merely tropical forms of *Ganoderma applanatum*.

——**colossum** (Fr.) Bres. (Sacc. **VI**:138 as *Polyporus colossus* Fr.; **300**:368, figs.; **288**, No. **38**:5; **70, IV**:3, figs.). On logs, Hooghly, Bengal (Bose); India (J. H. Irani).

——**lucidum** (Leyss.) Karst. (Sacc. **VI**:157 as *Fomes lucidus* (Leyss.) Fr.; **260**:151; **391**:141; **70, II**:142, figs.; **263**:325; **288**, No. **27**:1, No. **31**:3, No. **40**:3, No. **60**:5; **97**, figs.; **92; 113** (6):53; **467**:155; **372**:47; **287**:1295; **331; 495**:91; **393**:3; **506**:33; **57,** after Nos. 337 and 417, as *Polyporus lucidus* Fr.; **196**:122; **467**:155 as *Ganoderma resinaceum* (Boud.)). On trunks of old trees, Darjeeling and Sone River, Bihar (Hooker f.); Royal Botanic Garden, Calcutta (Kurz); on dead roots of *Casuarina*, Botanic Garden, Saharanpur (Gollan); on trunks, Victoria Garden, Bombay, Adheri, and Salsette (Blatter); Calcutta and Hooghly, Bengal (Bose); on *Areca catechu,* which it apparently kills, Sylhet (Butler); Kulna; on *Guazuma* sp., Pusa (Hafiz Khan); on *Acacia melanoxylon*, Nilgiris, and *Pongamia glabra*, Bangalore (McRae); on tea stumps, Assam and Bengal (Tunstall); on *Morus* sp. cult., Maymyo, Burma (Rhind); on wood, Sukkur (J. H. Irani); India (H. E. Houghton; P. L. Dey); Nicobar Islands. Specimens of a form of this species were sent by Mr. R. S. Hole, Dehra Dun, to Massee, and named in litt. *Polystictus egregius* Massee, n. sp. (a nomen nudum). Specimens, possibly duplicates of those sent Massee, were sent to Pusa by Hole, and referred from Pusa to Miss Wakefield, who regards them as belonging to *G. lucidum*. See also *G. amboinensis*.

——**subtornatum** Murr. (Sacc. **XXI**:300; **70, IV**:2, fig.; **299**:269). On logs, Darjeeling (Bose).

Gloeoporus conchoides Mont. (Sacc. **VI**:403; **70, VIII**:28 as *Polyporus conchoides* Mont.). On a dead tree, Chittagong (Bose).

——**corrugatus** Berk. in Herb. (**177**:105; Sacc. **XI**:103). On trunks, Nilgiris.

Grandinia granulosa Fr. (Sacc. **VI**:501; **196**:126). Pellowa, Toukyeghat, Burma (Kurz).

Guepinia spathularia (Schw.) Fr. (Sacc. **VI**:807; **287**:1263; **196**:127, figs., as *G. ramosa* Currey; **263**:323). On old wood, Arracan and Howrah (Kurz);

Royal Botanic Garden, Calcutta (Butler); Botanic Garden, Saharanpur (Gollan); India (Bose). Massee (**314**:6) states that *G. ramosa* is identical with *G. fissa* Berk., and Bresadola (**84**:234) states that *G. fissa* is a form of *G. spathularia*.

Hebeloma catervarium Lév. (**286**:113 as *Agaricus catervarius* Lév.; Sacc. **V**:804). On trunks, Nilgiris (Perrottet). A specimen was not found at Paris.

——**thomasianum** Cke (**172**:7 ; Sacc.**IX**:102). Growing in a cactus hedge, Belgaum (Mrs. Thomas).

Helicobasidium purpureum (Tul.) Pat. (Sacc. **XIV**:1175 as *Rhizoctonia violacea* Tul.; **326**:110 ; **409**:139). On stems and roots of *Medicago sativa*, Hosur, Salem District, Madras (McRae). The connection with *Helicobasidium* is reported by Buddin and Wakefield (Trans. Brit. Myc. Soc., XII, pp. 116-140, 4 pls., 1927).

Hexagonia aculeata Mont. (Sacc. **VI**:358 ; **286**:143). Nilgiris (Perrottet). Although Lloyd (**293**:9) states that this species is known only from one collection made in French Guiana, the Paris Herbarium has also two pilei so labelled from the Nilgiris, collected by Perrottet in 1840.

——**apiaria** Pers. (Sacc. **VI**:358 ; **70, IV**:5, figs.; **435**:74 ; **287**:1125 ; **273**:200, fig., as *Polyporus wightii* Klotzsch ; **57**, after No. 445, as *Hexagonia wightii* Klotzsch ; **293**:6, figs.). India (Wight); Sone River, Bihar (Hooker f.); Hooghly, Bengal (Bose); Madras (**435**); India (Maruda Rajan). See *H. sinensis* below.

——**burchelli** Berk. in Lloyd (**288**, No. **61**:7 ; **287**:1069 ; **70**, **VII**:34 ; **372**:56). India (G. H. Cave); on dead trunk of *Diospyros embryopteris*, Faridpur, Bengal (Bose).

——**discopoda** Pat. & Har. (Sacc. **XI**:98 ; **79**:140). India (Bose). Stated by Petch (**372**:56) to be the common Ceylon *Hexagonia*; according to Lloyd (**288**, No. **28**:3) a form of *H. tenuis* (q. v.). Cultured by Bose (**81**).

——**kurzii** Currey (**196**:126, figs.; Sacc. **VI**:360 ; **70, VII**:35). On trunks, Mutlah, Lower Bengal (Kurz); Jalpaiguri, Bengal (Bose).

——**levis** Berk. in Herb. (**175**:103 ; Sacc. **XI**:99). On trunks, Andaman Islands. Lloyd (**293**:17) remarks that the species is based on nondescript material, but that he judges that it belongs to the section *Ungulaformis*.

——**scutigera** Fr. (Sacc. **VI**:362 ; **472**:150). Reported for India, but Theissen (l.c.) states that he doubts that *H. scutigera* occurs in India, for it is a form of *H. variegata*, which is an exclusively American plant. (See also Lloyd, **293**:13).

——**sinensis** Fr. (Sacc. **VI**:357 ; **273**:201 ; **51**:382 as *Polyporus (Favolus) sinensis* Fr. ; **239**:489 as *Trametes sinensis* Fr.). Recorded by Klotzsch and Berkeley from India (Wight), and considered synonymous with *P., wightii* Kl., but Lloyd (**293**:45) states that African specimens from Klotzsch are *H. hirta*.

[**Hexagonia**] **subtenuis** (Berk. in Herb.) Lloyd (**293**:26, 41; Sacc. **XXI**:348; **70**, **I**:114; **287**:502, fig.; **288**, No. **31**:3; **175**:103 as *H. tenuis* var. *subtenuis* Berk. in Herb.). Nilgiris; Bombay (Kirtikar). Lloyd (l.c.) discusses this species, which he considers distinct from *H. tenuis*.

——**sulcata** Berk. (Sacc. **VI**:364; **70**, **VII**:35). On dead wood, Jalpaiguri, Bengal (Bose).

——**tenuis** Hooker (Sacc. **VI**:366; **51**:382 as *Polyporus* (*Favolus*) *tenuis* Hook.; **57**, after No. 445, and **196**:125 as *H. tenuis* Fr. and as *H. polygramma* Mont.; **260**:152 as form *minor*; **263**:327; **84**:231; **288**, No. **28**:2, No. **60**:6; **293**:22, figs.; **467**:157; **287**:1126, 1295; **285**:199; **349**:22). On branches and trunks of trees, Parasnath and Sone River in Bihar, East Nepal, and Darjeeling (Hooker f.); Pellowa, Toukyeghat, Burma (Kurz); Botanic Garden, Saharanpur (Gollan); Bassein and Khandala (Blatter); Bilaspur, Central Provinces (Marten); Bengal (Hutchings); India (P. D. Master; P. L. Dey); South India (D. Maruda Rajan); Nicobar Islands. Lloyd and Bresadola regard *H. polygramma* as a large pored form of *H. tenuis*. Lloyd (**288**, No. **28**:3) received a *Hexagonia* from Bengal (Hutchings) which he considered probably the same as Fries' *H. tricolor*, which is, from its description (no type exists), close to or identical with *H. discopoda*, which is a form of *H. tenuis*. An Indian specimen of *H. polygramma*, perhaps from the Nilgiris, is at Paris.

——**umbrinella** Fr. Sacc. **VI**:365; **288**, No. **49**:4). India (Cave). Lloyd (**293**:26) considers this species to be close to *H. tenuis*.

Hirneola ampla (Pers.) Fr. (Sacc. **VI**:765; **284**:179, figs., as *Exidia nobilis* Lév.; **285**:218). On dead wood, Mussoorie (Jacquemont). The specimen at Paris is placed under *Hirneola nobilis* (Lév.) Fr., and bears a note by Bresadola: " vix dubia-*Hirneola nigra* (Sw.) Fr." but he (**84**:235) published it as a synonym of *H. ampla*.

——**auricula-judae** (L.) Berk. (Sacc. **VI**:766; **196**:128; **288**, No. **28**:2, No. **47**:2, No. **69**:5; **287**:1295; **494**:117; **495**:87; **497**:37; **352**; **467**:154 as *Auricularia sambucina* Mart.). On trunks, etc., Nakawa, Toukyeghat, Burma (Kurz); Khandala, Bombay (Blatter); Bengal (Hutchings); Sonamarg, Kashmir (R. R. Stewart); India (Bashambar; Cave; P. L. Dey); on tea stems, which it injures (Tunstall). Petch (**367**, **II**:414) discussed this and related tropical species.

——**auriformis** (Schw.) Fr. (Sacc. **VI**:765; **57**, after No. 347, as *Exidia protracta* Lév.; **84**:234). On trunks of living trees, Sikkim (Hooker f.).

——**cochleata** Fr. (Sacc. **VI**:765; **57**, after Nos. 347 and 453, as *Exidia hispidula* Berk.; **84**:233). Darjeeling; Kosderah, Sone River (Hooker f.).

——**crassa** Lloyd? (**287**:1275, figs., 1266). India (D. Maruda Rajan). Considered by Lloyd to be a dark, thick form of the next.

[**Hirneola**] **delicata** (Fr.) Bres. (**84**:231; Sacc. **VI**:407 as *Laschia tremellosa* Fr. ; **57**, after No. 453 ; **287**:784, fig.). In woods, Lebong, Darjeeling, 5,000 ft. (Hooker f.).

——**nigra** (Swartz) Fr. (Sacc. **VI**:768 ; **196**:128 as *H. auricula canis* Fr.). On logs, North Yomah, Burma (Kurz).

——**polytricha** Mont. (Sacc. **VI**:766 ; **319, III**:152 ; **288**, No. **60**:6 ; **350**:154 as *Exidia polytricha* Mont. ; **263**:323 as *Auricularia polytricha* Mont. ; **467**:154 ; **287**:1295 ; **68**:147). On wood, India (Bélanger) ; Belgaum (Hobson) ; Poona (Woodrow) ; Arnigadh, Mussoorie (Gollan) ; Khauheri Caves, Bombay (Blatter) ; India (P. D. Master ; P. L. Dey) ; Dastikop, Dharwar (Sedgwick) ; on dead twigs of *Tectona grandis*, Nidungayam, Malabar (Fischer).

Hydnum aitchesoni Berk. (**65**:137 ; Sacc. **VI**:454). Gulmarg, Kashmir, 8,500 ft. (Aitcheson). Esculent ; vernacular name " Ryle gŭb ".

——**analogum** Berk. in Herb. (**178**:1 ; Sacc. **XI**:108). On rotten wood, Nilgiris.

——**auriscalpium** L. (Sacc. **VI**:445 ; **57**, after No. 398). On fir-cones, Myrong, Khasi Hills (Hooker f.).

——**coralloides** Scop. (Sacc. **VI**:446 ; **57**, after No. 343 ; **319, III**:152). In crevices on old tree trunks, Darjeeling, 7,500 ft. (Hooker f.) ; Chitral Relief Expedition (Duthie).

——**delicatulum** Klotzsch in Fr. (**239**:515 ; Sacc. **VI**:458). On trunks, India.

——**delicatum** Klotzsch in Berk. (**51**:395 ; Sacc. **VI**:470 ; **349**:23). On rotten *Jatropha curcas*, Madras (Wight) ; on dead trunks, Kunda, Nilgiris (Perrottet). A good specimen of the latter is in Herb. Montagne, marked " verum exc. Berkeley ".

——**erinaceus** Bull. (Sacc. **VI**:449 ; **57**, after No. 448). Abundant on dead wood, Sikkim, 7,500 ft. (Hooker f.).

——**flabelliforme** Berk. (Sacc. **VI**:457 ; **57**, after Nos. 344 and 448). On dead wood, Darjeeling, 7-8,000 ft., and Lebong (Hooker f.).

——**gilvum** Berk. (**57**, No. 344 ; Sacc. **VI**:459 ; **367**, **III**:276). On dead trunks, Darjeeling (Hooker f.). A specimen was sent by Berkeley to Montagne.

——**gleadowii** Massee (**319**, **II**:166 ; Sacc. **XVI**:175 ; **242**). On dead wood, Dehra Dun (Gleadow). The species name and collector were published in error as " gleadonii " and " Gleadon ". Miss Wakefield, Kew, informs us that this is not a *Hydnum*, but a *Polystictus*, probably *P. leoninus*.

——**lachnodontium** Berk. (**178**:2 ; Sacc. **XI**:108). On logs, etc., Nilgiris.

——**olidum** Berk. (Sacc. **VI**:443 ; **288**, No. **65**; 4, 8). India (Kashyap).

——**pulcherrimum** Berk. & Curt. (Sacc. **VI**:452 ; **287**:1069). India (Bose).

——**repandum** L. (Sacc. **VI**:435 ; **263**:325). On the earth, Arnigadh, Mussoorie (Gollan).

——**rufescens** Pers. (Sacc. **VI**:436 ; **288**, No. **65**:4). India (Kashyap).

[**Hydnum**] **thwaitesii** Berk. & Broome (**67**:58 ; Sacc. **VI**:433). Nilgiris (E. S. Berkeley).

——**udum** Fr. (Sacc. **VI**:469 ; **196**:126). Mutlah, Lower Bengal (Kurz).

——**vespertilio** Berk. (**57**, No. 448 ; Sacc. **VI**:442). On the ground, Nunklow, Khasi Hills (Hooker f.).

——**zonatum** Batsch (Sacc. **VI**:441 ; **57**, after No. 447). On the ground, Nunklow, Khasi Hills (Hooker f.).

Hygrophorus fulvus Berk. (**57**, No. 388 ; Sacc. **V**:420). In pine woods, Sikkim, 11,000 ft. (Hooker f.).

——**hobsoni** Berk. (**66**:39 ; Sacc. **V**:390). Central India (Hobson).

——**miniatus** Fr. (Sacc. **V**:413 ; **57**, after No. 386). In pine woods, Lachen, Sikkim, 10,000 ft. (Hooker f.).

——**pomona** Berk. (**57**, No. 387 ; Sacc. **V**:420). On clay banks, Moflong, Khasi Hills (Hooker f.).

Hymenochaete cacao Berk. (**57**, No. 452 as *Stereum cacao* Berk. ; Sacc. **VI**:592 ; **147**:146 ; **287**:955, 1266). On dead timber, Khasi Hills (Hooker f.) ; India (Cave ; D. Maruda Rajan).

——**c[illegible]teri** Berk. in Herb. in Cooke (**147**:149 ; Sacc. **VI**:603). Bombay. Cooke [illegible]ublished this as an imperfect, excluded species.]

——**depallens** Berk. & Curt. (Sacc. **VI**:596 ; **147**:147 ; **379**:276 ; **263**:324). On dead branches, Botanic Garden, Saharanpur (Gollan). This species was published in Berkeley & Broome (**67**:68) but marked "*H. depallens* B. & C." and is usually so cited, although Petch (**379**) gives Berkeley and Broome as the authors.

——**leonina** Berk. & Curt. (Sacc. **VI**:597 ; **319**, **I**:114). On dead bark, Jaunsar (Gamble).

——**mougeotii** (Fr.) Cke (Sacc. **VI**:595 ; **57**, after No. 452, as *Stereum mougeotii* Fr. ; **196**:127 as *Corticium mougeotii* Fr.). On wood, Yangma Valley, East Nepal, and Singalelah, Sikkim, 10,000 ft. (Hooker f.) ; on dead trees, Phallut, Sikkim, 11-12,000 ft. (Kurz).

——**nigricans** (Lév.) Patouill. (Sacc. **XXI**:389 ; **467**:155 as *H. strigosa* Berk. & Broome ; **84**:233 ; **379**:274). On bark, Khandala, Bombay (Blatter).

——**rheicolor** (Mont.) Lév. (**286**:151 ; Sacc. **VI**:591 ; **349**:23 as *Stereum rheicolor* Mont. ; **147**:145 ; **379**:273). On dead trunks, Gudalur, Nilgiris. The type is in Herb. Mont., and there are also three Nilgiri specimens in the general Herbarium at Paris.

——**rubiginosa** (Schrad.) Lév. (Sacc. **VI**:589 ; **147**:145). Recorded as occurring in India, but no definite reference noted. No Indian specimens were found in the collections of Léveillé at Paris.

——**tenuissima** Berk. (Sacc. **VI**:593 ; **147**:146 ; **288**, No. 65·:4 ; **287**:1263). India (Kashyap ; Bose). According to Bresadola (**84**:233) the type of this species is *H. rheicolor*.

[**Hymenochaete**] **villosa** (Lév.) Bres. (Sacc. **XXI**:389 ; **287**:1263). India (Bose).

Hypholoma appendiculatum (Bull.) Sacc. (Sacc. **V**:1039 ; **263**:330). On the ground, Botanic Garden, Saharanpur (Gollan).

——**atrichum** Berk. (**57**, No. 295 ; Sacc. **V**:1035). On dead timber and soil impregnated with charcoal, Darjeeling, 7-8,000 ft. (Hooker f.).

——**castanophyllum** Berk. (**57**, No. 296 ; Sacc. **V**:1035). On the ground, Jallapahar, Darjeeling (Hooker f.).

——**condensum** Berk. (**57**, No. 297 ; Sacc. **V**:1042). On the ground, Darjeeling (Hooker f.).

——**fasciculare** (Huds.) Fr. (Sacc. **V**:1029 ; **57**, after Nos. 292 and 403). Abundant on dead wood, Darjeeling, 7-8,000 ft. (Hooker f.) ; Simla (Thomson).

——**hemisodes** Berk. (**57**, No. 294 ; Sacc. **V**:1035). On earth banks, Darjeeling, 7,600 ft. (Hooker f.).

——**sublateritium** (Schaeff.) Fr. (Sacc. **V**:1028 ; **57**, after No. 292, as *Agaricus sublateritius* Fr.). On dead wood, Darjeeling, 8,000 ft. (Hooker f.).

——**velutinum** (Pers.) Fr. (Sacc. **V**:1034 ; **57**, after No. 293, as *Agaricus velutinus* Fr.). On earthy banks, Darjeeling, 7,500 ft. (Hooker f.).

Inocybe echinata (Roth) Cke (Sacc. **V**:773 ; **263**:332). On the ground, Botanic .Garden, Saharanpur (Gollan). Rea (British Basidiomycetæ, p. 206) refers this species to *Lepiota h ematosperma* (Bull.) Boud.

——**holophlebia** Berk. (**176**:104 ; Sacc. **XI**:52). On the ground, Masulipatam.

Irpex canescens Fr. (Sacc. **VI**:485 ; **467**:155). On wood, Khandala, Bombay (Blatter).

——**consors** Berk. (Sacc. **VI**:486 ; **287**:1069 as "*Irpex concors*"). India (Bose).

——**flavus** Klotzsch (Sacc. **VI**:486 ; **57**, after No. 449 ; **196**:126 ; **263**:325 ; **467**:155 ; **288**, No. **28**:7 and No. **42**:11 as *Polystictus flavus* Junghuhn ; **287**:1295). On old wood, Sone River, Bihar (Hooker f.) ; Arracan (Kurz) ; Botanic Garden, Saharanpur (Gollan) ; Khandala and Salsette, Bombay (Blatter) ; Bengal (Hutchings) ; India (J. Ray ; Kirtikar ; P. L. Dey). Lloyd (**287**:903) prefers the name *Polystictus flavus*. The fungus has both *Irpex* and *Polystictus* forms. Berkeley (l.c.) records the variety *orbicularis* Jungh. on old bamboo, Ganges (Hooker f.).

——**pallescens** Fr. (Sacc. **VI**:487 ; **196**:126). Yomah, Burma (Kurz).

——**vellereus** Berk. & Broome (Sacc. **VI**:489 ; **467**:155). On dead wood, Khandala, Bombay (Blatter).

——**zonatus** Berk. (**57**, No. 449 ; Sacc. **VI**:485). On dead wood, Sikkim and eastern Nepal (Hooker f.).

Kordyana indica Gäumann (**244**:264 ; **455**:280 as *K. tradescantiae* (Pat.) Rac. ; Sacc. **XVI**:199). On leaves of *Commelina* sp., Pusa (Subramaniam). Gäumann points out that this is *K. tradescantiae* of Raciborski, not of Patouillard.

Lachnocladium hookeri Berk. (**57**, No. 399 ; Sacc. **VI**:738). Khasi Hills (Hooker f.).

Lactarius deliciosus Fr. (Sacc. **V**:438 ; **57**, after No. 388). Lachen, Sikkim, 11,000 ft. (Hooker f.).

——**princeps** Berk. (**57**, No. 389, and after No. 404 ; Sacc. **V**:448). In woods, Kullung and Myrong, Khasi Hills, 6,000 ft. (Hooker f.).

——**stramineus** Berk. (**57**, No. 404, not in Sacc.). On the ground, Pomrang, Khasi Hills, 5,000 ft. (Hooker f.).

——**subdulcis** Fr. (Sacc. **V**:450 ; **57**, after No. 389). Identified from drawings without notes.

——**vellereus** Fr. (Sacc. **V**:437 ; **57**, after No. 388). Fir woods, Sikkim, 10,000 ft. (Hooker f.). As Berkeley notes under No. 404, he first confused the specimens of *L. stramineus* with those of *L. vellereus.*

Laschia intestinalis (Berk.) Bres. (**84**:234 ; **57**, No. 342 as *Favolus intestinalis* Berk. ; Sacc. **VI**:400 ; **287**:708, 785). Darjeeling (Hooker f.). Lloyd (l.c.) proposed the genus *Poroauricularia* for this fungus, which he regards as more an *Auricularia* than a Polypore.

——**lamellosa** Berk. (**57**, No. 454 ; not in Sacc. ; not *L. lamellosa* Pat., Sacc. **IX**: 205). Lebong, Darjeeling (Hooker f.).

——**subvelutina** Berk. (**57**, No. 343 ; Sacc. **VI**:410). On tree trunks, Darjeeling, 5-8,000 ft. (Hooker f.). Lloyd (**287**:839) states that this species "rests on a single, inadequate specimen, but not a *Laschia.* I would not say without cutting it whether it is a *Campanella* or an *Auricularia* ".

Lentinus alopecinus Fr. (**239**:392 ; Sacc. **V**:589). Recorded as apparently on the ground, India.

——**blepharodes** Berk. (Sacc. **V**:577 ; **288**, No. **47**:10). Botanic Garden, Saharanpur.

——**candidus** Graff (Sacc. **XXIII**:167 ; **287**:1069). India (Bose). Lloyd (l.c.) considers it a pale form of *L. sajor-caju,* except that the spores are different.

——**capronatus** Fr. (Sacc. **V**:575 ; **196**:119). Myodwine, Burma (Kurz).

——**coadunatus** Hooker f. (**57**, No. 323 ; Sacc. **V**:601). On dead wood, Darjeeling, 7,500 ft. (Hooker f.). Lloyd (**288**, No. **47**:13) is of the opinion that this species and *L. curreyanus* are the same as *L. subnudus.*

——**connatus** Berk. (Sacc. **XXI**:116 ; **71**: 347, fig.). On dead wood, Howrah District, Bengal (Bose). See *L. javanicus* below.

——**curreyanus** Sacc. & Cub. (Sacc. **V**:586 ; **70**, **I**:111 ; **196**:120, figs., as *L. caespitosus* Currey, not Berk.). Burma (Kurz) ; Calcutta (Bose).

——**descendens** Fr. (Sacc. **V**:587 ; **196**:119). Toukyeghat, Burma (Kurz).

——**exilis** Klotzsch (Sacc. **V**:606 ; **148**:98). Andaman Islands (Kurz).

——**glabratus** Mont. (Sacc. **V**:605 ; **196**:120). North Rajmahal Hills, North Bengal (Kurz).

[**Lentinus**] **hepaticus** Berk. (**57,** No. 324 ; Sacc. **V**:603). On tree trunks, Darjeeling, 7,500 ft. (Hooker f.).

——**hookerianus** Berk. (**57,** No. 322 ; Sacc. **V**:573). On dead wood, Darjeeling, 6-9,000 ft. (Hooker f.).

——**inquinans** Berk. (**57,** No. 407, figs. ; Sacc. **V**:583 ; **196**:120). On dead wood, Mai Valley, East Nepal, 5,000 ft., and Changachelling, Sikkim, 7,000 ft. (Hooker f.) ; near Rangoon (Kurz).

—**javanicus** Lév. (Sacc. **V**:599 ; **286**:120 as *L. decaisneanus* Lév.). On trunks, Bombay (Polydore Roux). The type of *L. decaisneanus* at Paris has a note by Bresadola " est une forme três developpée de *Lentinus javanicus* " and that *L. cretaceus* and *L. infundibuliformis* are also synonyms. Lloyd (**288,** No. **47**:13) considers *L. javanicus* a synonym of *L. connatus*, and Petch (**371**:147) finds *L. infundibuliformis* a synonym of *L. connatus*, but Bresadola (**84**:222) seems to consider *L. javanicus* distinct from *L. connatus*.

——**lecomtei** Fr. (Sacc. **V**:572 ; **57,** after Nos. 321 and 405 ; **65**:137 ; **371**:151). On wood, Tonglo, Sikkim, 6-8,000 ft., and East Nepal (Hooker f.) ; Gulmarg, Kashmir (Aitcheson). Edible ; vernacular name " Silry ".

——**melanophyllus** Lév. (Sacc. **V**:575 ; **260**:152). On *Shorea robusta*, Bilaspur, Central Provinces (Marten).

——**molliceps** Fr. (**240**:38 ; Sacc. **V**:603). Nicobar Islands (Kamphoevener).

——**nepalensis** Berk. (**57,** No. 405 ; Sacc. **V**:573). On dead wood, Nangki, East Nepal, 9,000 ft. (Hooker f.).

——**nicobarensis** Reichhardt (**391**:146, figs. ; Sacc. **V**:599). On rotten trunks, Nicobar Islands.

——**omphalomorphus** Mont. (Sacc. **V**:589 ; **242** ; **196**:120 as *L. furfurosus* Fr.). Yomah, Burma (Kurz). Bresadola (**84**:234) states that *L. omphalomorphus* is a species of *Omphalia*, and Lloyd (**288,** No. **47**:11) states that Currey's record of *L. furfurosus* refers to *L. praerigidus*.

——**pergameneus** Lév. (**286**:117 ; Sacc. **V**:600). The type at Paris is marked " herb. de Candolle, Indes ? " There is one other specimen there, from Cambodia.

——**polychrous** Lév. (Sacc. **V**:590 ; **196**:120, figs., as *L. kurzianus* Currey ; **84**:222). Yomah Range, Burma (Kurz). According to Lloyd (**288,** No. **47**:11 ; **287**:955) the next species is the commoner in India and Ceylon.

——**praerigidus** Berk. (**57,** No. 406, figs. ; Sacc. **V**:587 ; **70, I**:110 ; **288,** No. **28**:3 ; **287**:955, 1069, 1295). On dead wood, Sone River, Bihar (Hooker f.) ; Bengal (Hutchings) ; India (Cave ; Kashyap ; Bose). See *L. omphalomorphus*. Lloyd (**288,** No. **47**:11) considers *L. kurzianus* to be a synonym of this species rather than of the preceding, but the two species are very similar.

——**revelatus** Berk. (Sacc. **V**:592 ; **148**:98 ; **287**:1069). Andaman Islands (Kurz) ; India (Bose).

[Lentinus] sajor-caju Fr. (**239**:393 ; Sacc. **V**:598 ; **71**:348, fig. ; **288,** No. **47**:11 ; **287**:1069 ; **288,** Nos. **28**:2 and **42**:12 as *L. dactyliophorus* Lév. ; **196**:120, and 121, as *L. irregularis* Currey and *L. exilis* Fr.). India (Sundevall) ; South Andaman Island, and in pine forests from Theemeechu to Bookee, Karen Country, Burma (Kurz) ; on dead wood, Hooghli District, Bengal (Bose) ; Toukyeghat, Nakawa, and Sittang, Burma (Kurz) ; Calcutta (Bose) ; Bengal (Hutchings) ; India (J. Ray).

——**squarrosulsus** Mont. (**349**:21 ; Sacc. **V**:585). On dead trunks, Nilgiris (Perrottet). Type not found at Paris.

——**strigosus** Fr. (Sacc. **V**:573 ; **352**). Sonamarg, Kashmir (R. R. Stewart).

——**subdulcis** Berk. (**57,** No. 325 ; Sacc. **V**:611). On dead wood, Darjeeling, 7-8,000 ft. (Hooker f.).

——**subnudus** Berk. (Sacc. **V**:583 ; **71**:348, fig. ; **288,** Nos. **47**:~~13~~ and **49**:3 ; **467**:157 as *L.* aff. *subnudus*). On dead branches and logs, common in Bengal (Bose) ; India (Cave) ; Khandala, Bombay (Blatter). See also *L. coadunatus* and *L. curreyanus*.

——**tigrinus** Bulliard (Sacc. **V**:580 ; **288,** No. **47**:13). India (S. N. Ratnagar). " Seems to be this species of Europe " (Lloyd).

——**velutinus** Fr. (Sacc. **V**:589 ; **196**:120 ; **288,** No. **47**:10). Yomah, Burma (Kurz) ; Bengal (Hutchings).

——**villosus** Klotzsch (Sacc. **V**:574 ; **288,** No. **47**:9). Lloyd (l.c.) notes that there are specimens of this species from India at Kew.

Lenzites acuta Berk. (Sacc. **V**:643 ; **57,** after No. 410 ; **287**:1186, 1295). On wood, Nunklow, Khasi Hills (Hooker f.) ; India (Bose ; P. L. Dey). See next entry.

——**adusta** Massee (**319, XI**:250 ; Sacc. **XXII**:1506 (name only) ; **287**:1072 ; **70, IX**:41). On wood, Bengal (Hutchings) ; Sylhet (Bose). Lloyd (l.c.) considers this to be a form of *L. acuta,* but differing in having a white context. Bose (l.c.) gives *L. beckleri* as a synonym but Massee said " allied to *L. beckleri* ", and Lloyd (**287**:1000, fig.) considers the latter a good species.

——**albida** Fr. (Sacc. **V**:637 ; **196**:121). Toukyeghat, Burma (Kurz).

——**alutacea** Cke (Sacc. **V**:649 ; **287**:1010). Calcutta (Bose). Lloyd (l.c.) considers it " too close to *L. flavida* " and Bresadola (**84**:221) regards *L. alutacea* as a lenzitoid form of *Elmerina vespacea.* See also L. *murina.*

——**applanata** Fr. (Sacc. **V**:644 ; **57,** after No. 408). Lebong, Darjeeling, 6,000 ft. (Hooker f.). Petch (**372**:31) gives this name as a synonym of *L. repanda.*

——**beckleri** Berk. (Sacc. **V**:645 ; **319, XI**:250). " Also an Indian species " according to Massee. See *L. adusta.*

——**betulina** (L.) Fr. (Sacc. **V**:638 ; **57,** after No. 327 ; **263**:328 ; **287**:852, fig. ; **70, II**:138, figs.). On dead timber, Darjeeling, 3,000 ft. (Hooker f.) ; Arnigadh, Mussoorie (Gollan) ; India (G. H. Cave) ; Darjeeling (Bose).

[**Lenzites**] **eximia** Berk. & Curt. (**57**, No. 410 ; Sacc. **V**:648). On dead wood, Darjeeling, 7,500 ft. (Hooker f.).

——"**flavida**" (**288**, No. **60**:6 ; **287**:1295). India (P. D. Master ; P. L. Dey). Presumably Lloyd referred to a Lenzitoid form of *Daedalea flavida*.

——"**imbricata** Fr." (**57**, after No. 411). Darjeeling (Hooker f.). Presumably Berkeley referred to *Polyporus imbricatus* Fr., but he states that the specimens were old and uncertain.

—**malaccensis** Sacc. & Cub. (Sacc. **V**:645 ; **70**, **IX**:40). On old trunks and stumps, Assam, Darjeeling, and North Burma (Bose).

—**murina** Lév. (Sacc. **V**:642 ; **287**:952). India (Bose). Lloyd (**287**:1010) records that he found a later sending of the same number to be *L. alutacea*, and is uncertain as to the possible identity of the two collections.

——**repanda** (Mont.) Fr. (Sacc. **V**:650 ; **57**, after Nos. 327 and 408 ; **148**:98 ; **70**, **III**:4, figs. ; **467**:157 ; **288**, No. **27**:3 ; **287**:1069 ; **57**, after No. 410, as *L. pallida* Berk. ; **196**:121 as *L. palisota* Fr. ; **287**:1125 as *Daedalea repanda* Mont.). Hot valleys of Sikkim Himalaya on dead tree trunks, 2-5,000 ft., Khasi Hills, and East Nepal, 6,000 ft. (Hooker f.) ; Andaman Islands, and Seven Pagodas and Toukyeghat, Burma (Kurz) ; Dehra Dun and Simla (Blatter) ; Hooghly, Bengal (Bose) ; Bengal (Hutchings) ; India (Maruda Rajan). See *L. applanata*.

——**rugulosa** Berk. (**57**, No. 328 ; Sacc. **V**:649). On trunks, Darjeeling (?Hooker f.).

——**sepiaria** (Wulf.) Fr. (Sacc. **V**:639 ; **70**, **IX**:40). On wood palings of a bridge, Shillong (Bose).

——**striata** Swartz (Sacc. **V**:643 ; **287**:1295). India (P. L. Dey).

——**subferruginea** Berk. (**57**, No. 411 ; Sacc. **V**:643 ; **288**, No. **65**:4 and No. **42**:3 as "*Lenzites subferruginosus*"). On dead wood, Moflong, Khasi Hills (Hooker f.) ; India (Kashyap ; Bashambar).

Lepiota alliciens Berk. (**61**:20 ; **168**:105 ; Sacc. **IX**:7). On the roof of a house, Masulipatam.

——**altissima** Massee (**319**, **I**:114 ; Sacc. **XVI**:5 ; **524**:363, figs.). In open pastures near Poona (Woodrow). Petch (**367**, **I**:47) states that this species seems to differ from *L. dolichaula* Berk. & Broome only in the size of the spores.

——**anax** Berk. (**57**, No. 362 ; Sacc. **V**:71). On clay banks and amongst grass, Nunklow, Khasi Hills (Hooker f.).

——**badhami** Berk. (Sacc. **V**:35 ; **68**:151). Thana, Bombay (Blatter).

——**beckleri** Berk. (Sacc. **V**:56 ; **319**, **III**:151). On the ground, Poona (Woodrow).

——**cepaestipes** Sowerby (Sacc. **V**:43 ; **319**, **III**:151 ; **71**:350). On the ground, India (Gardener) ; Poona (Woodrow) ; on rotten wood indoors, Jessore and Calcutta (Bose).

[**Lepiota cepaestipes**] var. **lutea** With. (**263**:335). On the ground, Botanic Garden, Saharanpur (Gollan).

——**clypeolaria** (Bull.) Fr. (Sacc. **V**:36; **263**:335). A form, on the ground, Botanic Garden, Saharanpur (Gollan).

——**cristata** (Alb. & Schw.) Fr. (Sacc. **V**:39; **263**:335). On the ground, Botanic Garden, Saharanpur (Gollan).

——**deliciolum** Berk. (**57**, No. 244; Sacc. **V**:44). In the hollows of dead trees, Darjeeling, 8,000 ft. (Hooker f.).

——**erminea** Fr. (Sacc. **V**:40; **70**, **I**:112). In grassy places, Calcutta (Bose).

——**excoriata** (Schaeff.) Fr. (Sacc. **V**:31; **145**:13; **263**:336; **319**, **III**:151 as *L. excoriata* Karsten; **57**, after No. 243, as *Agaricus excoriatus* Fr.). In hot valleys on the bare earth, Darjeeling (Hooker f.); Punjab (Aitcheson); Poona (Woodrow); on the ground, Botanic Garden, Saharanpur (Gollan).

——**flavophylla** Massee (**319**, **XIV**:253; Sacc. **XXIII**:18). Calcutta (Burkill).

——? **hispida** (Lasch) Fr. (Sacc. **V**:36; **263**:335). On the ground, Botanic Garden, Saharanpur (Gollan).

——**holosericea** Fr. (Sacc. **V**:42; **263**:335). On the ground, Botanic Garden, Saharanpur (Gollan).

——**implana** Berk. (**57**, No. 363; Sacc. **V**:41). On dry, stony hills, Moflong, Khasi Hills (Hooker f.).

——**longicaudata** P. Henn. (**260**:153; Sacc. **XVI**:6; **263**:336). On the ground, Kalsia (Blandford); Botanic Garden, Saharanpur (Gollan).

——**malleus** Berk. (**61**:20; **168**:105; Sacc. **IX**:3). On the ground, Masulipatam.

——**mammosa** P. Henn. (**263**:335; Sacc. **XVII**:7). On the ground, Botanic Garden, Saharanpur (Gollan).

——**mastoidea** Fr. (Sacc. **V**:33; **70**, **II**:136, figs.; **74**:643). Hooghly District, Bengal (Bose). Edible.

——**meleagris** (Sow.) Fr. (Sacc. **V**:36; **263**:335). On the ground, Botanic Garden, Saharanpur (Gollan).

——**microspora** Massee (**319**, **IV**:92; Sacc. **XXI**:19). On the ground, Narcondam, Andaman Islands (Rogers).

——**mimica** Massee (**319**, **XIV**:253; Sacc. **XXIII**:11). In decaying mown grass, Calcutta Maidan (Burkill).

——**montosa** Berk. (**57**, No. 401; Sacc. **V**:41). On the ground, Sikkim (Hooker f.).

——**procera** (Scop.) Sacc. (Sacc. **V**:27; **263**:336). On the ground under *Ficus carica*, Botanic Garden, Saharanpur (Gollan).

——**punicea** Massee (**319**, **XIV**:253; Sacc. **XXIII**:17). Calcutta (Burkill).

——**rhacodes** Vittad., var. **puellaris** Fr. (Sacc. **V**:29; **260**:153). On the ground, Kalsia Hills (Blandford).

——**seminuda** (Lasch) Juel (Sacc. **V**:50; **263**:335). On the ground, Botanic Garden, Saharanpur (Gollan).

[Lepiota] sericea Massee (**319, XIV**:254 ; Sacc. **XXIII**:10). In a plant pot and amongst grass, Calcutta (Burkill).

——**sistrata** Fr. (Sacc. **V**:50 ; **263**:335). On the ground, Botanic Garden, Saharanpur (Gollan).

Lloydella bicolor (Pers.) v. Hoehn. & Lits. (Sacc. **VI**:565 as *Stereum bicolor* (Pers.) Fr. ; **57**, after No. 346 ; **472**:146). On dead wood, Darjeeling, 7,000 ft. (? Hooker f.).

——**membranacea** (Fr.) Bres. (Sacc. **VI**:576 as *Stereum membranaceum* Fr. ; **288**, No. **46**:5 ; **196**:127 as *S. papyrinum* Mont. ; **84**:232). Timeokee, Burma (Kurz) ; Bengal (Hutchings). Petch (**379**:263) states that Ceylon specimens recorded as *S. papyrinum* do not agree with the diagnosis of *L. membranacea.*

Lopharia mirabilis (Berk. & Broome) Patouill. (Sacc. **VI**:496 as *Radulum ;* **263**:325). On dead twigs, Botanic Garden, Saharanpur (Gollan). Lloyd **288**, No. **53**:9) prefers to return this species to *Radulum.* He considers it (**288**, No. **59**:4) a synonym of *R. neilgherrense* (q.v.).

Marasmius ?androsaceus Fr. (Sacc. **V**:543 ; **263**:329). On fallen bamboo on the ground, Botanic Garden, Saharanpur (Gollan).

——**burmensis** Cke (**149**:122 ; Sacc. **V**:532). On twigs, Moulmein, Burma (Parish).

——**caperatus** Berk. (**57**, No. 321 ; Sacc. **IX**:66). On twigs of live and dead bushes, Tonglo, Sikkim, 10,000 ft. (Hooker f.).

——**consocius** Berk. (**57**, No. 320 ; not in Sacc.). On dead twigs, Darjeeling, 8,000 ft. (Hooker f.).

——**?curreyi** Berk. & Broome (Sacc. **V**:556 ; **263**:329). On dead leaves, Botanic Garden, Saharanpur (Gollan).

——**equicrinus** Mueller (Sacc. **V**:553 ; **287**:1295 ; **497**:38 ; **504**). India (P. L. Dey) ; on tea, Cachar (Tunstall). Petch (**370**:61) referred certain Indian specimens to this species, since they possessed similar mycelial threads.

——**erythropus** Fr. (Sacc. **V**: 520 ; **57**, after No. 319). On the ground, Darjeeling (?Hooker f.).

——**graminum** (Lib.) Berk. (Sacc. **V**:542 ; **260**:152 ; **263**:329). On dead grass, Botanic Garden, Saharanpur (Gollan).

——**haematodes** Berk. (**57**, No. 390 ; Sacc. **V**:568). On pine twigs, Sikkim, 11,000 ft. (Hooker f.).

——**hookeri** Berk (**57**, No. 391 ; Sacc. **V**:514). In copse-wood, on moss, Khasi Hills, 5,000 ft. (Hooker f.).

——**iridescens** Berk. (**57**, No. 319 ; not in Sacc.). On mossy banks, Sinchul, Sikkim Himalaya, 8,000 ft. (Hooker f.).

——**korthalsii** Fr. (**240**:30 ; Sacc. **V**:518). On chips, Nicobar Islands (Kamphoevener).

——**?languidus** (Lasch) Fr. (Sacc. **V**:527 ; **260**:152). On dead stems, Botanic Garden, Saharanpur (Gollan).

[**Marasmius**] **parishii** Cke (**149**:122 ; Sacc. **V**:545). On grass, palm petioles, etc., Burma (?Parish).

——**proletarius** Berk. & Curt. (Sacc. **V**:555). Reported from Narcondam, Andaman Islands (C. G. Rogers).

——**pulcher** (Berk. & Broome) Petch (**377**:19 ; Sacc. **VI**:673 as *Cyphella ;* **497**:38). On tea, Assam (Tunstall). This is the common thread blight of tea in Ceylon, and is, according to Petch (**377**) the species with which Massee found his *Stilbum nanum* (q.v.) associated, and wrongly thought the latter to be its fructification. *M. pulcher* is apparently not parasitic, and the identity of the true parasitic thread blight of tea in India is still unknown. (See **111**:456, figs., and *Corticium repens*).

——**ramealis** (Bull.) Fr. (Sacc. **V**:531 ; **263**:328). On dead roots, Botanic Garden, Saharanpur (Gollan).

——**rotula** (Scop.) Fr. (Sacc. **V**:541 ; **57,** after No. 391 ; **263**:329). On leaves of maple, etc., in pine woods, Sikkim, 11,000 ft. (Hooker f.) ; on dead grass stems, Botanic Garden, Saharanpur (Gollan). Berkeley notes that the Sikkim form has adnate gills, not attached to a collar.

——**sacchari** Wakker (Sacc. **XIV**:115 ; **392**:4). On exotic varieties of *Saccharum officinarum,* Mandalay (Rhind).

——**spaniophyllus** Berk. (Sacc. **V**:568 ; **467**:157). On dead twigs, Khandala, Bombay (Blatter).

——**subomphalodes** P. Henn. (**263**:329 ; Sacc. **XVII**:41). On dead grass, Botanic Garden, Saharanpur (Gollan).

——**urens** Fr. (Sacc. **V**:504 ; **263**:328). On the ground, Arnigadh, Mussoorie (Gollan).

Merulius corium Fr. (Sacc. **VI**:413). Reported in Saccardo as occurring in India, but a more definite reference was not found.

——**lignosus** Berk. (**57,** No. 447 ; Sacc. **VI**:420). On dead wood, Darjeeling (Hooker f.).

——**pseudolachrymans** P. Henn. (**263**:328 ; Sacc. **XVII**:145). On tree roots, Botanic Garden, Saharanpur (Gollan).

——**similis** Berk. & Broome (Sacc. **VI**:420 ; **70, IV**:5, figs.). On exposed bamboo roots, Hooghly District, Bengal (Bose).

Mycena arata Berk. (**57,** No. 268 ; Sacc. **V**:269). On roots of trees, amongst moss, Sinchul, Sikkim Himalaya, 8,500-9,000 ft. (Hooker f.).

——**bicrenata** Berk. (**57,** No. 269 ; Sacc. **V**:278). On rotten wood, Jallapahar, Darjeeling (Hooker f.).

——**broomeiana** Berk. (**57,** No. 265 ; Sacc. **V**:281). On dead wood, Darjeeling (Hooker f.).

——**colligata** Berk. (**57,** No. 367 ; Sacc. **V**:271). In pine woods, Sikkim, 11,000 ft. (Hooker f.).

[Mycena] conocephala P. Henn. (**263**:334; Sacc. **XVII**:20). On the ground, Botanic Garden, Saharanpur (Gollan).

——**dentosa** Berk. (**57**, No. 370; Sacc. **V**:282). In pine woods, Sikkim, 11,000 ft. (Hooker f.).

——**discors** Berk. (**57**, No. 368; Sacc. **V**:264). On wood in pine forest, Sikkim, 11,000 ft. (Hooker f.).

——**epipterygia** (Scop.) Fr. (Sacc. **V**:294; **57**, after No. 372, as *Agaricus epipterygius* Scop.). In pine woods, Sikkim, 11,000 ft. (Hooker f.)

——**flavo-miniata** Berk. (**57**, No. 372; Sacc. **V**:289). On sticks in pine woods, Sikkim, 11,000 ft. (Hooker f.).

——**galericulata** (Scop.) Fr. (Sacc. **V**:268; **57**, after No. 366, as *Agaricus galericulatus* Scop.). In pine woods, Sikkim, 11,000 ft. (Hooker f.).

——?**gypsea** Fr. (Sacc. **V**:260; **263**:334). On roots, Botanic Garden, Saharanpur (Gollan).

——**incommiscibilis** Berk. (**57**, No. 369; Sacc. **V**:281). In pine woods, Sikkim, 11,000 ft. (Hooker f.).

——**macrothela** Berk. (**57**, No. 373; Sacc. **V**:294). Amongst moss in woods, Myrong, Khasi Hills (Hooker f.).

——**manipularis** Berk. (**57**, No. 274; Sacc. **V**:272). On trunks and stumps, Sinchul, Himalaya, 8,500 ft. (Hooker f.).

——**myriadea** Berk. (**57**, No. 266; Sacc. **V**: 271). On dead trunks, Darjeeling, 8,000 ft. (Hooker f.).

——**nubigena** Berk. (**57**, No. 267; Sacc. **V**:269). On old timber, Darjeeling, 7,500 ft. (Hooker f.).

——? **plicosa** Fr. (Sacc. **V**:279; **263**:334). On the ground, Botanic Garden, Saharanpur (Gollan).

——**prasia** Berk. (**57**, No. 275; Sacc. **V**:264). On the ground, top of Tonglo, Sikkim, 10,000 ft. (Hooker f.).

——**puberula** Berk. (**57**, No. 371; Sacc. **V**:284). In pine woods, Sikkim, 11,000 ft. (Hooker f.).

——**pura** (Pers.) Fr. (Sacc. **V**:256; **57**, after No. 366, as *Agaricus purus* Pers.). In pino woods, Sikkim, 11,000 ft. (Hooker f.).

——**rubiaetincta** Berk. (**57**, No. 270; Sacc. **V**:291). On tree trunks, Darjeeling, 7,500 ft. (Hooker f.).

——**rufata** Berk. (**57**, No. 273; Sacc. **V**:272). On tree trunks, Darjeeling, 8,000 ft. (Hooker f.).

——**rufo-picta** Berk. (**57**, No. 276; Sacc. **V**:294). On dead wood, Darjeeling, 8,000 ft. (Hooker f.).

——**russulina** Berk. (**57**, No. 272; Sacc. **V**:272). On tree trunks, Darjeeling, 8,600 ft. (Hooker f.).

——**xanthophylla** Berk. (**57**, No. 271; Sacc. **V**:272). On roots of trees, Darjeeling, 8,000 ft. (Hooker f.).

Naucoria cerodes Fr. (Sacc. **V**:836 ; **57,** after No. 401). Leh and Valley of the Indus, Kashmir (Thomson).

——? **conspersa** (Pers.) Fr. (Sacc. **V:**854 ; **263**:332). On the ground, Botanic Garden, Saharanpur (Gollan).

——**descendens** Berk. (**57,** No. 381 ; Sacc. **V**:849). Amongst moss in pine woods, Sikkim, 11,000 ft. (Hooker f.).

——**fusispora** P. Henn. (**263**:332 ; Sacc. **XVII:**72). On the ground, Arnigadh, Mussoorie (Gollan).

——**khasiensis** Berk. (**57,** No. 402 ; Sacc. **V**:837). Amongst grass, Kala Pani, Khasi Hills, 5,000 ft. (Hooker f.).

——**micromegala** Berk. (**57,** No. 380 ; Sacc. **V:**833). On dead wood, Myrong, Khasi Hills (Hooker f.).

——**pediades** Fr. (Sacc. **V**:844 ; **263**:332). On the ground, Botanic Garden, Saharanpur (Gollan).

——**scrupea** Berk. (**57,** No. 290 ; Sacc. **V**:851). On moist earth, Darjeeling, 7,500 ft. (Hooker f.).

——**semiorbicularis** Bull. (Sacc. **V:**844). India is included in the distribution as listed in Saccardo.

Nolanea nana Massee (**319, VI:**122 ; Sacc. **XXIII:**220). On walls, Calcutta (Burkill).

Omphalia calycinoides P. Henn. (**263**:334 ; Sacc. **XVII**:21). On the ground or on roots, Botanic Garden, Saharanpur (Gollan).

——**campanella** (Batsch) Fr. (Sacc. **V**:327 ; **352**). Sonamarg, Kashmir (R. R. Stewart).

——**fuliginosa** Massee (**319, VI**:122 ; Sacc. **XXIII:**109). On a wall, Calcutta (Burkill).

——**hydrogramma** (Bull.) Fr. (Sacc. **V**:309 ; **248**:15). Central Provinces and Berar (Graham).

——**oedipus** Massee (**319, VI:**122 ; Sacc. **XXIII**:105). On the ground at a base of a wall, Calcutta (Burkill).

——**radiatilis** Berk. (**57,** No. 375 ; Sacc. **V**:323). In pine woods, Sikkim, 11,000 ft. (Hooker f.).

——**ranunculina** Berk. (**57,** No. 374 ; Sacc. **V**:325). On turf, etc., Lachen, Sikkim, 14-16,000 ft. (Hooker f.).

——**rogersi** Massee (**319, IV**:92 ; Sacc. **XXI:**60). On the ground, Narcondam, Andaman Islands (Rogers).

——? **rustica** Fr. (Sacc. **V**:316 ; **260**:153). On the ground, Botanic Garden, Saharanpur (Gollan).

——**umbellifera** (L.) Fr. (Sacc. **V**:321 ; **57,** after No. 373, as *Agaricus umbelliferus* L.). In pine woods, Sikkim, 11,000 ft. (Hooker f.).

Panaeolus campanulatus (L.) Fr. (Sacc. **V**:1121 ; **71**:353 ; **260**:152 as *Chalymotta campanulata* (L.) Karst. ; **263**:329). On the ground, Botanic Garden, Saharanpur (Gollan) ; on dung, Hooghly District, Bengal (Bose).

——**cyanescens** Berk. & Broome (Sacc. **V**:1123 ; **71**:352, fig. ; **70**, II). Common in Bengal on dung heaps (Bose). Cultures of this fungus are also reported by Bose (**72**).

——**papilionaceus** (Bull.) Fr. (Sacc. **V**:1122 ; **57**, after No. 383, as *Agaricus papilionaceus* Bull.). On the ground, Jheels, Plains of Eastern Bengal (Hooker f.).

Panus conchatus Fr. (Sacc. **V**:615 ; **57**, after No. 326). Darjeeling (Hooker f.).

——**monticola** Berk. (**57**, No. 326 ; not in Sacc.). On the ground, probably attached to wood, Tonglo, Sikkim (Hooker f.).

——**ochraceus** Massee (**319**, **IV**:92 ; not in Sacc.). On dead wood, Dehra Dun (Butler).

Paxillus chrysites Berk. (**57**, No. 314 ; Sacc. **V**:986). On dead wood, Darjeeling, 7,500 ft. (Hooker f.).

——**pinguis** Hooker f. (**57**, No. 316 ; Sacc. **V**:987). On earth and mossy banks, Darjeeling, 7,500 ft. (Hooker f.).

——**sulphureus** Berk. (**57**, No. 315 ; Sacc. **V**:986). On dead wood and on the ground, Darjeeling, 7,500 ft. (Hooker f.).

Peniophora habgallae (Berk. & Broome) Cke (Sacc. **VI**:641 ; **288**, No. **58**:4, 7, as *Matula poroniaeformis* (Berk. & Broome) Massee ; **287**:390, figs.). " Maduré", India through Rev. C. Torrend. Petch (**380**) has solved the relationships of the *Matula*, formerly placed in the Gasteromycetes.

——**levigata** (Fr.) Massee (Sacc. **VI**:628 as *Corticium levigatum* Fr. ; **196**:127). Yomah, Burma (Kurz).

Pholiota aurivella (Batsch) Fr. (Sacc. **V**:748 ; **57**, after No. 288, as *Agaricus aurivellus* Batsch). On tree stumps, Darjeeling, 7,000 ft., and Tonglo, Sikkim, 5,000 ft. (Hooker f.).

——**examinans** Berk. (**57**, No. 287 ; Sacc. **V**:742). On dead wood, Darjeeling (Hooker f.).

——**gollani** P. Henn. (**263**:333 ; Sacc. **XVII**:62). On a living trunk of *Tamarindus indica*, Botanic Garden, Saharanpur (Gollan).

——**granuloso-verrucosa** P. Henn. (**263**:333 ; Sacc. **XVII**:64). On the ground, Botanic Garden, Saharanpur (Gollan).

——**indica** Massee (**319**, **III**:151 ; Sacc. **XVII**:63). On the ground, Poona (Woodrow).

——**microspora** Berk. (**57**, No. 288 ; Sacc. **V**:742). On dead wood, Darjeeling (Hooker f.).

——**praecox** (Pers.) Fr. (Sacc. **V**:738 ; **263**:333 as *P. candicans* (Schaeff.) Schroet.). On the ground, Botanic Garden, Saharanpur (Gollan).

Pilacre orientalis Berk. & Broome (**67**:101 ; Sacc. **IV**:580). On dead wood, Nilgiris (E. S. Berkeley).

[**Pilacre**] **tephrospora** Berk. & Broome (**67**:101 ; Sacc. **IV**:580). On dead leaves, Nilgiris (E. S. Berkeley).

Pleurotus anserinus Berk. (**57,** No. 279 ; Sacc. **V**:362). On dead wood, Jallapahar, Darjeeling, 7,500 ft. (Hooker f.).

——**cretaceus** Massee (**319, II**:165 ; Sacc. **XVI**:36 ; **248**:15). On wood, Peshawar (Watt), edible, sold in the Bazaar as " Dhingri " ; Central Provinces (Graham).

——**dryinus** (Pers.) Fr. (Sacc. **V**:340 ; **57,** after No. 401). On trees, near Avantipura, Kashmir (Thomson).

——**eöus** Berk. (**57,** No. 280 ; Sacc. **V**:361). In the hottest valleys, on dead tree trunks in open places, Sikkim (Hooker f.).

——**fimbriatus** Bolt. (Sacc. **V**:344 ; **248**:15). Central Provinces and Berar (Graham).

——**flabellatus** Berk. & Broome (Sacc. **V**:369 ; **71**:349, figs.). On dead wood or on the ground, Hooghly District, Bengal (Bose).

——**hapalosclerus** Berk. (**57,** No. 277 ; Sacc. **V**:351). On tree trunks, Darjeeling, 7-8,000 ft. (Hooker f.). The species name is spelled " *apalosclerus* " in the original description.

——**membranaceus** Massee (**319, III**:151 ; Sacc. **XVII**:24). On trunks, Poona (Woodrow).

——**ninguidus** Berk. (**57,** No. 281 ; Sacc. **V**:361). In hot valleys on dead timber, Sikkim (Hooker f.).

——**ostreatus** (Jacq.) Fr. (Sacc. **V**:349 ; **352**). Sonamarg, Kashmir (R. R. Stewart).

——**petaloides** (Bull.) Fr. (Sacc. **V** : 361 ; **57,** after No. 401). On dead wood, East Nepal (Hooker f.).

——**placentodes** Berk. (**57,** No. 376 ; Sacc. **V**:359). On birch wood, Sikkim, 11,000 ft. (Hooker f.).

——**platypus** Cke & Massee (**169**:121 ; Sacc. **IX**:47). On trunks, Nepal.

——**salignus** (Pers.) Fr. (Sacc. **V**:359 ; **57,** after No. 401). Sikkim (Hooker f.).

——**sapidus** Kalchbr. (Sacc. **V**:348 ; **248**:15).Central Provinces and Berar (Graham).

——**subpalmatus** Fr. (Sacc. **V**:343 ; **263**:334). On the ground, perhaps on roots, Arnigadh, Mussoorie (Gollan).

——**verrucarius** Berk. (**57,** No. 278 ; Sacc. **V**:351). On dead wood, Darjeeling, 7-8,000 ft. (Hooker f.).

Pluteus chrysoprasius Berk. (**57,** No. 283 ; Sacc. **V**:678). On burnt roots of *Abies*, base of Mount Tonglo, Sikkim, 7,000 ft. (Hooker f.).

——**cuspidatus** Berk. (**57,** No. 377 ; Sacc. **V**:677). On the ground, Khasi Hills (Hooker f.).

——**palumbinus** Berk. (**57,** No. 282 ; Sacc. **V**:677). On living tree trunks, Darjeeling, 7,500 ft. (Hooker f.).

Polyporus acervatus Lloyd (**287**:1126). South India (D. Maruda Rajan).

[Polyporus] adustus (Willd.) Fr. (Sacc. **VI**:125 ; **286**:134 ; **57,** after No. 421 ; **288,** No. **33**:4, No. **34**:2, No. **69**:5 ; **300**:328 ; **287**:955 ; **352** as "*Bjerkandera adusta*"; **70, VIII**:30 as *P. dissitus* Berk. & Broome ; **84**:224). Nilgiris (Perrottet, a good specimen at Paris) ; Sikkim, 7,000 ft. (Hooker f.) ; Sonamarg, Kashmir (R. R. Stewart) ; India (Cave) ; on dead date palm trunk, Kadambagachi, Bengal (Bose).

——**albellus** Peck (**287**:1126 ; **296**:130). South India (D. Maruda Rajan).

——**anthelminticus** Berk. (**60**:753 ; Sacc. **VI**:79 ; **70, VIII**:30). At the base of bamboo stems, Burma ; on dead root of bamboo, Bengal (Bose). Vernacular names "wa-mo", "ihan-mo". Said by Berkeley to be near *P. rufescens* (q. v.).

——**aquosus** P. Henn. (Sacc. **XVII**:112 ; **296**:130). India (Cave). Lloyd (**288,** No. **32**:1) states that this is the same as *P. lenzitoides* Berk., but Bresadola' (**84**:235) considers it to be quite different from "*Polyporus lenziteus* Berk." presumably meaning *P. lenzitoides* Berk.

——**arcularius** (Batsch) Fr. (Sacc. **VI**:67 ; **349**:22 ; **263**:326 ; **287**:1126 ; **70, IX**:36). India (? Perrottet) ; on dead twigs, Arnigadh, Mussoorie (Gollan) ; South India (D. Maruda Rajan) ; on the ground and on a fallen branch, Barkuda Island, Orissa (Bose, who reports it as *P. agariceus* Berk., with *P. arcularius* given as a synonym. These two names are usually given as synonyms, although Bresadola (**82**:222) states that *P. agariceus* Berk. is a synonym of *P. boucheanus* Klot.). See *P. cremoricolor* and *P. squamoso-maculatus*.

——**bambusicola** P. Henn. (**263**:326 ; Sacc. **XVII**:111). On roots of bamboo, Botanic Garden, Saharanpur (Gollan).

——**bicolor** Jungh. (Sacc. **VI**:207 as *Fomes* ; **196**:124 as *P. anebus* Berk. ; **70, V**:20, figs. ; **287**:1010 ; **84**:222). On wood, Toukyeghat, Pellowa, Burma (Kurz) ; Bengal, Assam, and Madras (Bose). Lloyd (**300**:338) also considers *P. anebus* to be the same as *P. bicolor*.

——**bosei** Bres. (**85**:55). On trunks, Bakarganj, Calcutta District (Bose).

——**brumalis** (Pers.) Fr. (Sacc. **VI**:63 ; **70, IX**:35). On a prostrate log, Chilka Lake, Orissa (Bose).

——**calcuttensis** Bose (**80**:179, figs.). In a cavity in the trunk of *Caesalpinia pulcherrima*, Calcutta (Bose).

——**campbelli** Berk. (**57,** No. 488 ; Sacc. **VI**:61 ; **240**:32). On the ground, Poona (J. D. Campbell). Lloyd (**295**:89) states that this species is "only known from a thin section. Probably could not be determined on comparison."

——**chocolatus** Bose (**77** :226, fig. ; **287**:1186, 1194, fig.). On the ground, Coimbatore, Madras (Bose).

——"**clementsii**" (**287**:1069). India (Bose). We are uncertain as to the interpretation of Lloyd's reference. *P. clementiae* Murr. is stated by Bresadola (**84**:67) to be a synonym of *P. vernicipes* Berk. *P. clemensiae* (Murr.) Sacc. & Trott. (Sacc. **XXI**:272) may be intended.

[Polyporus] **cremoricolor** Berk. (**57,** No. 334; Sacc. **VI**:71; **263**:326). On decayed wood, Darjeeling, 7,000 ft. (Hooker f.); on dead twigs, Arnigadh, Mussoorie (Gollan). Lloyd (**296**:176) includes this as close to, if not the same as, *P. arcularius.*

——**crispus** (Pers.) Fr. (Sacc. **VI**:125; **57,** after No. 338). On dead wood, Darjeeling, 7-8,000 ft., and Tonglo, Sikkim, 8,000 ft. (Hooker f.).

——**cubensis** Mont. (Sacc. **VI**:146; **467**:155). On wood, Dehra Dun (Blatter).

——**curtisii** Berk. (Sacc. **VI**:158 as *Fomes ;* **288,** No. **42**:3). India (Prof. D. E. Bashambar).

——**digitalis** Berk. (**57,** No. 422; Sacc. **VI**:123). On dead wood, Darjeeling, 7,500 ft. (Hooker f.). Lloyd (**300**:377) states that the types are very poor, but look much like *P. adustus.*

——**elatinus** Berk. (**57,** No. 424; Sacc. **VI**:141; **300**:295). On pine, East Nepal and Sikkim (Hooker f.).

——**emerici** (Berk. in Herb.) Lloyd (Sacc. **IX**:195 as *Trametes emerici* Berk. in Herb.; **288,** No. **40**:5; **70,III**:2, figs.; **296**:137). On trunks, Nilgiris; on dead wood, Hooghly, Bengal (Bose). Lloyd (**288,** l. c.) states that it is related to *P. gilvus.*

——**flammans** Berk. (**57,** No. 421; Sacc. **VI**:103). On dead wood, Darjeeling, 7,500 ft. (Hooker f.). The type was re-examined by Lloyd (**300**:378).

——**friabilis** Bose (**75**:300, fig.; **287**:1010, 1148, fig.). On the ground, Bengal and Orissa (Bose).

——**fumoso-olivaceus** Lloyd (**70, V**:20, figs.; **373, II**:284; **287**:1010). On tree trunks, Howrah, Bengal (Bose).

——**gilvus** Schwein. (Sacc. **VI**:121; **260**:151; **70, II**:138, figs.; **300**:346; **288,** No. **42**:12, No. **47**:2; **287**:1295; **252 ; 57,** after No. 339, as *P. isidioides* Berk.; **467**:155; **51**:393 as *P. cupreus* Berk.; **84**:224). On dead wood, Darjeeling (Hooker f.); Khandala, Bombay (Blatter); India (Wight; J. Ray; Bashambar; P. L. Dey); on a dead trunk of *Shorea robusta,* Kalsia (J. H. Blandford); a suspected parasite of *Dalbergia sissoo,* Dehra Dun (Hafiz Khan); Hooghly District, Bengal, and Darjeeling (Bose). See also *Fomes holosclerus* and *P. inamoenus.*

——**grammocephalus** Berk. (Sacc. **VI**:92; **148**:98; **260**:151; **467**:156; **372**:36; **70, IV**:1, figs.; **435**:71; **287**:1125, 1126). On stems in dense forest, Kalsia (J. H. Blandford); on dead trunks, Poona (Blatter); Hooghly, Bengal (Bose); on dead wood, Sidapur, Coorg (**435**); South India (Maruda Rajan); India (H. V. Ryan); Andaman Islands (Kurz).

——**guhae** Bose in Lloyd (**287**:1147, fig., 952, 1010). India (Bose).

[——**haematinus** Berk. in Herb. (Sacc. **VI**:149, a nomen nudum according to Lloyd, **300**:379). India.

——**hispidus** (Bull.) Fr. (Sacc. **VI**:129; **263**:326; **111**:21, figs.; **98**:14, fig.; **287**:1295, 1330, as *P. subhispidus*). On dead tree roots, Botanic Garden, Saharanpur

(Gollan); on living trunks of *Morus*, *Pyrus malus*, *Prunus armeniaca*, and plum trees, Kashmir (Butler); India (P. L. Dey).

[Polyporus] hobsoni (Berk.) Cke (**163**:20; Sacc. **VI**:135). On dead wood, Bombay. Lloyd (**300**:325, 380) thinks that this is *P. obtusus* Berk.

——**hookeri** Lloyd (**300**:384, fig.; Sacc. **XXIII**:366; **70, IX**:37; **57,** after No. 430, as a form of *P. scruposus* Fr.). Ratong River, 7,000 ft. (Hooker f.); on dead trunk, Darjeeling (Bose). Lloyd (l. c.) agrees with Berkeley that this is only an extreme form of *P. scruposus*.

——**ikenoi** ?Lloyd (**287**:1125). India (D. Maruda Rajan).

——**illicicola** P. Henn. (Sacc. **XVII**:110; **288,** No. **69**:5). India (G. H. Cave). Lloyd writes *P.* "*ilicicolor*", but presumably he means this species.

——**inamoenus** Mont. (**349**:22; Sacc. **VI**:191 as *Fomes*). Nilgiris (Perrottet). This specimen is in Herb. Montagne at Paris. Lloyd (**300**:381) considers it to be *P. gilvus*, but Bresadola (**84**:237) is not of the same opinion.

——**intybaceus** Fr. (Sacc. **VI**:96; **57,** after No. 338). On dead wood, Darjeeling, 7-8,000 ft. (Hooker f.). A doubtful record; Saccardo (**VI**:280) refers an earlier reference of Berkeley to this species to *Polystictus cichoriaceus* Berk.

——**luteo-umbrinus** Romell (Sacc. **XVI**:149; **70, VIII**:29; **319, XI**:250 as *P. indicus* Massee; **84**:67; **296**:162, fig.). On the ground, probably attached to buried wood or roots, Baroda (B. Cavanagh); on dead *Heritiera minor* tree, Madras (Bose).

——**luzonensis** Murr. (Sacc. **XXI**:283 as *Fomes*; **73**:130, figs.; **287**:1069). On dead branch of *Ficus bengalensis* and on logs, Bengal (Bose).

——**manilaensis** Lloyd (**287**:1131, 1126). South India (D. Maruda Rajan).

——**medullaris** Berk. (**57,** No. 425; Sacc. **VI**:140). On rotten wood, Bihar (Hooker f.). Lloyd (**299**:282) states that no type exists, but suggests that it may have been a *Ganoderma*.

——**miniatus** Jungh. (Sacc. **VI**:90; **57,** No. 338 as *P. rubricus* Berk.; **84**:227). Tonglo, Sikkim, 7,000 ft. (Hooker f.). Lloyd (**288,** No. **37**:3) considers *P. miniatus* to be a thin form of *P. sulphureus*.

— **molliculus** Bres. (**83**:33; Sacc. **XXIII**:352). On trunks, "India or., comm. C. G. Lloyd, No. 149".

— **montanus** (Quél.) Ferry (Sacc. **XXI**:260; **288,** No. **49**:4). India (Cave).

——**nigrocrustosus** Lloyd (**300**:373; Sacc. **XXIII**:366; **288,** No. **38**:8 erroneously referred to *Fomes nigrolaccatus*). Dodabetta, Nilgiris (H. V. Ryan).

——**nodipes** Berk. (**57,** No. 415; Sacc. **VI**:57). On the ground, Khasi Hills (Hooker f.). Lloyd (**295**:90) states that no type exists.

——**ochroleucus** Berk. (Sacc. **VI**:145; **70, IV**:1, figs.; **300**:311, figs.). On dead wood, Calcutta (Bose).

——**oerstedii** Fr. (Sacc. **VI**:178 as *Fomes*; **300**:370, fig.; **288,** No. **53**:4 as "*Ganodermus oerstedii*"). India (W. E. Houghton).

[**Polyporus**] **ostreiformis** Berk. (Sacc. **VI**:110 ; **70, IV**:1, figs.; **300**:307). Common on trunks, Bengal (Bose).

——**patouillardii** Rick (Sacc. **XXI**:311 as *Polystictus* ; **287**:1186, 1207, 1266). India, associated with a *Ptychogaster* (Bose) ; India (D. Maruda Rajan).

——**picipes** Fr. (Sacc. **VI**:83 ; **196**:122 ; **288**, No. **38**:8 ; **352**). Royal Botanic Garden, Calcutta, and Sikkim (Kurz) ; Sonamarg, Kashmir (R. R. Stewart) ; India (H. V. Ryan).

——**platyporus** Berk. (**57**, No. 337 ; Sacc. **VI**:83). On dead timber, Darjeeling, 8,000 ft. (Hooker f.). Lloyd (**296**:162) found the type very scanty, but he thinks it is the same as *P. rufescens* form *heteroporus*.

——**plorans** (Patouill.) Sacc. & D. Sacc. (Sacc. **XVII**:110 ; **288**, No. **42**:3). India (Prof. D. E. Bashambar). Lloyd (**300**:364) was in some doubt as to the identity of the specimens.

——**resinosus** (Schrad.) Fr. (Sacc. **VI**:137 ; **263**:325 ; **300**:384). On dead roots, Botanic Garden, Saharanpur (Gollan).

——**rhodophaeus** Lév. (Sacc. **VI**:175 as *Fomes* ; **435**:71 ; **287**:1266, 1290 ; **70, VIII**:29 as *P. semilaccatus* Berk. ; **84** :277 ; **435** :70 ; **287** :1126 ; **196** :124, fig., as *P. cinereo-fuscus* Currey ; **84**:223). Nakawa, Toukyeghat, Burma (Kurz) ; on dead wood, Sidapur, Coorg, and Dhoni forests, Madras (**435**) ; South India (D. Maruda Rajan) ; on dead *Heritiera minor*, Sundribuns and Madras (Bose). Lloyd (**296**:129 ; **300**:385) also considered *P. semilaccatus* as a synonym, but later (**287**:1290) preferred to use the latter name for forms with dark pores.

——**rubidus** Berk. (Sacc. **VI**:137 ; **196**:123, figs. ; **287**:955 ; **296**:133). Burma (Kurz) ; India (Cave). Lloyd (**296**:191) is uncertain whether this is distinct from *Polystictus modestus* (q. v.).

——**rufescens** Fr. (Sacc. **VI**:78 ; **57**, after No. 335). On dead wood, Darjeeling and Jallapahar, 3-5,000 ft. (Hooker f.). See also *P. anthelminticus*.

——**rugosus** Nees (Sacc. **VI**:152 as *Fomes ;* **57**, after No. 415 ; **372**:4 ; **296**:110). Khasi Hills (Hooker f.).

——**rugulosus** Lév. (Sacc. **VI**:168 as *Fomes ;* **287**:952, 1126, 1266 as *P. rigidus* Lév. ; **435**:70). India (Bose) ; South India (D. Maruda Rajan) ; on dead wood in Dhoni forests, Malabar (**435**). Lloyd (**287**:1078 ; **300**:337) considers *P. zonalis* as well as *P. rugulosus* to be the same as *P. rigidus ;* we have followed Bresadola (**84**:227).

——**rutilans** (Pers.) Fr. (Sacc. **VI**:119 ; **287**:1186). India (Bose).

——**saharanpurensis** P. Henn. (**263**:325 ; Sacc. **XVII**:108). On the ground near tree roots, Botanic Garden, Saharanpur (Gollan). Lloyd (**296**:162) did not find the type, but considers the description to indicate *P. schweinitzii*.

——**satpoorensis** Beck (**50**:146, fig. ; Sacc. **XI**:85). On rotten trunks, Satpoor Range. " The illustration is an excellent picture of *Polystictus leoninus* " (Lloyd, **300**:385).

[**Polyporus**]**schweinitzii** Fr. (Sacc. **VI**:76 ; **263**:326 as *P. sistotremoides* (Alb. & Schw.) Schroet.). On the ground, Botanic Garden, Saharanpur (Gollan). Recorded also by Saccardo as from the Himalaya. See *P. saharanpurensis* and *P. tabulaeformis*.

——**scopulosus** Berk. (**57**, No. 429 ; Sacc. **VI**:201 as *Fomes ;* **296**:128, fig., 190 ; **287**:952, 996 ; **391**:139, figs., as *Trametes rhizophorae* Reichardt). On dead wood, Darjeeling (Hooker f.) ; in Bengal (Hutchings) ; India (Bose) ; on *Rhizophora* trunks, near Car-Nicobar, Nicobar Islands. Lloyd (**296**:128) states that this is a lignicolus *Polyporus*, or might be classed as a *Trametes*.

——**scruposus** Fr. (Sacc. **VI**:121 ; **57**, after No. 430 ; **288**, No. **49**:3 ; **196**:124 ; **263**:325 as *P. gilvus* Schw. var. *scruposus* Fr.). Iwa River, East Nepal and Sone River, Bihar (Hooker f.) ; Martaban Hills (Kurz) ; on dead trunks, Arnigadh, Mussoorie (Gollan) ; India (Cave). Lloyd (**300**:347) considers this to be only an excessively rough form of *P. gilvus*. See also *P. hookeri*.

——**secernibilis** Berk. (Sacc. **VI**:102 ; **70, III**:3, figs., **288**, No. **49**:4). On dead wood, Darjeeling, 7,000 ft., and Hooghly, Bengal (Bose) ; India (Cave). Lloyd (**300**:329) places this species near *P. adustus*.

——**shoreae** Wakefield (Sacc. **XXIII**:358, authors erroneously listed as " Wakefield et Grove " ; **70, V**:21, figs. ; **111**:98, fig. ; **113** (7):53 ; **113** (8):53 ; **113** (9):64 ; **247**). On living trunks of *Shorea robusta*, Bengal (Shaw ; Bose ; McCrie ; Hole ; Glasson).

——**squamoso-maculatus** Sacc. (Sacc. **IX**:161 ; **57**, No. 336 as *P. maculatus* Berk., not Peck). On trunks of living trees, Darjeeling, 8,000 ft. (Hooker f.). Lloyd (**296**:176) lists this as close to, or identical with, *P. arcularius*.

——**squamosus** (Huds.) Fr. (Sacc. **VI**:79 ; **57**, after Nos. 336 and 416 ; **260**:151). On dead wood, Darjeeling, 7,500-8,000 ft. (Hooker f.) ; Panzi, North-western Himalaya (Marten).

——**subvirgatus** Lloyd (**296**:172 ; **288**, No. **32**:1). India (Cave).

——**sulphureus** (Bull.) Fr. (Sacc. **VI**:104 ; **57**, after No. 338 ; **352** as " *Lactiporus sulphureus* "). Tonglo, Sikkim (Hooker f.) ; Sonamarg, Kashmir (R. R. Stewart). See also *P. miniatus*.

——**tabulaeformis** Berk. (**57**, No. 78 and after No. 415 ; Sacc. **VI**:76 misprinted " *tubulaeformis* "). On dead wood, Darjeeling, 7,500 ft. (Hooker f.). Lloyd (**296** :162) considers that this name, as applying to the type from the United States of America, refers to *P. schweinitzii*.

——**turboformis** Lloyd (**296**:194, fig. ; Sacc. **XXIII**:371). Baroda (Krumbiegel). Lloyd does not give the habitat, but Saccardo says " ad truncos ".

——**umbellatus** Fr. (Sacc. **VI**:95 ; **352** as " *Grifola umbellata* "). Sonamarg, Kashmir, " a favourite edible species " (R. R. Stewart).

——**umbilicatus** Berk. (**57**, No. 335 ; Sacc. **VI**:68 ; **467**:156). On dead wood, Tonglo and Sinchul, Sikkim Himalaya, 8,000 ft. (Hooker f. ; part of the material also in Herb. Montagne); St. Xavier's College, Bombay (Blatter). Lloyd (**296**: 171) states: "Close to *brumalis* but more smooth and *rigid.*"

[**Polyporus**]**ungulatus** Berk. var. **hobsoni** Sacc. (**62**:165; Sacc. **VI**:142). Bombay (Hobson).

——**vallatus** Berk. (**57**, No. 419; Sacc. **VI**:77; **84**:235; **296**:162). On the ground, Khasi Hills (Hooker f.).

——**varius** Fr. (Sacc. **VI**:84; **288**, No. **49**:4). India (Cave).

——**versiformis** Berk. (**57**, No. 417; Sacc. **VI**:81). On dead wood, Sikkim (Hooker f.). Lloyd (**296**:188) records that this species is based on two small specimens, one of which seems to be *P. melanopus*, the other apparently different.

——**vivax** Berk. (**57**, No. 423; Sacc. **IX**:171 and **XI**:84). On dead wood, East Nepal (Hooker f.). Lloyd (**296**:148) states that *P.* "*virax*" (error for *vivax*) seems to be the same as *P. liebmanni* Fr.

——**zonalis** Berk. (Sacc. **VI**:145; **57**, after No. 339; **260**:151; **70**, **III**:3, figs.; **300**: 336; **287**:1069). On dead wood, Darjeeling, 7,000 ft. (Hooker f.); Hooghly, Howrah, and Darjeeling (Bose); on dead trunk of *Cedrela toona*, Kalsia (J. H. Blandford). See *P. rugulosus*.

Polystictus aethiops Cke (**148**:99; Sacc. **VI**:284). On bark, India. Lloyd (**300**: 375) notes that the type specimens cannot be accepted as representing a species.

——**affinis** Nees (Sacc. **VI**:219; **196**:122; **472**:152; **288**, No. **28**:2, No. **60**:6; **294**: 53, 57; **296**:142, fig.; **287**:1126; **435**:73; **70**, **VIII**:31, figs.). Yomah Range and Karen Country, Burma (Kurz); on dead wood, Coorg (**435**); Bengal (Hutchings); India (P. D. Master); South India (D. Maruda Rajan); on fallen trunk of *Terminalia arjuna*, Midnapur, Bengal (Bose); also recorded for Nicobar Islands. See *P. squamaeformis*.

——**asper** Jungh. (Sacc. **VI**:224; **196**:124, figs., erroneously as *Polyporus xerophyllaceus* Berk.; **163**:21 as *Fomes curreyi* Berk. in Herb.; **148**:98 as *P. curreyi* Berk.; **299**:279 as "*Trametes strigata*"; **84**:224, 227). On old logs, Botanic Garden, Calcutta and Andaman Islands (Kurz).

——**badius** Berk. (Sacc. **VI**:281; **288**, No. **28**:3; **70**, **IV**:4, figs., as *Trametes badius* Berk.; **287**:1010. On old logs, Bengal (Hutchings); Assam and Bengal (Bose). Lloyd (**288**, No. **28**:3) states that the Indian specimens are more like the type at Kew than Philippine specimens similarly named.

——**beharensis** Berk. (**57**, No. 438; Sacc. **VI**:289). On dead wood, Sone River, Bihar (Hooker f.). Lloyd (**287**:1145) notes that *P. glabro-rigens* may be the same as *P. beharensis*, but the type of the latter is too poor to decide.

——**berkeleyi** Bres. (Sacc. **VI**:369 as *Hexagonia pergamenea* Berk. & Broome, not *Polystictus pergamenus* Fr.; **70**, **V**:22, fig.; **287**:1010). On dead wood, Hooghly District, Bengal (Bose). Bose states that this is more a *Hexagonia* than a *Polystictus*.

[Polystictus] caperatus Berk. (Sacc. **VI**:282 ; **57,** after No. 432 ; **70, IV**:2 ; **287**:1032, fig. ; **287**:1126 as *P. phocinus* Berk. ; **84**:226). On dead wood, Parasnath, Bihar (Hooker f.) ; Hooghly, Bengal (Bose) ; South India (D. Maruda Rajan).

——**cervino-gilvus** Jungh. (Sacc. **VI**:288 ; **70, VIII**:27 ; **288,** No. **33**:2 as *P. dermatodes* Lév. ; **67**:51 as *P. peradeniyae* Berk. & Broome ; **84**:224, 226). On fallen branches, Nicobar Islands ; Nilgiris ; India (Kirtikar) ; Chittagong (Bose). See also *P. zeylanicus.*

——**cichoriaceus** Berk. (Sacc. **VI**:280 ; **287**:955 ; **294**:62). India (Cave). See *Polyporus intybaceus* and *P. setiporus.*

——**cineraceus** Lév. (**286**:139 ; Sacc. **VI**:261). On trunks, Nilgiris (Perrottet). The type was not found at Paris, but a specimen marked *P. cineraceus*, not from India, is in Herb. Montagne with a note that it is perhaps a form of *P. adustus.* Lloyd (**296**:146) did not find a type.

——**cinerescens** Schwein. (Sacc. **VI**:285 ; **196**:124). On wood, Yomah, Burma (Kurz).

——**cingulatus** Fr. (Sacc. **VI**:268). Recorded in Saccardo as occurring in India.

——**cinnabarinus** (Jacq.) Fr. (Sacc. **VI**:245 ; **57,** after No. 430 ; **196**:124 ; **352** as " *Pycnoporus cinnabarinus* " ; **51**:387 as *Polyporus cristula* Klotzsch ; **70, IV**:3, fig., as *Trametes cinnabarina* (Jacq.) Fr.). On dead wood, East Nepal (Hooker f.) ; on old logs, Royal Botanic Garden, Calcutta (Kurz) ; Sonamarg, Kashmir (R. R. Stewart) ; on rotten bamboos (in India ?; Wight) ; on trunks, Howrah (Bose). Lloyd (**296**:144) considers *P. cinnabarinus* to be limited to temperate regions, and *P. sanguineus* the corresponding species of the tropics.

——**cinnamomeus** Jacq. (Sacc. **VI**:210 ; **287**:1295 ; **472**:152 ; **296**:164). India (P. L. Dey). See also *P. oblectans* var. *lahorensis* Lloyd (**288,** No. **65,** Note 526). India (Kashyap).

——**coriaceus** Lév. (**286**:137 ; Sacc. **VI**:270). On trunks, Nilgiris (Perrottet). Type not found at Paris.

——**cumingii** Berk. (Sacc. **VI**:209 ; *P. garckeanus* P. Henn. in Sacc. **IX**:181 ; **84**:224). On trunks, Mergui, ? Burma (or ? Chile) (T. Philippi). See *P. nilgheriensis.*

——**elongatus** Berk. (Sacc. **VI**:231 ; **57,** after Nos. 339 and 433 ; **288**, No. **49**:4, 11). On wood, forming large masses, Jallapahar and Lebong (7,000 ft.), Darjeeling, and a velvety form, Parasnath (Hooker f.) ; Darjeeling (Cave). " Merely the tropical form of the common *Polystictus pergamenus* of temperate America" (Lloyd, **288,** Note 125).

——**feei** Fr. (Sacc. **VI**:244 ; **196**:124). Yomah, Burma (Kurz). Lloyd (**288,** No. **32**:3 ; **299**:226) considers this a tropical American plant similar to *Trametes carnea.*

——**fibula** Fr. (Sacc. **VI**:239 ; **70, VIII**:31 as *P. vellereus* Berk. ; **84**:228). On root of dead bamboo, Bengal (Bose).

[**Polystictus**] **flabelliformis** Klotzsch (Sacc. **VI**:216 ; **57,** after Nos. 337 and 420 ; **51**:386 ; **196**:122 ; **288,** No. **27**:3). On dead wood, Darjeeling and Lebong, 4-8,000 ft. (Hooker f.) ; Yomah Range, Burma (Kurz) ; Bengal (Hutchings) ; Ootacamund (Perrottet).

——**flavidus** Berk. (**57,** No. 432 ; Sacc. **VI**:278). On dead wood, East Nepal (Hooker f.).

——**floccosus** Jungh. (Sacc. **VI**:250 ; **288,** No. **32**:3, No. **60**:6 ; **287**:1266 ; **435**:72). India (Kirtikar ; P. D. Master ; D. Maruda Rajan) ; on dead wood in Dhoni forests, Madras (**435**). Lloyd (**287**:1036) decided later that the specimens sent by Kirtikar and Master were *P. zeylanicus.*

——**floridanus** Berk. (Sacc. **VI**:251 ; **467**:156). A form on wood, Dehra Dun (Blatter).

——**gallo-pavonis** Berk. & Broome (Sacc. **VI**:234 ; **288,** No. **27**:3). Bengal (Hutchings). Lloyd (**288,** No. **65**:16) notes that this species is intermediate between *Polyporus* and *Polystictus.*

——**gaudichaudii** Lév. (Sacc. **VI**:233 ; **288,** No. **49**:4 ; **296**:134, fig.). India (Cave). Bose (**70, VIII**:28) includes this name as a synonym of *P. thwaitesii,* and the latter name is considered by Bresadola (**84**:227) to be a synonym of *P. menziesii* (q. v.).

——**gleadowii** Massee (**319, III**:152 ; Sacc. **XVII**:130). On dead wood, Dehra Dun (Gamble). Bresadola (**85**:29) says this is *P. tephroleucus.*

——**gollani** P. Henn. (**263**:327 ; Sacc. **XVII**:130). On trunks of *Terminalia tomentosa,* Siwalik Range (Gollan).

——**gratus** Berk. (**57,** No. 436 ; Sacc. **VI**:264). On the ground, doubtless attached to wood, Himalaya (Hooker f.). Lloyd (**300**:379) says the type seems the same as *F. floriformis* Quél.

——**hasskarlii** Lév. (Sacc. **VI**:280). Recorded in Saccardo as occurring in Nilgiris, but no other reference found. No Indian specimens were found at Paris.

——**hirsutus** Fr. (Sacc. **VI**:257 ; **57,** after No. 339 ; **288,** No. **28**:2, No. **31**:2). On dead wood, Darjeeling, 4,000 ft. (Hooker f. ; part of this collection sent by Berkeley to Montagne) ; Bengal (Hutchings) ; India (Cave). See *P. steinheilianus.*

——**hutchingsii** Lloyd (**287**:1316, fig.). Bengal (Hutchings).

——**hypothejus** Kalchbr. (Sacc. **VI**:263 ; **70, VIII**:31, fig.). Bengal (Bose). Previously known from Australia.

——**inquinatus** Lév. (**286**:140 ; Sacc. **VI**:270 ; **288,** No. **60**:7). On trunks, Nilgiris (Perrottet) ; North-west Himalaya, 6,500 ft. (W. T. Saxton). The type at Paris consists of one pileus in good condition.

——**lanatus** Fr. (**239**:490 ; Sacc. **VI**:274). India.

——**leoninus** Klotzsch (**274**:486 ; **51**:390 ; Sacc. **VI**:235 ; **467**:156 ; **287**:1125 ; **70, II**:140, figs. ; **57,** after No. 422, as *P. funalis* Fr. ; **196**:123 ; **263**:327 ; **288,** No. **37**:3). On trunks, India (Wight) ; Sone River (Hooker f.) ; Mutlah,

Bengal (Kurz); Hooghly District, Bengal (Bose); Botanic Garden, Saharanpur (Gollan); Anand, Bombay (Blatter); India (D. Maruda Rajan). See also *P. satpoorensis* and *Hydnum gleadowii.*

[**Polystictus**] **licnoides** Mont. (Sacc. **VI**:281; **57,** after No. 339; **70, III**:2, figs.; **288,** No. **34**:2). Sinchul, Sikkim, 8,000 ft. (Hooker f.); on dead wood, Darjeeling (Bose); India (Cave). Lloyd (**300**:349) considers it a pronounced, tropical form of *P. gilvus.*

——**luteus** Blume & Nees (Sacc. **VI**:218). Recorded by Saccardo as occurring in the Nicobar Islands, presumably from Cooke's reference regarding the distribution of this species, given in the report of the " Challenger " expedition 1885. No Indian specimen was found at Kew, nor a more definite record of a collection from the Nicobar Islands.

——**malaiensis** Cke (Sacc. **VI**:227 as *P.* " *malacensis* " Cke; **287**:1125 as " *Polyporus maliensis* "). South India (D. Maruda Rajan).

——**membranaceus** (Swartz) Berk. (Sacc. **VI**:287; **349**:22 as *Polyporus flabellum* Mont.; **84**:224). Ootacamund, Nilgiris (Perrottet). There are two specimens in the general Herbarium, Paris.

——**menziesii** Berk. (Sacc. **VI**:226; **288,** No. **69**:5; **70, VIII**:28, figs., as *P. thwaitesii* Berk. & Broome; **84**:227). India (Cave); on dead timber and log of *Excoecaria agallocha,* Calcutta, and Sundribuns (Bose). See *P. gaudichaudii.*

——**meyenii** Klotzsch (Sacc. **VI**:261; **287**:1126, and 1186 as " *Trametes meyenii*"). South India (D. Maruda Rajan); India (Bose). See *T. obstinata.*

——**modestus** Kunze (Sacc. **VI**:226; **196**:122). Yomah Range, Burma (Kurz). See *Polyporus rubidus.*

——**nepalensis** Berk. (**57,** No. 434; Sacc. **VI**:228; **57,** No. 485 as *Polyporus corium* Berk.; **84**:223). On dead wood, Nangki, East Nepal, 10,000 ft., and Sikkim, 12,000 ft. (Hooker f.). Types discussed by Lloyd (**296**:145; **300**:377; **287**:951).

——**nilgheriensis** Mont. (**349**:22; Sacc. **VI**:231; **57,** after Nos. 339 and 433; **70, V**:20, figs.; **300**:357, figs.; **287**:1010). On old trunks, Nilgiris (Perrottet); East Nepal and Darjeeling (Hooker f.); Darjeeling (Bose). Lloyd (**300**:357) considers this species to be a lignescent *Polyporus,* resembling a *Fomes.* He (l.c.:379) calls *P. garckeanus* P. Henn. a synonym, but we have followed Bresadola who includes *P. garckeanus* with *P. cumingii* (q.v.). Type specimens of *P. nilgheriensis* are at Paris, also specimens marked " Sikkim " which were probably sent to Montagne by Berkeley.

——**oblectans** Berk. (Sacc. **VI**:211; **57,** after Nos. 335 and 415; **263**:327). On dead wood and on the ground, Sikkim, 7,500 ft., and Churra, Moflong, and Nunklow, Khasi Hills (Hooker f.); Siwalik Range (Gollan). Lloyd (**296**:164) considers this an Australian species related to *P. cinnamomeus.*

——**occidentalis** Klotzsch (Sacc. **VI**:274; **196**:125; **263**:327; **70, III**:3, figs.; **287**:952, 1069, 1126; **467**:156, forms *resupinata* and *tenuis,* the latter perhaps *P. gibberulosum*). On old wood, Royal Botanic Garden, Calcutta (Kurz); Arnigadh, Mussoorie (Gollan); Bandra and Bombay (Blatter); Bengal and Assam (Bose); South India (D. Maruda Rajan).

[**Polystictus**] **ozonioides** Berk. (**57**, No. 339; Sacc. **VI**:236; **57**, after No. 443, as *Trametes ozonioides* Berk.). On dead wood, Darjeeling, and on charred wood, Lebong, 6,000 ft. (Hooker f.).

——**parishii** Berk. in Herb. (**163**:51; Sacc. **VI**:263; **288**, No. **69**:5, 11). On trunks, Moulmein, Burma; India (G. H. Cave). A form of *P. versicolor*, according to Lloyd (l.c.).

——**pectunculus** Lév. (**286**:138; Sacc. **VI**:261). On trunks, Nilgiris (Perrottet). The type is in good condition at Paris. No other specimens of this species were found there.

——**perennis** (L.) Fr. (Sacc. **VI**:210; **70, IX**:38, figs.; **352** as "*Coltricia perennis*"). On the ground, Shillong (Bose); Sonamarg, Kashmir (R. R. Stewart).

——**pergamenus** Fr. (Sacc. **VI**:242; **288**, No. **38**:8). India (H. V. Ryan). Bresadola (**84**:226) considers *P. pergamenus* Fr. to be "*P. prolificans* Fr., juvenilis." See *P. elongatus*.

——**persoonii** Fr. (Sacc. **VI**:272; **196**:123; **467**:155; **70, III**:4, figs.; **288**, No. **28**:2, No. **34**:2, No. **61**:8, No. **65**:4, 8; **287**:1126; **274**:481 as *Daedalea sanguinea* Klotzsch; **51**:381; **57**, after No. 443). Nakawa, Toukyeghat, Burma (Kurz); on dead wood, India (Wight); East Nepal (Hooker f.); Andheri, Bombay (Blatter); Bengal and Assam (Bose); Bengal (Hutchings); South India (D. Maruda Rajan); India (Cave; Kashyap). Lloyd (**288**, No. **42**:10) notes that when thin it is a *Polystictus* and when thick a *Trametes*.

——**pictilis** Berk. (**57**, No. 433; Sacc. **VI**:255). On dead *Betula*, East Nepal, 12,000 ft. (Hooker f.).

——**pinsitus** Fr. (Sacc. **VI**:262; **196**:124; **286**:136; **349**:22 as *Polyporus setosus* Mont.). Nilgiris (Perrottet); Nattoung Range, Burma (Kurz); India (Jacquemont). Lloyd (**287**, Polyporoid Issues:27) considers *P. pinsitus* to be confined to America, and the specimens so referred from India to be "abundantly different." Montagne did not include *P. setosus* in his Sylloge, nor is it in Saccardo, and from the fact that the type is included at Paris in the folder with *P. pinsitus*, we infer that Montagne later considered it the same as specimens referred to the latter species.

——**polyzonus** (Pers.) Fr. (Sacc. **VI**:278 and **X**:188; **288**, No. **60**:7). North-Western Himalaya, 6,500 ft. (W. T. Saxton).

——**proteus** Berk. (Sacc. **VI**:250; **70, V**:21, figs.; **287**:1010, 1069). On dead bark of logs, Calcutta (Bose).

——**pterygodes** Fr. (Sacc. **VI**:222; **287**:1186; **70, IX**:38, figs.). Sylhet (Bose). Lloyd (**294**:56) regards this as perhaps a sessile form of *P. xanthopus*.

——**russogramme** Berk. (Sacc. **VI**:229; **467**:156). On dead wood, Khandala, Bombay (Blatter). Lloyd (**296**:147) considers the type inadequate.

[Polystictus] sacer Fr. (Sacc. **VI**:213 ; **111**:18, fig.; **70, IX**:37, figs.). On the ground, growing from sclerotia, Assam (Butler) ; Chittagong (Bose). Lloyd (**287**: 1037) considers *P. rhinocerotis* Cke as a small-pored form of this species.

——**sanguineus** (L.) Mey. (Sacc. **VI**:229 ; **349**:22 ; **57,** after No. 337 ; **196**:122 ; **260**:151 ; **467**:156 ; **70, I**:114 ; **288,** No. **27**:3, No. **34**:2, No. **42**:11 ; **287**: 1295). Nilgiris (Perrottet: specimen in Herb. Montagne) ; on dead wood, Darjeeling, 7,000 ft. (Hooker f.); Mutlah, Lower Bengal, Royal Botanic Garden, Calcutta, and Tonglo, Sikkim (Kurz); Bilaspur (Marten); Dehra Dun (Blatter) ; Calcutta (Bose); Bengal (Hutchings); India (Cave; P. L. Dey ; J. Ray). See note under *P. cinnabarinus*.

——**sarawacensis** Berk. (Sacc. **VI**:306 as *Poria ;* **68**:150). Khandala, Bombay (Blatter).

——**sarbadhikarii** Bose (**75**:301, figs.; **79**:138). On dead trunk of *Tamarindus indica*, Hooghly District, Bengal (Bose).

——**setiporus** Berk. (**57,** No. 446 as *Favolus setiporus* Berk.; Sacc. **VI**:402 ; **472** : 153). On dead wood, Nunklow, Khasi Hills (Hooker f.). Lloyd (**294**:60) considers this to be a form of *P. cichoriaceus* Berk.

——**splitgerberi** Mont. (Sacc. **VI**:234 ; **196**:122). Natoung Hills, Burma, 6-7,000 ft. (Kurz). A doubtful record, for Lloyd (**300**:355, 386) says that this is the same as *Polyporus rheicolor* Berk., which is confined to the American tropics.

——**squamaeformis** Berk. (**57,** No. 420 ; Sacc. **VI**:221 ; **391**:140). On dead wood, Khasi Hills (Hooker f.); Nicobar Islands. Lloyd (**294**:57) remarks **:** " No type exists, but specimens so determined by Cooke are small forms of *affinis*."

——**steinheilianus** Berk. & Lév. (not in Sacc.; **70, VIII**:32, figs.). On a dead branch, Puri, Orissa (Bose). Identified by Bresadola. He first (**84**:227) gave this as a synonym of *Trametes rigida* Mont. & Berk., but later (**84**:68) seems to consider it a good species. Bose (l.c.) states " to me it seems a resupinate form of *Polystictus hirsutus*."

——**stupeus** Berk. (Sacc. **VI**:236). On trunks, India. Lloyd (**294**:65, fig.) states that there is a specimen in Cooke's herbarium marked " India, Herb. Griffith."

——**suboccidentalis** Sacc. & Syd. (Sacc. **XIV**:188 ; **287**:952, 1000, fig., 1010). India (Bose).

——**tabacinus** Mont. (Sacc. **VI**:280 ; **472**:153 ; **70, II**:141, figs.; **288,** No. **69**:5 ; **287**:1126 ; **435**:74). On dead wood in very moist and shady places, Coorg (**435**) ; Darjeeling (Bose); India (G. H. Cave); South India (D. Maruda Rajan).

—— **tephroleucus** Berk. (**57,** No. 442 as *Trametes tephroleuca* Berk.; Sacc. **VI**:275). On dead wood, Nangki, East Nepal, 10,000 ft. (Hooker f.). See *P. gleadowii.*

——**tomentosus** Fr. (Sacc. **VI**:208 ; **260**:152). On wood of *Shorea robusta*, Kalsia (J. H. Blandford).

[**Polystictus**] **velutinus** Fr. (Sacc. **VI**:258; **70, IX**:39, figs.). On dead *Alnus nepalensis*, Calcutta and Pashok, Darjeeling (Bose).

——**versatilis** Berk. (Sacc. **VI**:244; **260**:151; **472**:153; **70, III**:1, figs., as *Trametes versatilis* Berk.; **288,** No. **28**:3). On dead trunks, Bilaspur, Central Provinces (Marten); Hooghly District, Bengal (Bose); Bengal (Hutchings).

——**versicolor** (L.) Fr. (Sacc. **VI**:253; **349**:22; **57,** after Nos. 339 and 433; **196**: 124; **263**:327; **288,** No. **60**:7, No. **65**:4; **287**:1295; **70, II**:140, figs.). Nilgiris (Perrottet: specimens at Paris); on wood, Darjeeling, 7,500 ft., Tonglo, Sikkim, 8,000 ft., and Nangki, East Nepal, 9,000 ft. (Hooker f.); Sikkim Himalaya (Kurz); Arnigadh, Mussoorie (Gollan); N. W. Himalaya, 6,500 ft. (W. T. Saxton); India (Kashyap; P. L. Dey); a "new variety" (unnamed), Darjeeling 7,000 ft., and Hooghly, Bengal (Bose).

——**villosus** Massee (**319, IV**:93; Sacc. **XXI**:311; not *P. villosus* (Swartz) Fr., Sacc. **VI**:238). On dead branches, Dehra Dun and Mysore (Butler).

——**virgineus** Schwein. (Sacc. **VI**:224; **260**:152). On dead wood, Bilaspur, Central Provinces (Marten).

——**vittatus** Berk. (Sacc. **VI**:268; **287**:1069 as "*Trametes vittatus*"). India (Bose).

——**xanthopus** Fr. (Sacc. **VI**:215; **349**:22; **467**:156; **57,** after Nos. 335 and 415; **391**:140; **288,** No. **27**:3, No. 31:2, No. **49**:3, No. **60**:6; **294**:51; **435**:73; **288**:955; **57,** No. 416 as *Polyporus florideus* Berk.; **196**:121; **196**:122, fig., as *Polyporus crassipes* Currey). India (Bélanger) and Nilgiris (Perrottet: good specimens of both collections are in Herb. Montagne); Darjeeling, 7,500 ft., Pimhabania, banks of Sone River, Bihar, and East Nepal (Hooker f.) Burma and South Andaman Island (Kurz); Simla (Blatter); Coorg (**435**); Bengal (Hutchings); India (Cave; P. D. Master); Nicobar Islands. See *P. pterygodes.*

——**xeranticus** Berk. (**57,** No. 431; Sacc. **VI**:279). On charred wood, Lebong, Darjeeling (Hooker f.).

——**zeylanicus** Berk. (Sacc. **VI**:271; **467**:156; **435**:72; **287**:1036, 1266). On trunks, Khandala, Bombay (Blatter); Dhoni Forests, Malabar (**435**); India (D. Maruda Rajan). Petch (**367, VII**:129-133) considers *P. cervino-gilvus* to be the same, but Lloyd (**287**:1163) and Bresadola do not agree. See also *P. floccosus.*

——**zonatus** Fr. (Sacc. **VI**:280; **467**:156 as *P. zonatus* (Koen.) Berk.). On branches, Bombay (Blatter).

Poria arenaria Klotzsch (**274**:487; **51**:394; Sacc. **VI**:331). On sandy soil, India (Wight).

——**carteri** Berk. (**163**:25; Sacc. **VI**:309). On trunks, Bombay (Carter).

——**cerea** Berk. (**57,** No. 437; Sacc. **VI**:320). On dead wood, high valleys of East Nepal (Hooker f.).

[**Poria**] **diversispora** Berk. & Broome (Sacc. **VI**:324 ; **70, III**:1, figs.). Common, usually on old bamboo, Bengal (Bose).

——**gallo-grisea** Berk. (**163**:25 ; Sacc. **VI**:306). On rotten wood, Nilgiris.

——**hypobrunnea** Petch (**372**:51 ; Sacc. **XXIII**:419 ; **497**:38). A common cau[se] of die-back of tea stems, Cachar (Tunstall).

——**hypolateritia** Berk. (**163**:24 ; Sacc. **VI**:297 ; **372**:50 ; **287**:375). On wood, Nilgiri Hills (E. S. B[erkeley ?]). Injurious to tea and coffee in Ceylon, and stated by Petch (**368**:158) to occur also on tea in southern India.

——**membranicincta** Berk. (Sacc. **VI**:315 ; **70, VIII**:32, fig.). On decomposed wood, Khulna, Bengal (Bose).

[——**porriginosa** Berk. (**163**:26 ; Sacc. **VI**:318). Bombay. Bresadola (**84**:229) states "=*Ascomyces ?* indeterminabilis."]

——**ravenalae** Berk. & Broome (Sacc. **VI**:307 ; **70, IV**:3, fig.). Common on dead petioles of palms, Calcutta (Bose).

Psathyra calvescens Berk. (**57**, No. 310 ; Sacc. **V**:1064, 1067). On mossy earth, Darjeeling, 7,500 ft. (Hooker f.).

——**flavo-grisea** Berk. (**57**, No. 300 ; Sacc. **V**:1063). On dead wood, in tufts, Darjeeling, 7-8,000 ft. (Hooker f.).

——**nana** Massee (**319, III**:152 ; Sacc. **XVII**:85 in error as " *Agaricus nanus* (Mass.) Sacc. & D. Sacc."). On the ground, Poona (Woodrow).

——**nassa** Berk. (**57**, No. 299 ; Sacc. **V**:1061). On dead wood, Darjeeling, 7,000 ft. (Hooker f.).

——**obtusata** Fr. (Sacc. **V**:1066 ; **263**:329). On the ground, Botanic Garden, Saharanpur (Gollan).

Psathyrella discolor Berk. (**57**, No. 311 ; Sacc. **V**:1132). On the ground and on dead timber, Darjeeling, 7,500 ft. (Hooker f.).

——**gracilis** Fr. (Sacc. **V**:1127 ; **57**, after No. 403 ; **260**:152). On the ground, Kursar in the Nubra Valley, Kashmir (Thomson) ; Botanic Garden, Saharanpur (Gollan).

—— ? **hydrophora** (Bull.) Sacc. (Sacc. **V**:1129 ; **263**:329). On the ground, Botanic Garden, Saharanpur (Gollan).

——**prona** Fr. (Sacc. **V**:1131 ; **260**:152 ; **263**:329). On the ground, Botanic Garden, Saharanpur (Gollan).

Psilocybe caespiticia Berk. (**57**, No. 298 ; Sacc. **V**:1053). On clay banks, Darjeeling, 7,500 ft. (Hooker f.).

——**tristis** P. Henn. (**263**:330 ; Sacc. **XVII**:89). On the ground, Botanic Garden, Saharanpur (Gollan).

Pterula himalayensis (Massee) Lloyd (**287**:867 ; **319, I**:114 as *Lachnocladium himalayense* Massee ; Sacc. **XVI**:211). On the ground in a pine forest, Phallaloong Ridge, Sikkim, 10,000 ft. (Gamble).

——**penicellata** Berk. in Lloyd (**287**:863, figs., and 1295). India (P. L. Dey).

Radulum emerici Berk. (**178**:3 ; Sacc. **XI**:111). On logs, Nilgiris.

[**Radulum**] **neilgherrense** Berk. in Herb. (**178**:3 ; Sacc. **XI**:111 ; **288**, No. **59**:4). On logs, Nilgiris. Lloyd (l.c.) thinks that this and the preceding are identical, and also *R. mirabile* Berk. (*Lopharia mirabilis*, q.v.).

——**spongiosum** Berk. (**57**, No. 450 ; Sacc. **XI**:111 ; **303**:2, figs.). On dead wood, East Nepal (Hooker f.).

Russula alutacea Fr. (Sacc. **V**:479 ; **65**:137). In woods, Gulmarg, Kashmir (Aitcheson).

——**cinnabarina** Hooker f. (**57**, No. 318 ; not in Sacc.). On clay banks, Darjeeling, 7,500 ft. (Hooker f.).

——**emetica** Fr. (Sacc. **V**:469 ; **57**, after Nos. 318 and 389). Khasi Hills and on clay banks, Darjeeling (Hooker f.). Berkeley remarks " Found with ***Russula rubra***."

——**furcata** Fr. (Sacc. **V**:456 ; **57**, after No. 316). On clay banks, Sinchul, Sikkim Himalaya, 8,500 ft. (Hooker f.).

——**grossa** Berk. (**57**, No. 317 ; not in Sacc.). Darjeeling (Hooker f.).

——**lepida** Fr. (Sacc. **V**:461 ; **57**, after No. 318). On clay banks, Darjeeling, 8,000 ft. (Hooker f.).

——**sanguinea** (Bull.) Fr. (Sacc. **V**:457 ; **57**, after No. 389 ; **263**:328). In pine woods, Lachen, Sikkim Himalaya 11,000 ft. (Hooker f.); on the ground, Arnigadh, Mussoorie (Gollan).

Schizophyllum commune Fr. (Sacc. **V**:655 ; **57**, after Nos. 326 and 408 ; **196**:121 ; **263**:328 ; **467**:157 ; **70**, **I**:109 ; **287**:1295 ; **288**, No. **41**:1). On dead wood, Darjeeling, 7,000 ft., and Parasnath, 3,000 ft. (Hooker f.); Calcutta, Sikkim, South Andaman Island, and Toukyeghat, Burma (Kurz); Arnigadh, Mussoorie (Gollan); Poona (Blatter); Calcutta (Bose); India (P. L. Dey).

——**flabellare** Fr. (Sacc. **V**:655 ; **349**:21). Nilgiris (Perrottet). A small, much branched specimen is in Herb. Montagne.

Sebacina alutacea Wakefield (**519**, **XXVI**:162). Encrusting the bases of young saplings of *Shorea robusta*, Ganjam, Madras (Minchin).

Septobasidium pteruloides (Mont.) Patouill. (**362**:337 ; **381**:279 ; **350**:152 as *Hydnum ? pteruloides* Mont.; not in Sacc.). On dead bark, Madura (Bélanger). Specimens are in Herb. Montagne. Petch (**381**) gives notes on the morphology and taxonomy of this species.

Stereum adustum Lév. (Sacc. **VI**:562 ; **196**:127 ; **391**:137). Southern Yomah, Burma (Kurz); on old *Pandanus*, Nicobar Islands.

[——**alternum** Lloyd (nomen nudum, **287**:1263). India (Bose).]

——**coalescens** Lloyd (**287**:1338, fig., 1295). Almora, Kumaon Himalaya (S. D. Joshi).

[——**curreyi** Sacc. (Sacc. **VI**:557 ; **196**:127, fig., as *S. cyathiforme* Currey, not Fr.; **313**:168 as *S. crucibuliforme* Massee). On wood, Karen Hills, Burma, 5-6,000 ft. (Kurz). Lloyd (**302**:14) states that this was based on the empty cups of a *Nidula*.]

[Stereum] elegans Meyer (Sacc. **VI**:553 ; **196**:126 ; **263**:323 ; **467**:154 ; **379**:260 ; **288,** No. **46**:2 ; **297**:24, fig.; **287**:1295, 1336, fig.). Lower Burma (Kurz); on tree roots, Botanic Garden, Saharanpur (Gollan); on wood, Khandala, Bombay (Blatter); India (P. L. Dey); India (collector unknown).

——**endocrinum** Berk. (**57,** No. 451 ; Sacc. **VI**:569). On dead branches, Yangma Valley, East Nepal (Hooker f.).

——**floriforme** Bres. (Sacc. **XXIII**:508 ; **288,** No. **46**:2 ; **297**:24). India (Gammie).

——**hirsutum** (Willd.) Fr. (Sacc. **VI**:563 ; **57,** after Nos. 346 and 451 ; **263**:323 ; **467**:154 ; **288,** No. **60**:7, No. **69**:5 ; **352**). On dead wood, Darjeeling, 7,000 ft., and Lebong, (Hooker f.); Arnigadh, Mussoorie (Gollan); Khandala, Bombay (Blatter); N. W. Himalaya, 6,500 ft. (W. T. Saxton); India (Cave); Sonamarg, Kashmir (R. R. Stewart).

——**lobatum** Fr. (Sacc. **VI**:568 ; **57,** after No. 451 ; **196**:126 ; **467**:154 ; **379**:262 ; **288,** No. **38**:8, No. **47**:3 ; **242**). On dead wood, Churra, Khasi Hills (Hooker f.); South Andaman Island and Toukyeghat, Burma (Kurz); Khandala, Bombay (Blatter); India (H. V. Ryan ; Cave). Lloyd (**288,** No. 38:8, No. **39**:3, No. **46**:3) regards this as a luxuriant tropical form of *S. fasciatum.*

——**malabarense** Lloyd ? (**297**:39). Apparently on the ground, Malabar, erroneously referred by massee (**313**:162) to *S. elegans.*

——**medicum** Currey (**196**:127, figs.; Sacc. **VI**:582 ; **313**:202). On wood, Sikkim 5-6,000 ft. (Kurz). Used medicinally by the Lepchas.

——**nitidulum** Berk. (Sacc. **VI**:552 ; **260**:151 ; **263**:323 ; **319, III**:152 ; **379**:259). On stump and roots of *Casuarina*, Botanic Garden, Saharanpur (Gollan); Kalsia (J. H. Blandford); on rotten wood, Poona (Woodrow).

——**ostrea** Nees (Sacc. **VI**:571 ; **349**:23 ; **196**:126). Nilgiris (Perrottet, specimens in Herb. Montagne); Sikkim Himalaya, 4-6,000 ft. (Kurz).

——**papyraceum** Massee (**319, IV**:94 ; Sacc. **XXI**:387). On dead wood, Wahjain, Khasi Hills (Butler).

——**princeps** Jungh. (Sacc. **VI**:570 ; **288,** No. **49**:4, No. **69**:5, 11 ; **287**:923, fig.; **57,** No. 453 as *S. scytale ;* **84**:232 ; **196**:127, figs.; **313**:171). On dead wood, Khasi Hills (Hooker f. and Thomson); Western Himalaya (Strachey); Karen Hills and Yomah, Burma (Kurz); India (Cave). Causes a pocket rot in timbers, resembling that caused by *S. frustulosum.*

——**purpureum** Pers. (Sacc. **VI**:563 ; **57,** after No. 346 ; **263**:323). On dead wood, Darjeeling, 7-8,000 ft. (Hooker f.); Arnigadh, Mussoorie (Gollan).

——**rimosum** Berk. (**57,** No. 346 ; Sacc. **VI**:568 ; **379**:265). On vegetable soil, old trees, etc., Darjeeling, 7,500 ft. (Hooker f.).

——**schomburgkii** Berk. (Sacc. **VI**:568 ; **287**:1069). India (Bose).

——**spadiceum** Fr. (Sacc. **VI**:564 ; **57,** after Nos. 346 and 451). On dead wood, Darjeeling, 7,000 ft., and Lachen, Sikkim, 8,000 ft. (Hooker f.).

Strobilomyces indicus Lloyd (**287**:1331, fig., 1295). India (P. L. Dey).

[**Strobilomyces**] **montosus** Berk. (**57,** No. 333 ; Sacc. **VI**:50). On dead wood and on the ground, Jallapahar, Darjeeling, 7,500 ft. (Hooker f.).

——**nigricans** Berk. (**57,** No. 398, fig.; Sacc. **VI**:50). In woods, Kala Pani, Khasi Hills, 5,000 ft. (Hooker f.).

——**polypyramis** Hooker f. (**57,** No. 332 ; Sacc. **XI**:81). Jallapahar, Darjeeling, 7,500 ft. (Hooker f.).

——**strobilaceus** (Scop.) Berk. (Sacc. **VI**:49 ; **352**). Sonamarg, Kashmir (R. R. Stewart).

Stropharia aureo-fulva Berk. (**57,** No. 292 as *Agaricus* (*Psalliota*) *aureofulvus* Berk.; Sacc. **V**:1015). On dead wood, Jallapahar, Darjeeling, 7-8,000 ft. (Hooker f.).

——**aurivella** Massee (**319, XIV**:255 ; Sacc. **XXIII**:306). Among grass, Calcutta Maidan (Burkill).

——? **crocopepla** Berk. & Broome (Sacc. **V**:1017 ; **260**:153). On the ground, Kalsia (J. H. Blandford).

——**gollani** P. Henn. (**260**:152 ; Sacc. **XVI**:120). On shady ground, Saharanpur (Gollan).

——**mephistopheles** Cke (**172**:7 ; Sacc. **IX**:139). On the ground, Belgaum (Hobson).

——**merdaria** Karst. (Sacc. **V**:1020 ; **319, III**:151 ; **263**:330). On dung, Poona (Woodrow); Botanic Garden, Saharanpur (Gollan).

——**psathyroidea** P. Henn. (**263**:330 ; Sacc. **XVII**:86). On the ground, Botanic Garden, Saharanpur (Gollan).

——**pygmaea** P. Henn. (**260**:152 ; Sacc. **XVI**:121). On the ground, Saharanpur (Gollan).

——**semiglobata** (Batsch) Quél. (Sacc. **V**:1022 ; **57,** after No. 383, as *Agaricus semiglobatus* Batsch). On the ground, Myrong, Khasi Hills, 6,000 ft. (Hooker f.).

Thelephora ? aurantiaca Pers. (Sacc. **VI**:526 ; **263**:324). On the ground, Botanic Garden, Saharanpur (Gollan).

——**gelatinosa** Saut. (Sacc. **VI**:541 ; **287**:955). India (Cave).

——**palmata** (Scop.) Fr. (Sacc. **VI**:529 ; **57,** after No. 450 ; **196**:126 ; **352**). On the ground, Nunklow, Khasi Hills (Hooker f.); on bamboo stems, Royal Botanic Garden, Calcutta (Kurz); Sonamarg, Kashmir (R. R. Stewart).

——**pusilla** Currey (**196**:126, fig.; Sacc. **VI**:532). Sikkim, 2,000-2,500 ft. (Kurz).

——**sowerbyi** Berk. & Broome (Sacc. **VI**:522 ; **263** :324). On tree roots and dead trunks, Botanic Garden, Saharanpur (Gollan).

——**sparassoides** P. Henn. (**263**:324 ; Sacc. **XVII**:162). On dead branches, Arnigadh, Mussoorie (Gollan).

Trametes acu-punctata Berk. (Sacc. **VI**:279 as a syn.; **287**:943, fig., 1266, as *Polystictus luteo-olivaceus* Berk. & Broome ; **84**:67 ; **435**:72). India (D. Maruda Rajan); on dead wood in Dhoni forests, Malabar (**435**).

[**Trametes carteri**], Berk. in Herb. (Sacc. **IX**:196 ; **287,** No. **49**:4). On trunks, Bombay (? Carter); India (Cave).

——**cincta** Bose (**76**:173, fig.; **287**:1069). On trunk of *Artocarpus integrifolia*, Darjeeling (Bose).

——**cingulata** Berk. (**57,** No. 441 ; Sacc. **XXI**:866 ; **196**:125 ; **70, II**:142, figs.; **288,** No. **42**:12 ; **287**:1125 ; **70, I,** fig.; **70, VII**:27 as *T. picta* Berk. & Broome ; **84**:229 ; **287**:1069). On dead wood, Sone River, Bihar (Hooker f.); Yomah, Burma, and Royal Botanic Garden, Calcutta (Kurz); Sibpur, Bengal, and on dead bark of living coconut tree, Bengal (Bose); India (J. Ray ; D. Maruda Rajan). Saccardo (**VI**:199) first changed the name to *Fomes virginiae* Manc. & Sacc., because there is an earlier *Polyporus cingulatus* Fr.

——**colliculosa** Berk. (Sacc. **VI**:349 ; **57,** after No. 441). On dead wood, Darjeeling (Hooker f.).

——**crenulata** Berk. (**57,** No. 440 ; Sacc. **VI**:336 ; **70, IV**:5, fig.; **287**:952). On dead wood, Darjeeling (Hooker f.); Howrah, Bengal (Bose).

——**curreyi** Cke (Sacc. **VI**:356 ; **196**:124, fig., as *T. umbrina* Currey, not Fr.). Nakawa, Toukyeghat, Burma (Kurz).

——**devexa** Berk. (Sacc. **VI**:341 ; **70, VII**:27 ; **287**:1010). On a dead tree, Cox's Bazar, Bengal (Bose). Bresadola (**84**:229) states that *Trametes* " *devexa* Bres."=*Polyporus occidentalis* Klotz. f. *obesa*.

——**elegans** (Spreng.) Fr. (Sacc. **VI**:335 ; **350**:146 as *Daedalea elegans* Spreng.). The specimen in Herb. Montagne is in good condition, and is marked " Pegu, ex Bélanger."

——**floccosa** Bres. (Sacc. **XIV**:192 ; **70, IV**:4, figs.; **287**:1010). On bark at base of a tree trunk, Calcutta, (Bose).

——**fuscella** Lév. (Sacc. **VI**:347 ; **73**:131, figs.; **287**:1069). On rotten wood, Bengal (Bose).

——**gibbosa** (Pers.) Fr. (Sacc. **VI**:337 ; **288,** No. **69**:5, as *Daedalea gibbosa*). India (Cave).

——**hololeuca** (Kalchbr.) ? Lloyd (Sacc. **VI**:241 as *Polystictus* ; **288,** No. **28**:2). Bengal (Hutchings). Lloyd states that it is close to *T. muelleri*, and not a *Polystictus*. We are not certain that Lloyd was the first to make the *Trametes* combination.

——**hookeri** Berk. (**57,** No. 439 ; Sacc. **VI**:336). On dead wood, Darjeeling (Hooker f.).

——**immutata** Berk. (**57,** No. 443 ; Sacc. **VI**:343). On dead wood, Khasi Hills (Hooker f.).

——**incerta** (Currey) Cke (**163**:56 ; Sacc. **VI**:352 ; **196**:123, figs., as *Polyporus incertus* Currey ; **175**:102 as *T. burchelli* Berk. in Herb.; **287**:899). On wood, Burma (Kurz).

[Trametes] **karii** Bose in Lloyd (**287**:1148, fig.; 952, 1010). India (Bose). Lloyd twice spelled the name " karie " and once " karii "; but we assume that Prof. Bose intended to name the plant for Mr. Kar.

——**lactinea** Berk. (Sacc. **VI**:343 ; **70, III**:1, figs.; **288,** No. **45**:2, No. **60**:7 ; **287**: 955). On dead wood, Howrah District, Bengal (Bose); India (Irani, Cave); N. W. Himalaya, 6,500 ft. (W. T. Saxton).

——**muelleri** Berk. (Sacc. **VI**:339 ; **288,** No. **27**:3 ; **70, V**:24, figs.). On wood, Howrah District, Bengal (Bose); Bengal (Hutchings).

——**obstinata** Cke (**288,** No. **49**:3). India (Cave). Bresadola (**84**:229) considers this to be a form of *Polystictus meyenii* (q.v.).

——**pini** (Brot.) Fr. (Sacc. **VI**:345 ; **251,** figs.; **299**:275, figs.; **416** ; **265**:194 ; **439**; **111**:97). On *Pinus excelsa* near Simla (Hole); on *Cedrus libani* var. *deodara*, Kashmir (Singh); Punjab (Suri); on roots and base of trunk of *Pinus longifolia*, Himalaya (Hafiz Khan).

——**plebeja** (Berk.) Lloyd (**299**:227 ; **56**:179 as *Polyporus plebeius* Berk. var. α). Himalaya. See note with *Fomes semitostus* above.

——**serpens** Fr. (Sacc. **VI**:355 ; **263**:327 ; **70, IV**:4, fig.). On dead trunks, Botanic Garden, Saharanpur (Gollan); on dead bark, Hooghly, Bengal (Bose).

——**sycomori** P. Henn. (Sacc. **XI**:96 ; **287**:1126). South India (D. Maruda Rajan).

——**versiformis** Berk. & Broome (Sacc. **VI**:339 ; **287**:1125). India (D. Maruda Rajan).

Tremella fuciformis Berk. (Sacc. **VI**:782 ; **288,** No. **28**:2 ; **287**:790, fig.). Bengal (Hutchings).

——**protensa** Berk. (**57,** No. 455 ; Sacc. **VI**:782). On trees, Darjeeling, 7,500 ft. (Hooker f.).

Tremellodon gelatinosum (Scop.) Pers. (Sacc. **VI**:479 ; **287**:1126). South India (D. Maruda Rajan).

Tricholoma cremoriceps Berk. (**57,** No. 252 ; Sacc. **V**:113). On tree trunks, Darjeeling, 7,500 ft. (Hooker f.).

——**giganteum** Massee (**319, XIV**:254 ; Sacc. **XXIII**:25). Shamnagar, near Calcutta (Burkill).

——**melaleucum** (Pers.) Fr. (Sacc. **V**:134 ; **352** as " *Melanoleuca melaleuca* "). Sonamarg, Kashmir (R. R. Stewart).

——**subpulverulentum** (Pers.) Fr. (Sacc. **V**:136 ; **57,** after No. 401). On the ground, near Sassar in Kashmir, 16,000 ft. (Thomson). Found also by Strachey.

Trogia belangeri (Mont.) Fr. (**239**:402 ; Sacc. **V**:635 ; **350**:145, fig., as *Agaricus* (*Crepidotus*) *belangeri* Mont.). On trunks, Mannentoddy, Western Ghats (Bélanger). Montagne later (**351**:121) stated " Verus *Xerotus* " and Lloyd (**287**:1227, fig.) writes it " *Xerotus belangeri.*"

——**königii** Fr. (**239**:402 ; Sacc. **V**:626). Recorded by Fries as " In India orientali. Koenig."

[Trogia] montagnei Fr. (Sacc. **V**:636 ; **350**:149, fig., as *Cantharellus aplorutis* Mont.). On fallen branches in forest, Trichinopoli Hills (Bélanger). Specimen not found at Paris.

Tubaria asperata P. Henn. (**263**:331 ; Sacc. **XVII**:76). On the ground, Botanic Garden, Saharanpur (Gollan).

——**furfuracea** (Pers.) W. G. Smith (Sacc. **V**:872 ; **263**:331). On the ground, Botanic Garden, Saharanpur (Gollan).

——**saharanpurensis** P. Henn. (**263**:332 ; Sacc. **XVII**:76). On the ground, Botanic Garden, Saharanpur (Gollan).

Typhula fuscipes (Pers.)Fr. (Sacc. **VI**:750 ; **196**:127). Sikkim, 7,000 ft., with *Polystictus versicolor* (Kurz).

Ulocolla foliacea (Pers.) Bref. (Sacc. **VI**:778 ; **196**:127 as *Tremella foliacea* Pers.; **57**, after No. 347, as *T. ferruginea* Smith). On mossy and rocky wet places, Tonglo, Sikkim, 10,000 ft. (Hooker f.); on shrubs, Sikkim, 10,000 ft. (Kurz).

Urobasidium rostratum Giesenhagen (**246**:139 ; Sacc. **XI**:131). On galls of *Taphrina cornu-cervi* on leaves of *Aspidium aristatum*, Nepal (Wallich).

Volvaria castanea Massee (**319, XIV**:254 ; Sacc. **XXIII**:179). Under the north side of a wall, Calcutta (Burkill).

——**delicatula** Massee (**319, XIV**:254 ; Sacc. **XXIII**:180). Under a wall, Calcutta (Burkill).

——**diplasia** Berk. & Broome (Sacc. **V**:658 ; **71**:350, fig.). On rotten straw heaps, Hooghly District, Bengal (Bose). Eaten by the villagers.

——**liliputiana** P. Henn. (**263**:333 ; Sacc. **XVII**:53). On the ground, Botanic Garden, Saharanpur (Gollan).

——**media** (Schum.) Fr. (Sacc. **V**:662 ; **263**:334). On the ground, Botanic Garden, Saharanpur (Gollan).

——**terastria** Berk. & Broome (Sacc. **V**:661 ; **70, II**:137, figs., as *V.* " *terastius* "; **74**:643). Very common on heaps of rotten straw, Bengal (Bose). Edible.

——**thwaitesii** Hooker f. (**57,** No. 286 ; Sacc. **V**:656). On dead wood, Darjeeling, 7,000 ft. (Hooker f.).

——**volvacea** (Bull.) Sacc. (Sacc. **V**:657 ; **263**:334). Arnigadh, Mussoorie, 5,500 ft. (Gollan).

——**woodrowiana** Massee (**319, II**:166 ; Sacc. **XVI**:69). On the ground, Poona (Woodrow).

Xerotus cantharelloides Berk. (**57,** No. 327 ; Sacc. **V**:630). On dead wood, Jallapahar, Darjeeling (Hooker f.).

——**lateritius** Berk. & Curt. (Sacc. **V**:634 ; **467**:157). On dead branches, Khandala, Bombay (Blatter). Theissen (l.c. and **472**:155, 159) quotes Lloyd as considering this species as probably identical with *Anthracophyllum nigritum* (q.v.).

[**Xerotus**] **perrottetii** Mont. (**351**:151 ; Sacc. **V**:631 ; **349**:22, erroneously referred to *X. berteri* Mont.). On branches, near Ootacamund, Nilgiris (Perrottet). A good specimen is in the general herbarium at Paris.

GASTEROMYCETES

Astraeus hygrometricus (Pers.) Morg. (Sacc. **VII**:90 as *Geaster hygrometricus* Pers.; **57**, after No. 455 ; **145**:13 ; **263**:338 ; **288,** No. **65**:4 ; **287**:1295). On the ground, Simla (Thomson) ; Saharanpur (Duthie) ; Arnigadh, Mussoorie (Gollan) ; India (Kashyap ; P. L. Dey).

Battarrea levispora Massee (**319, III**:152 ; Sacc. **XVII**:224). On the ground, Poona (Woodrow). Lloyd (**290**:7, fig.) considers it a form of *B. phalloides* (Dicks.) Pers.

Bovista argentea Berk. (**51**:400 ; Sacc. **VII**:102). On the ground, Madras (Wight). Lloyd (**287**:190) states that the type at Kew is *Lanopila bicolor*, but the description suggests *B. dealbata*.

——**brasiliensis** (Fr.) de Toni (Sacc. **VII**:100 ; **196**:128 as *Lycoperdon brasiliense* Fr.). On wood, Nakawa, Toukyeghat, Burma (Kurz).

——**plumbea** Pers. (Sacc. **VII**:96 ; **263**:337). On the ground, Arnigadh, Mussoorie (Gollan).

Bovistella aspera (Lév.) Lloyd (Sacc. **VII**:119 ; **287**:1295). India (P. L. Dey).

——**bovistoides** (Cke & Massee) Lloyd (**166**:26 as *Mycenastrum bovistoides* Cke & Massee ; Sacc. **VII**:489 as *Scleroderma bovistoides* (Cke & Massee) Sacc.; **287**:284, fig.; **288,** No. **65**:4). On the ground, amongst moss, Nilgiris (Herb. Berkeley); India (G. H. Cave ; Kashyap).

——**trachyspora** Lloyd (**287**:287, figs.; Sacc. **XXI**:483). Evidently in moss, Respanna Valley, Mussoorie (Gollan).

Calathiscus sepia Mont. (**349**:23 ; Sacc. **VII**:24). On tree trunks in a damp forest, near Ootacamund, Nilgiris (Perrottet). Lloyd (**292**:43, fig.) writes the name "*Anthurus calathiscus*", but states that nothing is known of this fungus except Perrottet's crude figure.

Calvatia gardneri Berk. (Sacc. **VII**:129 as *Lycoperdon ;* **288**, No. **15**:3, No. **34**:2). On the ground, Shillong, Khasi Hills (Butler) ; Bengal (Bose).

——**lilacina** (Mont. & Berk.) Morgan (Sacc. **VII**:126 as *Lycoperdon ;* **288**, No. **12**:2, No. **13**:2). Bombay (Kirtikar) ; Poona (Gammie).

Cauloglossum elatum Fr. (**237, III**:61 ; Sacc. **VII**:57). On the ground, India (Koenig).

Clathrus cancellatus Tournef. (Sacc. **VII**:19 ; **57**, after No. 455 ; **400**:154). On the ground, Myrong, Khasi Hills (Hooker f.) ; Sikkim (Remy). *C. ruber* (Mich.) Pers. is the same species.

Crucibulum vulgare Tul. (Sacc. **VII**:43 ; **286**:160). Nilgiris (Perrottet). The specimen is still at Paris.

Cyathus hookeri Berk. (**57**, No. 461 ; Sacc. **VII**:35). On wood and in moss, Khasi Hills (Hooker f.). Lloyd (**291**:28, fig.) regards it as close to *C. microsporus*, but Cunningham (**193**:65, fig.) considers it distinct.

——**intermedius** (Mont.) Tul. (Sacc. **VII**:35 ; **196**:128 ; **291**:23, fig.). Sibpur, near Calcutta (Kurz).

——**limbatus** Tul. (Sacc. **VII**:37 ; **288**, No. **15**:3, No. **17**:2 ; **291**:16, fig.; **374**:61 ; **144**:96 as *C. limbatus* Fr.). Belgaum (Hobson) ; in a flower pot, Royal Botanic Garden, Calcutta (Butler).

——**microsporus** Tul. (Sacc. **VII**:35 ; **288**, No. **15**:3 ; **291**:27, fig.). On wood and on the ground, Wahjain and Shillong, Khasi Hills (Butler).

——**poeppigii** Tul. (Sacc. **VII**:37 ; **263**:336 ; **374**:61 ; **288**, No. **17**:2 ; **291**:15, fig.). On the ground and on charred wood, Botanic Garden, Saharanpur (Gollan); on dead branches, Pusa (Butler).

——**stercoreus** (Schwein.) de Toni (Sacc. **VII**:40 ; **288**, No. **13**:2, No. **65**:4 ; **291**:20, fig.; **287**:1295). Poona (Gammie) ; India (Kashyap ; P. L. Dey).

Dictyophora indusiata (Ventenat) Pers. (Sacc. **VII**:3 as *D. phalloidea* Desvaux ; **467**:158 ; **374**:58 ; **57**, after No. 455, as *D. speciosa* Klotzsch ; **287**:332, 453 as *Phallus indusiatus* Vent.; **292**:18, fig.). Sikkim, and Churra, Khasi Hills (Hooker f.) ; on the ground, Khandala, Bombay (Blatter) ; North Bengal (Hutchings). Fischer in a recent paper (Ann. Myc., XXV:472, 1927) gives the authorities as above ; Dodge, in his translation of Gäumann's "Comparative morphology of Fungi" writes it "*D. indusiata* (Ventenat ex Pers.) E. Fischer".

——**nana** Berk. (**66**:39 ; Sacc. **VII**:7) Andaman Islands (Berkeley Jr.). Lloyd (**292**:78) lists this as a synonym of the preceding species.

Geaster englerianus P. Henn. (Sacc. **XI**:162 ; **288**, No. **19**:5). India (G. H. Krumbiegel). Lloyd (**287**:310, fig.) says "In the tropics *G. saccatus* takes a black form which has been called *englerianus*".

——**fimbriatus** Fr. (Sacc. **VII**:82 ; **319, III**:152). On the ground, ? Punjab (Duthie). Lloyd (**289**:23) thinks *G. fimbriatus* grows only in Europe.

——**lageniformis** Vittadini (Sacc. **VII**:86 ; **263**:337 ; **288**, No. **65**:4). On the ground, Arnigadh, Mussoorie (Gollan) ; India (Kashyap).

——**lilacinus** Massee (**319, II**:166 ; Sacc. **XVI**:237). On the ground, Dehra Dun (Gamble).

——**limbatus** Fr. (Sacc. **VII**:81 ; **57**, after No. 455). "Form minor", Simla (Thomson).

——**plicatus** Berk. (**51**:399 ; Sacc. **VII**:76 as *G. tenuipes* Berk.; **289**:18, fig.). On the ground, Madras (Wight).

Gyrophragmium delilei Mont. (Sacc. **VII**:51 ; **352**). Sonamarg, Kashmir (R. R. Stewart).

Ithyphallus aurantiacus (Mont.) Ed. Fisch. (Sacc. **VII**:9 ; **349**:23 as *Phallus aurantiacus* Mont.; **61**:21 as *P. truncatus* Berk.). On the ground, Pondicherry (Perrottet); "Plains of India" (**61**). Perrottet's collections are in Herb. Montagne. Lloyd (**292**:14; see also **287**:458) makes this a synonym of *I. rubicundus* (Bosc) Ed. Fisch.

[**Ithyphallus**] **impudicus** (L.) Fr. (Sacc. **VII**:8). As Lloyd (**287**:328) notes, "there is a very small and very doubtful specimen so named" from Herb. Griffiths, India (? Churra), in Kew Herbarium.

Lanopila bicolor (Lév.) Patouill. (Sacc. **XVI**:240; **286**:162 as *Bovista bicolor* Lév.; **287**:190, fig.; **374**:67). On the ground, Bombay (Polydore Roux); reported in Saccardo also from the Nicobar Islands. Lloyd (**288**, No. **66**:8) considers it safe to place this species as a synonym of *L. wahlbergii* Fr. See also *Bovista argentea*.

Lasiosphaera fenzlii Reichardt (**391**:135, fig.; Sacc. **VII**:96 as *Eriosphaera fenzlii* Reich.; **287**:191, fig.). Locality unknown: probably India (Novara Expedition); on manure, India (Cave).

Lycoperdon alveolatum Lév. (**286**:163; Sacc. **VII**:120). On the ground, Nilgiris (Perrottet). Type at Paris in good condition.

——**berkeleyi** de Toni (Sacc. **VII**:124; **57**, No. 457 as *L. delicatum* Berk., not Berk. & Curt.). On the ground, Khasi Hills (Hooker f.). Lloyd (**287**:243) discusses the names, and see also Coker and Couch, Gasteromycetes of the Eastern United States and Canada.

——**elongatum** Berk. (**57**, No. 456; Sacc. **VII**:123; **263**:337). On mossy ground, Darjeeling, 7,500 ft., and East Nepal (Hooker f.); Arnigadh, Mussoorie (Gollan).

——**emodense** Berk. (**57**, No. 458; Sacc. **VII**:110). On the ground, Sikkim, 15,000 ft., and Phallut, East Nepal, 9,000 ft. (Hooker f.).

——**fucatum** Lév. (Sacc. **VII**:125; **57**, after No. 456). Khabili river, East Nepal, 5-6,000 ft. (Hooker f.). Capillitium and spores only seen by Berkeley.

——**gemmatum** Batsch (Sacc. **VII**:106; **57**, after No. 349; **196**:128; **287**:1266, 1295). On the ground, paths, clay banks, and decayed timbers, Jallapahar, Darjeeling, 7-8,000 ft. (Hooker f.); Sikkim Himalaya, 7-8,000 ft. (Kurz); India (D. Maruda Rajan; P. L. Dey).

——**giganteum** Batsch (Sacc. **VII**:109; **352**). Sonamarg, Kashmir (R. R. Stewart). This species is often called *Calvatia gigantea*.

——**hiemale** Bulliard (Sacc. **VII**:115; **57**, after No. 348, as *L. caelatum* Fr.). On the ground, Darjeeling, 7,000 ft. (Hooker f.).

——? **marginatum** Vittad. (Sacc. **VII**:127; **263**:337). On the ground, Botanic Garden, Saharanpur (Gollan).

——**microspermum** Berk. (**57**, No. 350; Sacc. **VII**:110). On the ground, Darjeeling (Hooker f.); Gauhati and Dauracherra, Assam (Butler). Lloyd (**289**:30) states that this appears to be the same as *L. pusillum*.

[Lycoperdon] "nigrum" (**288,** No. **31**:3 ; not in Sacc.). India (Kirtikar). We are unable to interpret the name applied by Lloyd. Perhaps he meant *L. nigrescens* Pers.

——**piriforme** Schaeff. (Sacc. **VII**:117 ; **288,** No. **17**:2, No. **35**:2 ; **57,** after No. 349 ; **374**:72). On dead wood, Sikkim, 8,000 ft. (Hooker f.) ; India (Butler) ; India, 8,000 ft. (Cave).

——**pusillum** Batsch (Sacc. **VII**:110 ; **57,** after No. 459 ; **196**:128 ; **288,** No. **17**:2). On the ground, Eastern Nepal (Hooker f.) ; in elephant grass jungles, Kayosoo, Lower Burma (Kurz) ; on the ground, Dehra Dun (Butler : specimens larger than those found in Europe).

——**saccatum** Vahl (Sacc. **VII**:128 ; **263**:336). On the ground, Arnigadh, Mussoorie (Gollan).

——**sericellum** Berk. (**57,** No. 349 ; Sacc. **VII**:116). On the ground, Darjeeling, 7,000 ft. (Hooker f.).

——**xanthospermum** Berk. (**57,** No. 459 ; Sacc. **VII**:111). On the ground, Moflong, Khasi Hills (Hooker f.).

Melanogaster durissimus Cke (**143**:94 ; Sacc. **VII**:167). A few inches below the surface of the ground, near Chakrata, Himalaya (Baden-Powell). Edible.

Mitremyces junghuhnii Schlect. & Muell. (Sacc. **VII**:69 ; **57,** after No. 460 ; **57,** No. 352 as *M. viridis* Berk.). On the ground and on dead timber, Tonglo, Sinchul, and Chola, Sikkim, 6-9,000 ft. (Hooker f.) ; Butan (Nuttal). Lloyd (**287**:241) found no difference between *M. viridis* and *M. junghuhnii.*

Nidula emodensis (Berk.) Lloyd (**291** : 12, fig.; **57,** No. 462 as *Cyathus emodensis* Berk.; Sacc. **VII**:40). On dead wood, Lachen, Sikkim, 12-13,000 ft. (Hooker f.).

Podaxon calyptratus Fr. (Sacc. **VII**:59 ; **145**:13). On the ground, Punjab (Aitcheson) ; on sandy soil, Chatrapur, Ganjam (Butler).

——**carcinomalis** (L.) Fr. (Sacc. **VII**:58 ; **61**:21). "Probably common in India" Berkeley.

——**emerici** Berk. in Herb. (**312**:75, figs.; Sacc. **IX**:267). On the ground, Masulipatam (Capt. Emeric Berkeley).

——**gollani** P. Henn. (**263**:338 ; Sacc. **XVII**:219). On the ground, Botanic Garden, Saharanpur (Gollan).

——**pistillaris** (L.) Fr. (Sacc. **VII**:59 ; **287**:170 ; **288,** No. **15**:3, No. **17**:2 ; **312**:74, figs., as *P. indica*). On the ground, India (Koenig) ; Nadiad, Bombay (Butler); Mirpur, Matholo, Sind ; Madras ; Punjab ; Afghanistan (Aitcheson). Lloyd (**288,** No. **17**:2) mentions that the Indian collections he examined agreed well with the type in the Linnaean Herbarium, which came originally from India. Described first by Linnaeus as *Lycoperdon pistillare.* (See p. i.)

Polygaster sampadarius (Rumph.) Fr. (**237,** **II**;295 ; Sacc. **VII**:146). In woods, India (Koenig).

Scleroderma aurantium Pers. (Sacc. **VII**:134 as *S. vulgare* Hornem.; **263** : 338 ; **288,** No. **2**:2, No. **17**:2 ; **289**:15). India (R. L. Proudlock ; Butler) ; on the ground, Arnigadh, Mussoorie (Gollan). Lloyd states that the Indian specimens he examined were just like those of Europe and America.

——**bovista** Fr. (Sacc. **VII**:135 ; **57,** after No. 351 ; **263**:338). On the ground, Sikkim (Hooker f.) ; Arnigadh, Mussoorie (Gollan).

——**cepa** (Vaill.) Pers. (Sacc. **VII**:135 ; **288,** No. **17**:2 ; **287**:1295). South India (Butler) ; India (P. L. Dey).

——**columnare** Berk. & Broome (Sacc. **VII**:144 as *Alveolaria* ? ; **288,** No. **17**:2). India (Cave). Lloyd states that this species is known only from India and Ceylon.

——**cookei** de Toni (Sacc. **VII**:140 ; **158**:6 as *Mycenastrum lycoperdioides* Cke, not Schwein.). Amongst moss, Nila Valley, Garhwal, Himalaya, 12,000 ft.

——**dictyosporum** Patouill. (Sacc. **XIV**:266 ; **288,** No. **15**:3). On the ground, Dehra Dun (Butler).

——**geaster** Fr. (Sacc. **VII**:138 ; **57,** after No. 459 ; **113** (19):54, as *Sclerangium polyrhizon* (Gmel.) Lév.). On clay banks, Nunklow, Khasi Hills, 4-5,000 ft. (Hooker f.) ; Shillong, edible. The late Abbé Bresadola determined the Shillong specimen as " *Sclerangium polirrhizon* ".

——**nitidum** Berk. (**57,** No. 460 ; Sacc. **VII**:138 ; **287**:1295, 1306, fig.). On the ground, Nangki, Eastern Nepal, 10,000 ft. (Hooker f.) ; India (P. L. Dey).

——**verrucosum** (Bull.) Pers. (Sacc. **VII**:136 ; **263**:338 ; **288,** No. **15**:3, No. **17**:2). On the ground, Arnigadh, Mussoorie (Gollan) ; Dehra Dun (Butler : recorded by Lloyd as " *S. caespitosum*, new form of *verrucosum* ").

Simblum periphragmoides Klotzsch (Sacc. **VII**:17 ; **287**:424 ; **292**:66 as *S. gracile* Berk.). Lloyd states that this is a very common species in Java, Ceylon, and India.

Sphaerobolus stellatus Tode (Sacc. **VII**:46 ; **291**:28, fig.; **263**:336 as *S. carpobolus* (L.) Schroet.). On dead moss, Botanic Garden, Saharanpur (Gollan).

Tylostoma bonianum Patouill. (Sacc. **XI**:159). Lloyd (**290**:14, fig.) states that he has seen what he takes to be this plant from India in the Kew Herbarium.

——**exasperatum** Mont. (Sacc. **VII**:64 ; **290**:12, 26, fig.). Lloyd states that he has seen specimens from India.

——**mussooriense** P. Henn. (**263**:337 ; Sacc. **XVII**:223 ; **374**:63 ; **290** :14, fig.). On the ground, Arnigadh, Mussoorie (Gollan).

——**pusillum** Berk. (Sacc. **VII**:64 ; **286**:165). On twigs, Madras. Lloyd (**290**:26) states that it is evidently very similar to *T. exasperatum*.

——**wightii** Berk. (**52**:157 ; Sacc. **VII**:62). On the ground, Madras (Herb. Hooker).

Xylopodium aitchesonii Cke & Massee (**167**:69 ; Sacc. **VII**:489). On the ground, Afghanistan frontier (Aitcheson). Lloyd (**289**:10) says there is no doubt *Xylopodium* is a synonym of *Phellorina.*

FUNGI IMPERFECTI.

HYPHOMYCETES (AND MYCELIA STERILIA).

Acrothecium lunatum Wakker (Sacc. **XIV**:1089 ; **113**(13):36 ; **341**:70 ; **311**:2). On leaves of *Andropogon sorghum*, *Setaria italica*, *Panicum frumentaceum*, *Eleusine coracana*, male inflorescence of *Zea mays*, and by inoculation on young leaves of *Pennisetum typhoideum*, Pusa (Mitra) ; isolated from paddy soil, Burma (Rhind).

——**penniseti** Mitra (**341**:70, figs.; **311**:5). On leaves and ears of *Pennisetum typhoideum* and by inoculation on male inflorescence of *Zea mays*, Pusa (Mitra).

Actinomyces bovis Harz (Sacc. **VIII**:928 as *Nocardia ;* **2**:50 ; **339**). Causing a human lung infection, Goa (Fr. de Mello).

——**scabies** (Thaxt.) Güssow (Sacc. **XXII**:1240 as *Oospora ;* **113**(17) :54). On tubers of *Solanum tuberosum*, Khasi Hills (McRae).

Alternaria brassicae (Berk.) Bolle, not Sacc. (Sacc. **IV**:526 as *Macrosporium ;* **111**:81, 300, figs.; **319**, **III**:153, as *Sporodesmium brassicaeae* Massee ; **311**:19). Parasitic on *Brassica campestris* var. *sarson*, Tirhoot (Watt) ; Pusa (Butler).

——**circinans** (Berk. & Curt.) Bolle (Sacc. **IV**:524 as var.; **311**:19). On *Brassica oleracea* (cabbage), Pusa (A. N. Khan).

——**crassa** (Sacc.) Rands (Sacc. **IV**:448 as *Cercospora*). This is perhaps the fungus referred to *A. solani* on *Datura stramonium*, Calcutta (Bal, **28**, **II**:7-9).

——**dianthi** Stev. & Hall (Sacc. **XXII**:1410 ; **119**). On *Hibiscus tiliaceus*, *Jasminum* sp., and *Calendula officinalis*, Lahore (Chaudhuri).

——**palandui** Ayyangar (**25**:13). On leaves of *Allium cepa*, Coimbatore (Ayyangar).

——**solani** (Ell. & Mart.) Jones & Grout (Sacc. **IV**:530 as *Macrosporium ;* **111**:288, fig.; **87** ; **89** ; **28**, **II**:8 ; **113**(17):55). On leaves of *Solanum tuberosum*, Farukhabad, United Provinces (Butler) ; and throughout northern India.

——**tenuis** Nees (Sacc. **IV**:545 ; **311**:22). On leaves of *Saccharum officinarum*, India (Mohendra).

——?**violae** Galloway & Dorsett (Sacc. **XVI**:1080 ; **28**, **II**:6-7, figs.). Recorded doubtfully by Bal on *Nicotiana plumbaginifolia* leaves, Bengal.

Anthromycopsis indica P. Henn. (**263**:342 ; Sacc. **XVIII**:652). On dead branches, Botanic Garden, Saharanpur (Gollan).

Aspergillus aguiari de Mello (**340**:91, figs.). In urine, Goa (de Mello).

——**albicans** de Mello (**340**:58, figs.). As a contamination of agar media, Goa (de Mello).

——**candidus** Link (Sacc. **IV**:66 ; **340**:58 ; **461**:145). Common as a contamination of culture media, Goa (de Mello) ; from soil, Madras (Thakur and Norris).

[**Aspergillus**] **castaneus** Patterson (Sacc. **XVI**:1029 as *Sterigmatocystis ;* **109**). On fruits of *Punica granatum*, India. Thom & Church (**485**:74, 174) state that it may be a strain of the *A. niger* group.

——**corolligenus** Massee (as *Sterigmatocystis*; **319**, **X**:5 ; Sacc. **XXII**:1259 ; **485**: 164). On corolla of *Impatiens* sp., Manipur (comm. Hooker f.).

——**ferrugineus** Cke (as *Sterigmatocystis* ; **144**:95 ; **160**:139, figs.; Sacc. **IV**:74). On pupa of Lepidoptera (Eri silk moth=*Attacus ricini*), Cachar (Moore). Thom & Church (**485**:217) would drop this species.

——**flavus** Link (Sacc. **IV**:69 ; **267**:141, 148 ; **461**:145 ; **311**:24). On roots of *Polygala arillata*, used for rice-beer brewing, Khasi Hills (**267**) ; from soil, Madras (Thakur and Norris) ; from paddy soil, Burma (Rhind).

——**fumigatus** Fresenius (Sacc. **IV**:65 ; **113**(7):56 ; **113**(9):69 ; **229**:63 ; **461**:145). In soil, Pusa (Shaw) ; Madras (Thakur and Norris) ; in jute bales affected by " heart damage ", Dacca (Finlow).

—— " **fuscus** " (**461**:145). From soil, Madras (Thakur and Norris). The author of the species was not given ; Thom & Church (**485**) record four species described as " *A. fuscus* ".

——**nidulans** (Eidam) Wint. (Sacc. **X**:524 as *Sterigmatocystis ;* **461**:145). From soil, Madras (Thakur and Norris).

——**niger** van Tiegh.(Sacc. **IV**:75 as *Sterigmatocystis ;* **267**:147 ; **113**(7):56 ; **340**:98 ; **461**:145 ; **496** ; **311**:25 ; **129**:90 as *Torula incarcerata* Cke). Within seed of *Gossypium*, Dharwar ; in soil, Pusa (Shaw) ; in rice-beer ferment, Ranchi (Hutchinson and Ayyar) ; in tea during fermentation (Tunstall) ; from paddy soil, Burma (Rhind) ; in soil, Madras (Thakur and Norris).

——**ortae** de Mello (**340**:88, figs.). Very frequent as a laboratory mould, Goa (de Mello).

——**phaeocephalus** Durieu & Mont. (Sacc. **IV**:76 as *Sterigmatocystis ;* **140**:118). On roots of *Asparagus racemosus*, Madras.

——**polychromus** de Mello (**338**:158 ; **340**:82, figs.; **485**:137). Goa (de Mello).

——**repens** (Cda) de Bary & Woronin (Sacc. **IV**:64 ; **485**:113 ; **461** :145). In soil, Madras (Thakur and Norris).

——**sulphureus** (Fres.) Wehmer (**485**:185 ; Sacc. **IV**:73 as *Sterigmatocystis ;* **338**: 158). Recorded as a laboratory mould, Goa (de Mello).

——**tamarii** Kita (**485**:194 ; **311**:25). From paddy soil, Burma (Rhind) ; India.

——**terreus** Thom (**485**:150 ; **311**:25). From paddy soil, Burma (Rhind).

——**ustilago** Beck (**50**:148, fig.; Sacc. **X**:526 as *Sterigmatocystis ;* **485**:177). In pericarp of *Phyllanthus emblica*, Satpoor Mountains.

Basidiella sphaerocarpa Cke (**140**:118 ; **160**:138, figs.; Sacc. **X**:698). In decayed roots of *Gloriosa superba*, Madras.

Bispora catenula (Lév.) Sacc. (Sacc. **IV**:344 ; **285**:71 as *Septonema catenula* **Lév.**). On leaves of *Quercus dealbata*, India (Jacquemont).

Botrytis cinerea Pers. (Sacc. **IV**:129 ; **461**:145). From soil, Madras (Thakur and Norris).

[——**vulgaris** Fr. (Sacc. **IV**:128 ; **89**:48). On *Primula*, Dehra Dun (Butler). Usually regarded as a synonym of the last.]

Campsotrichum cinnamomi Cda (**185**, **IV**:28, fig.; Sacc. **IV**:296). On leaves of *Cinnamomum* sp., Tenasserim (Helfer).

Cephalosporium acremonium Cda (Sacc. **IV**:56 ; **461**:145). From soil, Madras (Thakur and Norris).

——**lecanii** Zimm. (**375**, **VI**:153 ; **454**:392). On *Lecanium viride* on *Coffea*, South India ; on *L. hemisphaericum* on *Coffea*, Mysore (Lefroy).

——**sacchari** Butler (**115**:181, figs.; **111**:402, figs.). In culms of *Saccharum officinarum* throughout India.

Ceratophorum hypodermium (Niessl) Sacc. (Sacc. **IV**:397 ; **388**:146 as *Sporidesmium hypodermium* Niessl). Habitat not stated, Royal Botanic Garden, Calcutta (Kurz).

Cercospora ajrekari Syd. (**443**, **XII**:202). On leaves of *Jatropa nana*, Poona (Ajrekar).

——**annulata** Cke (**144**:95 ; Sacc. **IV**:475 ; **457**:262). On leaves of *Ficus hispida*, Calcutta (J. Scott) ; Pusa ; Mozufferpore ; Godavary ; of *F.* sp., Peshawar and Dehra Dun.

——**anthelmintica** Atkinson (Sacc. **X**:636 ; **457**:262). On leaves of *Chenopodium ambrosioides*, Wahjain, Assam (Butler) ; Peshawar.

——**apii** Fresen. (Sacc. **IV**:442 ; **111**:315, fig.; **457**:263). On leaves of *Apium graveolens*, Poona and Pusa.

——**asparagi** Sacc. (Sacc. **IV**:477 ; **457**:263). On leaves and stems of *Asparagus officinalis*, Solon (near Simla) and Pusa.

——**batatae** Zimm. (Sacc. **XVIII**:605 ; **457**:263). On leaves of *Ipomoea batatas*, Godavary and Pusa.

——**beticola** Sacc. (Sacc. **IV**:456 ; **457**:263). On leaves of *Beta vulgaris*, Pusa.

——**biophyti** Syd. (**457**:263). On leaves of *Biophytum* sp., Samalkota.

——**blumeae** Thuem. (Sacc. **IV**:445 ; **457**:263). On leaves of *Blumea* sp., Nagpur, Pusa, Dehra Dun, and Samalkota.

——**caladii** Cke (**144**:95 ; Sacc. **IV**:478). On drooping leaves of *Caladium* sp., Belgaum (Hobson);

——**calotropidis** Ell. & Ev. (Sacc. **XVI**:1072 ; **457**:263). On leaves of *Calotropis gigantea*, Pusa.

——**cannabina** Wakef. (**457**:264). On leaves of *Cannabis sativa*, Peshawar ; Godagiri, Bengal.

——**capsici** Heald & Wolf (**457**:264). On leaves of *Capsicum annuum*, Pusa and Cawnpore.

——**carthami** Sundar. & Ramak. (**438**:389, figs.). On leaves of *Carthamus tinctorius*, Coimbatore (Sundararaman and Ramakrishnan).

[Cercospora] **catappae** P. Henn. (Sacc. **XVIII**:598 ; **457**:264). On leaves of *Terminalia catappa*, Insein, Burma (Inayat).

— **cearae** Petch (Sacc. **XXII**:1421 ; **111**:516, fig.; **113**(3):54 ; **457**:264). On leaves of *Manihot piauhyensis*, Kamalpur, Assam.

— **cleomis** Ell. & Halst. (Sacc. **X**:621 ; **457**:264). On leaves of *Cleome* sp., Pusa. Saccardo spells the specific name " *cleomes* ".

— **cocculi** Syd. (**457**:264). On leaves of *Cocculus villosus*, Pusa (Inayat).

— **coffeicola** Berk. & Cke (Sacc. **IV**:472 ; **111**:485, figs.; **487**:703 ; **113**(9):67 ; **395** ; **457**:265). On leaves and berries of *Coffea arabica*, Mysore (Thomas) ; Burma (Rhind) ; Dauracherra, Sylhet (Butler).

— **concors** (Casp.) Sacc. (Sacc. **IV**:449 ; **111**:287, fig.). On leaves of *Solanum tuberosum*, Bengal, Pusa, and Poona (Butler).

— **cruenta** Sacc. (Sacc. **IV**:435 ; **111**:261, fig.; **457**:265). On leaves of *Phaseolus vulgaris*, Pusa ; of *P. mungo* var. *radiatus*, Yelwigi in Dharwar District, and Gilgit ; of *P. aconitifolius*, Jullundur, Punjab.

— **diodiae** Cke (Sacc. **IV**:441 ; **457**:265). On leaves of *Spermacoce hispida*, Erramacolla and Panora, Wynaad.

— **dioscoreae** Ell. & Mart. (Sacc. **IV**:479 ; **457**:265). On leaves of *Dioscorea* sp., Harwan, Kashmir (Butler) ; Nilphamari, Rangpur, Bengal.

— **dolichi** Ell. & Ev. (Sacc. **X**:622 ; **457**:265). On leaves of *Dolichos lablab* and *D. typica*, Pusa.

— **euphorbiae** Kellerm. & Swingle (not in Sacc.; **457**:265). On leaves of *Euphorbia* sp., Dehra Dun, Nadiad, and Surat ; of *E. tirucalli*, Pusa ; of *E. neriifolia*, Dohad, Bombay ; of *Pedilanthus tithymaloides*, Pusa. There is a *C. euphorbiae* Patouill. (Sacc. **XI**:629), but Kellerman and Swingle published the name earlier (Journ. Myc., V:76).

— **foeniculi** P. Magn. (**457**:266). On leaves, peduncles, and stems of *Foeniculum vulgare*, Harwan, Kashmir and Pusa.

— **gloriosae** Syd. (**457**:266). On leaves of *Gloriosa superba*, Pusa (Kar).

— **gossypina** Cke (Sacc. **IV**:441 ; **111**:369, fig.; **457**:266). On leaves of *Gossypium* sp., Pusa, Cawnpore, and Lyallpur.

— **henningsii** Allesch. (Sacc. **XIV**:1104 ; **111**:310, figs., as *Septogloeum manihotis* Zimm.; **311**:32). On *Manihot utilissima*, Travancore (Butler).

— **hibisci** Tracy & Earle (Sacc. **XIV**:1099 ; **311**:34 ; **457**:266). On *Hibiscus sabdariffa* and *H. cannabinus*, Mandalay (Rhind) ; on leaves of *H. esculentus*, Pusa and Nagpur.

— **ipomoeae** Wint. (Sacc. **X**:633 ; **457**:267). On leaves of *Ipomoea hederacea*, Dehra Dun and Pusa.

— **longipes** Butler (**93**:41, figs.; Sacc. **XXII**:1432; **111**:405, fig.; **393**:1 ; **457**:267). On leaves of *Saccharum officinarum* throughout Northern India and Burma.

[**Cercospora**] **menispermi** Ell. & Holw. (Sacc. **X**:618 ; **457**:267). On leaves of *Menispermum cordifolia*, Pusa.

—— **momordicae** McRae (**457**:267). On leaves of *Momordica charantia*, Pusa (Subramaniam).

—— **morindae** Syd. (**445**:490). On leaves of *Morinda tinctoria*, Coimbatore (McRae).

—— **neriella** Sacc. (Sacc. **IV**:473 ; **457**:267). On leaves of *Nerium oleander*, Janakpur, Bihar.

—— **nicotianae** Ell. & Ev. (Sacc. **XI**:628 ; **111**:341, figs.; **2**:26 ; **393**:5 ; **457**: 267). On leaves of *Nicotiana tabacum* throughout India and Burma.

—— **occidentalis** Cke (Sacc. **IV**:463 ; **457**:267). On leaves of *Cassia occidentalis*, Mozufferpore and Pusa.

—— **oryzae** Miyake (Sacc. **XXII**:1431 ; **105**:35). On *Oryza sativa*, Burma (Butler).

—— **pentaleuca** Syd. (**445**:490). On leaves of *Clitoria ternatea*, Coimbatore (McRae).

—— **penzigii** Sacc. (Sacc. **IV**:466, 808 ; **457**:268). On leaves of *Citrus* sp., Pusa.

—— **personata** (Berk. & Curt.) Ell. & Ev. (Sacc. **IV**:439 ; **89**:48 ; **110**, figs.; **111**:319, figs.; **395** ; **28**. **IV**, figs.; **457**:268). On leaves, petioles, and stems of *Arachis hypogaea* throughout India and Burma.

—— **punicae** P. Henn. (Sacc. **XXII**:1418 ; **457**:268). On leaves of *Punica granatum*, Orai, United Provinces.

—— **punjabensis** Syd. (**457**:268). On leaves of *Vallaris heynii*, Naganwari, Pathankot, Punjab (Mitter).

—— **rosicola** Thuem. (Sacc. **IV**:460 ; **457**:268). On leaves of *Rosa* sp., Srinagar, Almora, and Dehra Dun ; of *R. centifolia*, Tarnab near Peshawar, and Jullundur ; of *R. damascena*, Akbarpore, Peshawar.

—— **rubi** Sacc. (Sacc. **IV**:461 ; **457**:269). On leaves of *Rubus* sp., Verinag, Kashmir ; of *R. ellipticus*, Dehra Dun.

—— **sesbaniae** P. Henn. (Sacc. **XXII**:1419 ; **445**:490). On leaves of *Sesbania grandiflora*, Coimbatore (McRae) ; of *S. aegyptiaca*, Pusa.

—— **solanacea** Sacc. (Sacc. **IV**:449 ; **457**:269). On leaves of *Solanum nigrum*, Pusa ; of *S. melongena*, Renibennur in Bombay, and Pusa.

—— **strychni** Syd. (**457**:269). On leaves of *Strychnos nux-vomica*, Bhubaneshwar, Cuttack (Butler).

—— **subsessilis** Syd. (**457**:269). On leaves of *Melia azedarach*, Dehra Dun, Pusa, and Coimbatore.

—— **terminaliae** Syd. (**457**:270). On leaves of *Terminalia bellerica*, Nagpur (Pandit).

—— **ternateae** Petch (Sacc. **XXII**:1419 ; **457**:270). On leaves of *Clitoria ternatea*, Pusa, Chittagong, Khulna, Samalkota, and Orai, United Provinces.

—— **theae** Breda de Haan (Sacc. **XVIII**:598 ; **89**:47 ; **111**:116, fig.; **503**) On leaves of *Thea sinensis*, Assam (Butler) ; north east India (Tunstall).

[**Cercospora**] **trichosanthis** McRae (**457**:270). On leaves of *Trichosanthes anguina*, Pusa (Subramaniam).

——**ubi** Racib. (Sacc. **XVI**:1073 ; **457**:270). On leaves of *Dioscorea* sp., Chittagong.

——**vaginae** Krueger (Sacc. **XIV**:1106 ; **89**:47 ; **93**:43). On leaf sheaths of *Saccharum officinarum*. Reported from India but the records need confirmation.

——**vicoae** Syd. (**457**:270). On leaves of *Vicoa auriculata*, Dehra Dun (Butler).

——**viticola** (Ces.) Sacc. (Sacc. **IV**:458 ; **457**:271). On leaves of *Vitis vinifera*, Achibal, Kashmir (Butler).

——**woodfordiae** Syd. (**457**:271). On leaves of *Woodfordia floribunda*, Puttimari, Kamrup, Assam (Taslim).

Cercosporina ricinella (Sacc. & Berl.) Speg. (Sacc. **XXII**:1432 ; **111**:331, fig.). On leaves of *Ricinus communis*, Pusa (Butler).

Cerebella andropogonis-contorti Subram. (**423**:205, fig.). On ovaries of *Andropogon contortus*, Maymyo, Burma and Dumraon, Bihar (Butler). This and other species of *Cerebella* apparently grow upon forms of *Sphacelia* on the various hosts.

——**antidotale** Subram. (**423**:206, fig.). On ovaries of *Panicum antidotale*, Sangla Hill, Punjab (Cheema).

——**burmanensis** Subram. (**423**:205, fig.). On ovaries of *Panicum setigerum*, Mandalay (Shaw).

——**cenchroidis** Subram. (**423**:206, fig.). On ovaries of *Pennisetum cenchroides* and *Cenchrus biflorus*, Lahore (Cheema).

——**cynodontis** Syd. (**397** ; **423**:207, fig.). On ovaries of *Cynodon dactylon*, Mysore and Sylhet (Butler) ; Mysore (Barber) ; on *Panicum prostratum*, Bihar (Butler) ; Burma (Rhind) ; on *P. distachyum*, Bassein, Burma (Butler: stated to be slightly different from the type in that the majority of the spores are smooth.)

——**inquinans** (Berk. & Broome) Petch (Sacc. **VII**:508 as *Thecaphora* ; **423**:207, fig.). On *Paspalum scrobiculatum*, Bassein, Burma (Butler) ; on *P. longiflorum*, Sylhet (Butler) ; on *P. royleanum*, Ranchi (Butler) ; on *P. sanguinale* var. *ciliare*, Dacca (Kar) ; on *Panicum javanicum*, Ranchi (Butler).

——**nardi** Butler in Subram. (**423**:206, fig.). On glumes of *Andropogon nardus*, Mundanthorai, Tinnevelly (Barber).

——**sorghi-vulgaris** Subram. (**423**:206, fig.; **14** ; **397**). On ovaries of *Andropogon sorghum* (*Sorghum vulgare*), Rishivandayam, Madras (Barber) ; Poona (Ajrekar) ; Mandalay (Rhind) ; " common in Bombay and Madras " (Subramaniam).

Cladosporium buteacolum Cke (**133**:15, fig.; Sacc. **IV**:353 as *C.* " *butaecolum* "). On legumes of *Butea frondosa*, India (Hobson).

——**chodati** (Nechitsch) Sacc. (Sacc. **XVIII**:577 ; **355**:22, figs., as *Dematium chodati* Nechitsch ; **267**:150 ; **113**(6):58). In the ferment used for rice-beer brewing, Khasi Hills and elsewhere in north eastern India (Hutchinson and Ayyar). Placed by Berkhout in her new genus *Candida* (De Schimmelgeslachten *Monilia, Oidium, Oospora* en *Torula*. Thesis, Utrecht, 1923, p. 54).

[Cladosporium] delicatulum Cke (**133**:17 ; Sacc. **IV**:361 ; **144**:95). On dead leaves, Belgaum (Hobson).

——**fulvum** Cke (Sacc. **IV**:363 ; **89**:46 ; **3**:27). On leaves of *Lycopersicum esculentum*, ? Calcutta (Butler) ; Bombay (Ajrekar).

——**herbarum** (Pers.) Link (Sacc. **IV**:350 ; **144**:95 ; **111**:177, figs.; **363**). On leaves and legumes of *Acacia* sp., Belgaum (Hobson) ; common throughout India on withering plant material, and frequent as a laboratory contamination.

——**puccinioides** Cke (**133**:15, fig.; Sacc. **IV**:361). On the lower surface of living leaves, India (Hobson). Stated by Cooke to be " certainly intermediate between *Cladosporium* and *Helminthosporium* ".

——**scopiforme** Berk. (**57**, No. 471 ; Sacc. **IV**:358). On the lower surface of leaves of *Myristica* " *churra* ", Khasi Hills (Hooker f.).

——**subtile** Rabenh. (nomen nudum ; Sacc. **XI**:621). On pods of *Leucaena glauca*, Calcutta. Specimens were distributed under this name in Rabenhorst's Fungi Europaei, No. 2364.

Cladotrichum foliicola (Niessl) Ferro (Sacc. **XXII**:1365 ; **387**:176 as *Myxotrichum foliicolum* Niessl, nomen nudum). On leaves of *Bombax malabaricum*, Calcutta (Kurz).

——**glenosporoides** Sacc. (Sacc. **IV**:373 ; **132**:117, figs., as *Glenospora didyma* Cke, not *Cladotrichum didymum* (K. & Schm.) Sacc. On fading leaves, Kolapore (Hobson).

Clasterosporium maculatum Cke (**132**:117, fig.; Sacc. **IV**:392). On leaves of *Ficus cordifolia*, Kolapore (Hobson).

Coniosporium arundinis Sacc. (Sacc. **IV**:243 ; **319, II**:166). On *Thysanolaena agrostis*, Dehra Dun (Gamble).

——**donacis** (Niessl) Sacc. (Sacc. **IV**:243 ; **387**:176 as *Gymnosporium donacis* Niessl). On fading leaves of *Arundo donax*, Calcutta (Kurz).

Coniothecium chomatosporum Cda (Sacc. **IV**:510 ; **113**(10):79 ; **113**(11):73). On twigs and fruits of *Pyrus malus*, Kumaon (Shaw).

Epidermophyton inguinale Sabouraud (**6**). This variable species, with several synonyms, occurs in " tinea cruris " of man in India, commonly known as Dhobie itch (Acton and McGuire).

Exosporium palmivorum Sacc. (Sacc. **XVI**:1106 ; **89**:48). On leaves of *Phoenix humilis*, India (Butler).

Fusarium caeruleum (Lib.) Sacc. (Sacc. **IV**:705 ; **16**, figs. ; **514**). The cause of dry rot of *Solanum tuberosum* in western India is thought by Ajrekar and Kamat to be nearer this species than *F. trichothecioides*, as suggested by Nagpurkar and Kulkarni (Bombay Dept. Agric. Bull. 102, 1920), but Uppal (**514**) found both species to be present.

[**Fusarium**] **cubense** E. F. Sm. (**259**). Recorded on *Musa*, Bengal (Hector) and Madras.

—— **lini** Bolley (Sacc. **XVIII**:670 ; **5**:7 ; **113**(18)). On *Linum usitatissimum*, India (S. N. Sil ; McRae).

—— **oxysporum** Schlecht. (Sacc. **IV**:705 ; **514**). On *Solanum tuberosum*, Bombay (Uppal).

—— **pannosum** Massee (**319**, **I**:117 ; Sacc. **XVI**:1098). On living trunks of *Cornus macrophylla*, Punjab, 7,000 ft. (Aitcheson).

—— **radicicola** Wollenw. (**16**, figs. ; **514**). The cause of the wilt of *Solanum tuberosum* in western India is thought by Ajrekar and Kamat to be this species rather than *F. oxysporum* as suggested by Nagpurkar and Kulkarni (Bombay Dept. Agric. Bull. 102, 1920), and Uppal (**514**) found it to cause dry rot.

—— **trichothecioides** Wollenw. (**514**). Recorded by Uppal on *Solanum tuberosum*, Bombay.

—— **uncinatum** Wollenw. (**525**:54, pl. 237). On dry stems of *Cajanus indicus*, Pusa (Butler).

—— **vasinfectum** Atkinson (Sacc. **XXII**:1481 ; **113**(18) ; **335** ; **112**:273 ; **283**, figs. ; **113**(15):51 ; **101**:54, figs., as *F. udum* Butler ; **111**:9, 244, figs. ; **112**:273 ; **13** ; **15** ; **113**(5):64 ; **113**(6):54 ; **113**(15):56 ; **113**(16):44 ; **113**(17):45 ; **363**). On roots and stems of *Cajanus indicus*, *Gossypium* spp., and *Sesamum indicum* throughout India. Wollenweber (Phytopath., III:38) changed the name *F. udum* to *F. butleri* because of an earlier *Pionnotes udum* Berk., which becomes *F. udum*, since *Pionnotes* is usually merged in *Fusarium*. It is now considered, however, that the fungi causing wilt in the three crops mentioned above must be included in Atkinson's *F. vasinfectum*, described in 1892. A special study was made of this fungus on cotton in India by Kulkarni and Mundkur (**283**).

Fusicladium butleri Syd. (**443**, **XIV**:260). On leaves of *Jasminum arborescens*, Orai, United Provinces (Butler).

Glenospora uromycoides Sacc. (**403**:20). On living leaves of *Memecylon edule*, Matheran, Bombay (Ajrekar).

Gliocladium compactum Cke & Massee (**165**:16 ; Sacc. **X**:528). On paper from India.

Gonatobotryum dichotomum Cke & Massee (**165**:15 ; Sacc. **X**:579). On damp, decomposing, amylaceous substances from India.

Helminthosporium avenae Eidam (Sacc. **XXII**:1393 ; **89**:46 ; **111**:183, figs.). Common on *Avena sativa* in India.

—— **bambusae** Cke (**182**:91 ; Sacc. **X**:616). On *Bambusa spinosa*, Assam (Mann).

—— **gramineum** Rabenh. (Sacc. **X**:615 ; **89**:46 ; **111**:186, figs. ; **113**(13):36 ; **113**(16):41). On *Hordeum vulgare*, Pusa (Butler ; Mitra).

[**Helminthosporium**] **heveae** Petch (Sacc. **XXII**:1391 ; **344**). On *Hevea brasiliensis*, Andaman Islands (Mitra).

——**nodulosum** Berk. & Curt. (Sacc. **IV**:421 ; **111**:341, fig.). Common on *Eleusine coracana* in India (Butler).

——**obclavatum** Massee (**319, II**:166 ; Sacc. **XVI**:1063). On twigs of *Helicteres isora*, Siwalik Hills (Gamble). There is an earlier *H. obclavatum* Sacc. (see Sacc. **IV**:416).

——**oryzae** Breda de Haan (Sacc. **XXII**:1394 ; **514**). On *Oryza sativa*, Upper Sind (Uppal). Probably this fungus described and figured, but not identified, by Sundararaman (**427**) as occurring in the Godavari and Kistna Deltas. It is probably prevalent throughout India.

——**sacchari** Butler (**115**:204, figs. ; **111**:7, 388, figs. ; **113**(13):35). On living leaves of *Saccharum officinarum*, Pusa.

——**sativum** Pammel, King, & Bakke (**113**(15):55 ; **113**(16):41 ; **113**(21):63). On *Hordeum vulgare* and *Triticum vulgare*, Pusa (Mitra ; McRae). See *H. teres*.

——**teres** Sacc. (Sacc. **IV**:412 ; **111**:189, figs. ; **113**(13):35 ; **113**(16):41). On leaves of *Hordeum vulgare*, Pusa (Butler ; McRae). Drechsler (Jour. Agric. Res., XXIV, p. 658, 1923) remarks that the earlier account of this fungus in India (**111**) failed to distinguish between it and *H. sativum*.

——**turcicum** Passerini (Sacc. **IV**:420 ; **111**:201, fig. ; **113**(13):35 ; **342,** figs.). On leaves and inflorescences of *Zea mays* throughout India, and on leaves of *Andropogon sorghum*, Punjab (Mitra). Wheat, barley, oats, and sugarcane were infected by inoculation.

Hymenopsis cudraniae Massee (**319, II**:167 ; Sacc. **XVI**:1105 ; **264,** No. 405 as *Melasmia cudraniae* (Massee) v. Hoehn.). On living leaves of *Cudrania javanensis*, Dehra Dun (Gamble). Von Hoehnel (l. c.) considers the fungus to be a non-typical *Melasmia*. Petch (**373, II**:314) states that his *Phacodiscula cudraniae* is perhaps identical with *H. cudraniae*. *Sclerodiscus nitens* Patouill. (Journ. de Bot., IV:66, fig., 1890) is almost certainly the same fungus, and this is the earliest of these names.

Isaria elata (Kalchbr.) Sacc. & Trav. (Sacc. **XXII**:1442 ; **400**:154, fig., as *Institale elata* Kalchbr.). Substratum not stated, Sikkim Himalaya (J. Remy).

——**farinosa** (Dicks.) Fr. (Sacc. **IV**:584 ; **263**:342). On chrysalids, Arnigadh, Mussoorie (Gollan).

[——**stellata** Cke (**132**:116 ; **134**:4 ; Sacc. **IV**:586). Encrusting dead insects on under surfaces of mango leaves, Mysore. Lyle (The Entomologist, LI, No. 665, p. 227, 1918) thought the host possibly a species of *Praon*, but Petch (**378, II**:265), from an examination of the type, found the insect to be *Phenacoccus mangiferae*, and the supposed fungus to be the waxy projections common on leaf aphids in the tropics.]

Metarrhizium anisopliae Sor. (**11**3(16):49 ; **392**:6). On grubs of *Oryctes rhinoceros* Burma.

Monilia albicans Robin (**228**). Isolated from man, Bombay. The genus *Candida* Berkhout should probably be used for the species of *Monilia* recorded here.

——**krusei** Castellani (**228**). Isolated from human patients, and from monkeys, India.

——**psilosis** Ashford (**306** ; **228** as *M. ashfordi*). From human patients suffering from sprue, Bombay, and from monkeys.

Mylitta ? lapidescens Horaninow (Sacc. **VIII**:907 ; **58** ; **197**, figs. ; **111**:13 ; **262**). Travancore (Waring) ; Nilgiris (Warburg) ; on the ground, Nilgiris (Butler). Petch (**367, VII**:147) considers that the Ceylon sclerotia of this type should be referred to *M. ligulata* Cesati, and that the south Indian form is more likely to be the latter than *M. lapidescens*, which is Chinese. The perfect stage is not definitely known in either species, although Hennings (**262**:20) follows Schroeter in placing *M. lapidescens* doubtfully in the genus *Omphalia*. In the Nilgiris the fungus is called " little man's bread ".

Nigrospora sphaerica (Sacc.) Mason (Sacc. **XXII**:1490 as *Epicoccum hyalopes* Miyake ; **105**:35 ; **392**:2). On *Oryza sativa*, Burma (Butler ; Rhind). The synonymy of this fungus is discussed by E. W. Mason in Trans. Brit. Mycol. Soc., XII, pp. 152-165, 1927.

Oedocephalum aurantiacum Cke (**138**:147 ; Sacc. **IV**:48). On leaves of forest trees, associated with *Diplodia phyllostictae*, Mysore.

Oidium carneum Cke (**144**:94 ; Sacc. **IV**:42). " On leaves of Malvaceae, etc." Belgaum (Hobson).

——? **tingitaninum** Carter. Recorded by McRae (**326**:110, as *O. citri* Butler, a nomen nudum) on *Citrus aurantium*, Coonoor, Nilgiris. Petch (Phytopath., V:350 ; IX:266) points out that Berkeley (Gard. Chron., 1874, pp. 477-478) had early mentioned an *Oidium* on orange from India and Ceylon. Petch found the spores of the Ceylon mildew to differ considerably from the Californian as described by Carter.

Oospora citri-aurantii (Ferrar.) Sacc. & Syd. (Sacc. **XVI**:1024). Not uncommon in fruits of *Citrus* spp.

——**lactis** (Fres.) Sacc. (Sacc. **IV**:15 ; **461** as *Oidium lactis* Fres.). In soil, Madras (Thakur and Norris).

——**maydis** P. Henn. (**263**:341 ; Sacc. **XVIII**:497). On rotten panicles of *Zea mays*, Botanic Garden, Saharanpur (Gollan).

Penicillium digitatum Sacc. (Sacc. **IV**:78 ; **486**:245 ; **461**). In soil, Madras (Thakur and Norris).

——**glabrum** (Wehmer) Westling (Sacc. **XI**:594 as *Citromyces glaber* Wehmer ; **461**:145). In soil, Madras (Thakur and Norris).

——**glaucum** Link (Sacc. **IV**:78 ; **486**:2 ; **461**). In soil, Madras (Thakur and Norris).

——**oxalicum** Currie & Thom (**486**:247 ; **461**). In soil, Madras (Thakur and Norris).

[Penicillium] tenellum Cke (**141**:15 ; Sacc. **IV**:89). On fading leaves of *Symplocos*, Bengal. Thom (**486**:573) states that identification is impossible.

——**wortmanni** Kloecker (Sacc. **XVIII**:518 ; **486**:449). From humus from the Himalayas.

Piricularia oryzae Cavara (Sacc. **X**:563 ; **105** ; **330** ; **113**(14):44 ; **3**:24 ; **392** ; **434**). On *Oryza sativa* throughout India and Burma. Collections of *Piricularia* similar to that on rice have been obtained on *Eleusine coracana*, *Panicum repens*, *P. ramosum*, *Setaria italica*, *Paspalum* (*Panicum*) *sanguinale*, and *Triticum vulgare* in India, but whether or not they are distinct species is still undecided. Some of the forms are restricted in host range, but the form on rice will infect wheat, oats, and *Eleusine coracana* (McRae, **113**(14): 45). Nisikado (Ber. Ohara Inst. Landw. Forsch., I, 1917) separated the Japanese forms on *Panicum sanguinale* and *Setaria italica* as *P. grisea* and *P. setariae*, respectively.

Pityrosporium ovale Bizzozzero (Sacc. **XXII**:1336 as *Trichosporum* ; **7** as *Malassezia ovalis*). In cases of " pityriasis capitis " of human patients, Calcutta (Acton and Panja).

Ramularia areola Atkinson (**111**:368, fig.). On leaves of *Gossypium* spp., Madras, Pusa, and Bombay (Butler). Mason (**311**:36) finds that *Cercosporella gossypii* Speg. is an earlier name of this fungus.

——**viticis** Syd. (**445**:490). On leaves of *Vitex negundo*, Pollachi, Coimbatore (McRae).

Rhinocladium corticolum Massee (**319, III**:153 ; Sacc. **XVIII**:572 ; **28, IV**:7, figs.). On bark of *Mangifera indica*, Poona (Woodrow) ; Bengal (Bal and Banerjee).

Sclerographium aterrimum Berk. (**57,** No. 472, fig. ; Sacc. **IV**:632). Hypophyllous on *Indigofera ? atropurpurea*, India (? Hooker f.).

Sclerotium rolfsii Sacc. (Sacc. **XXII**:1500 ; **326**:110 ; **113**(9):63 ; **5**:8 ; **113**(19) ; **334** ; **363** ; **393** ; **12** ; **513** ; **413**:182, figs., erroneously as " *Rhizoctonia destruens* "). On *Solanum tuberosum*, *Arachis hypogaea*, *Piperbetle*, *Amorphophallus campanulatus*, *Delphinium* sp., *Dianthus* sp., and *Medicago sativa*, Bengal and Bombay (Shaw and Ajrekar) ; on potato, Ganjam (McRae) ; Western India (Ajrekar) ; on *Arachis hypogaea*, *Cicer arietinum*, *Lens esculenta*, and *Triticum vulgare*, Burma (Rhind) ; on *Eleusine coracana*, India (McRae) ; on *Sesbania grandiflora*, South Arcot (McRae) ; on *Piperbetle*, Bombay (Uppal).

——**oryzae** Cattaneo (Sacc. **XIV**:1153 ; **111**:231, figs. ; **105**:34 ; **410**, figs. ; **392** ; **393** ; **113**(5):57 ; **3**:24 ; **363**). In lower leaf sheaths and within the cavity of the stem of *Oryza sativa* throughout India and Burma.

Spegazzinia meliolae A. Zimm. (Sacc. **XVIII**:690 ; **403**:303). Parasitic on mycelium of *Meliola ambigua* or related species on dead leaves of *Holarrhena antidysenterica*, Dacca (Som).

Sphacelia sorghi McRae (**326**:109 ; **14** ; **113**(16):48 ; **399**). On inflorescences of *Andropogon sorghum*, Coimbatore and Tinnevelly (McRae) ; Burma (Robertson ; Rhind). Ajrekar (**14**) found unidentified *Sphacelia* stages also on *A. caricosus* var. *molicomous*, *Pennisetum alopecuros*, and *Ischaemum pilosum*. Species of *Cerebella* interfere with the development of the sclerotia. *S. sorghi* causes the sugary disease of sorghum mentioned by Ajrekar (**1**:7 ; **2**:18) as being common in the Bombay Presidency.

Sporodesmium polymorphum Cda (**185, I**:7, figs. ; Sacc. **IV**:501). On bark and wood, India.

Sporotrichum beurmanni Matruch. & Ramond (Sacc. **XXII**:1285 ; **2**:51). As a cause of human abscesses, Goa (Fr. de Mello).

——**roseum** Link (Sacc. **IV**:106 ; **461**:145). From soil, Madras (Thakur and Norris).

Stilbum erythrocephalum Dittm. (Sacc. **IV**:567 ; **196**:129). On dead bamboo stems, Royal Botanic Garden, Calcutta (Kurz).

——**inconspicuum** Currey (**196**:129 ; **167**:71 as *S. kurzianum* Cke ; Sacc. **X**:682). On branches, Sibpur, near Calcutta (Kurz). Currey's species was not entered in Saccardo, and Cooke copied the description with insignificant alterations, but for some reason gave it a new name.

——**lateritium** Berk. (Sacc. **IV**:571 ; **57**, after No. 470 ; **263**:341 as "*Stilbella lateritia*"). On bark, E. Nepal (Hooker f.) ; on dead bamboo, Botanic Garden, Saharanpur (Gollan).

——**nanum** Massee (**318**:112, fig. ; Sacc. **XVI**:1082 ; **111**:460 ; **89**:47 ; **367, III**:297). Recorded by Massee as on leaves and branches of living *Thea sinensis* and thought by him to be the fruiting stage of thread blight (see *Marasmius pulcher* above) but Petch (**367, III**:297, and **377**:2) notes that it is a saprophyte most probably identical with *Stilbella heveae* Zimm. and *S. theae* Bernard.

Trichoderma album Preuss (Sacc. **IV**:60 ; **461**). In soil, Madras (Thakur and Norris).

Trichophyton ? **rosaceum** Sabouraud (**2**:49). As a cause of human onychomycosis, Goa (Fr. de Mello).

——**viannai** De Mello (**337**:233, figs. ; **2**:50). As a cause of human skin disease, Goa (Fr. de Mello).

Trichosporium aterrimum Massee (**319, II**:167 ; Sacc. **XVI**:1052 ; **242**). On the bark of *Morus indica*, Changa Manga, Punjab (Gamble). There is an earlier *T. aterrimum* (Cda) Sacc. (see Sacc. IV:289).

——**purpureum** Massee (**319, II**:167 ; Sacc. **XVI**:1052). On decayed wood, Dehra Dun (Gamble).

——**vesiculosum** Butler (**90**:490. figs. ; Sacc. **XXII**:1356 ; **5**:8). On bark and wood of *Casuarina equisetifolia*, Chatrapur, Ganjam (Butler) ; on *C. muricata*, India (Dalia).

Tubercularia circinata Lév. (**285**:67 ; Sacc. **IV**:646). On leaves of *Hoya coronaria* (*H. wallichiana*), India.

——**maculicola** Sacc. (**403**:303). On dead leaves of *Capparis* sp., Pusa (Taslim).

Verticillium glaucum Bonord. (Sacc. **IV**:157 ; **461**). In soil, Madras (Thakur and Norris).

Volutella indica (Niessl) Sacc. (Sacc. **IV**:687 ; **387**:176 as *Psilonia indica* Niessl). On some kind of pod, Calcutta (Kurz).

Xenosporella berkeleyi (Curt.) Lindner (Sacc. **IV**:560 as *Helicosporium*). Stated by Lindner (Ann. Missouri Bot. Gard., XVI:319, 1929) to be " reported from India ".

SPHAEROPSIDALES AND MELANCONIALES.

Aschersonia badia Patouill. (Sacc. **XIV**:989 ; **196**:130, figs., in part, as *Hypocrea variabilis* Currey ; **375, II**:239, 250). On [insects on] living leaves of bamboo, Yomah, Burma (Kurz). The specimens consisted, according to Petch (l.c), of this species and *Hypocrella mollii* (q.v.).

Ascochyta citri Penzig (Sacc. **III**:390 ; **456**:195). On leaves of *Citrus medica*, Kumaon (Butler).

——**dioscoreae** Syd. (**456**:195). On leaves of *Dioscorea* sp., Dehra Dun (Butler).

——**gossypii** Syd. (**456**:194). On leaves of *Gossypium* sp., Kashmir (Inayat).

——**phaseolorum** Sacc. (Sacc. **III**:398 ; **456**:194). On leaves of *Phaseolus mungo* var. *radiatus* and *P. vulgaris*, Kashmir (Butler).

——**pisi** Lib. (Sacc. **III**:397 ; **113**(3):53 ; **456**:194 ; **111**:268, fig.). On leaves, branches, and pods of *Cicer arietinum*, Taru, near Peshawar (Roberston Brown) ; Punjab (Mohendra). The exact identity of the cause of gram blight in India, attributed to this fungus, is uncertain. It may be *Phyllosticta rabiei* (Pass.) Trotter. Petrak (Ann. Myc., XXII, p. 18, 1924) considers that *A. pisi* belongs to *Didymella pinodes* (Berk. & Blox.) Petrak.

——**rheea** (Cke) Grove (**249**:439 as *A. rheae* Grove ; **144**:93 as *Phoma rheea* Cke ; Sacc. **III**:140). On stems of *Boehmeria nivea*, Assam.

——**saccardiana** F. Tassi (Sacc. **XIV**:944 ; **456**:194). On fallen pods of *Albizzia lebbek*, Pusa (Butler).

Asterostomella balanseana (Karst. & Roum.) Theiss. (Sacc. **IX**:390 as *Asterina* ; **18**:214). See entry under *Asterina lawsoniae*.

Botryodiplodia diospyri P. Henn. (**263**:341 ; Sacc. **XVIII**:333). On dead branches of *Diospyros embryopteris*, Botanic Garden, Saharanpur (Gollan).

——**ficina** Syd. (**456**:202). On dead bark of *Ficus glomerata*, Pusa (Inayat).

——**hypoxyloidea** (Cke) Sacc. (Sacc. **III**:379 ; **144**:93 as *Diplodia hypoxyloidea* Cke). On bark of root of *Moringa*, Bengal.

——**manihoticola** Petrak (**384**:144 ; **456**:202 as *B. manihotis* Syd., not *B. manihotis* (P. Henn.) Petrak). On bark of *Manihot utilissima*, Pusa (Inayat).

——**nerii** Syd. (**456**:203). On branches of *Nerium odorum*, Pusa (Butler).

[**Botryodiplodia**] **persicae** Diedicke (**456**:202). On dead branches of *Prunus persica*, Pusa (Butler).

——**saccharina** Diedicke (**456**:203). On dead culms of *Saccharum officinarum*, Pusa (Butler).

——**theobromae** Patouill. (Sacc. **XXII**:1011 as *Lasiodiplodia theobromae* (Pat.) Griff. & Maubl. ; **456**:205 ; **89**:47 ; **93**:28 ; **111**:445, figs. ; **394**:11 ; **113**(17): 50 ; **494**:118 ; **495**:86 ; **493**:15, figs., as *Thyridaria tarda*, q.v. ; **476** ; **395** ; **344**; **421**). On culms of *Saccharum officinarum*, Bihar (Butler) ; on *Cinchona* spp., Lower Burma (McRae) ; on roots of *Thea sinensis*, Assam (Butler) ; North East India (Tunstall) ; on *Hevea brasiliensis* stems, Burma (Rhind) ; Andaman Islands (Mitra). N. E. Stevens (**421**) suggests that this fungus may be the conidial stage of *Physalospora rhodina* (Berk. & Curt.) Cke, q.v.

Camarosporium staurophragmium Tassi (**460**:20 ; Sacc. **XVI**:953). On dead pods and twigs of *Dalbergia sissoo*, India.

Catinula leucoxantha Massee (**319**, **I**:116 ; Sacc. **XVI**:993). On living leaves of *Leucas hyssopifolia*, Dehra Dun (Gamble).

Ceuthospora diospyri Wint. (**388**:148 ; Sacc. **III**:280). On leaves of *Diospyros embryopteris*, Royal Botanic Garden, Calcutta (Kurz).

Chaetomella atra Fckl (Sacc. **III**:321 ; **456**:194). On leaves of *Pennisetum* sp., Yelwal, Mysore (Butler). Petrak and Sydow (**384**:486) delete this genus, and state that this species will be reported upon later.

[——**furcata** Cke & Massee (**170**:43 ; Sacc. **X**:271). On coriaceous leaves, Sikkim. Petrak and Sydow (**384**:488) find that this is an old *Chaetomium*, perhaps *C. setosum*.]

Cicinnobolus cesatii de Bary (Sacc. **III**:216 ; **456**:190). A form on *Oidium* sp. on *Phaseolus mungo* var. *radiatus*, Pusa (Butler).

Colletotrichum agaves Cavara (Sacc. **XI**:570 ; **89**:44 ; **111**:374, fig. ; **411**, figs.). On living leaves of *Agave rigida* var. *sisalana*, Assam, Bengal, Cawnpore.

——**atramentarium** (Berk. & Broome) Taubenh. (Sacc. **III**:227 as *Vermicularia* ; **113**(8):54). On roots and lower stem of *Solanum tuberosum*, Pusa (Butler) ; Ranchi (Dobbs).

——**camelliae** Massee (Sacc. **XVI**:1007). See entry under *Glomerella cingulata*.

——**capsici** (Syd.) comb. nov. (**111**:352, figs., as *Vermicularia capsici* Syd. ; **208**, figs. ; **2**:25 ; **431**). On leaves, stems, and fruit of *Capsicum* spp., the following infected by inoculation : flowers and young fruit of *Carica papaya*, and fruit of *Vigna catjang*, *Dolichos lablab*, *Solanum melongena*, and *Citrus* sp., Bihar (Dastur) ; on *Capsicum*, *Solanum nigrum*, *S. xanthocarpum*, *Datura fastuosa*, *Hibiscus esculentus*, and *Canavalia ensiformis*, Madras (Sundararaman) ; on *Capsicum*, Bombay (Ajrekar). Sundararaman (**431**) considers that cross-inoculation experiments indicate that *C. curcumae* is the same species.

——**catechu** Diedicke (**456**:219). On leaves of *Areca catechu*, Chittagong (Sen).

[**Colletotrichum**] **clerodendri** Diedicke (**456**:220). On living leaves of *Clerodendron infortunatum*, Malda (Butler).

——**coffeanum** Noack (Sacc. **XVIII**:466). See entry under *Glomerella cingulata.*

——**curcumae** (Syd.) comb. nov. (**333** as *Vermicularia curcumae* Syd. ; **430**; **431** ; **326**:110). On leaves of *Curcuma longa*, Kistna, Coimbatore, and Kurnool (McRae) ; Godavari and elsewhere in Madras (Sundararaman). The following can be infected by artificial inoculation : *Brassica* spp., *Solanum melongena*, *Capsicum annuum*, and *Withania somnifera* (Sundararaman). See *C. capsici* above.

——**falcatum** Went (Sacc. **XI**:570 ; **89**:46 ; **93**:2, figs. ; **111**:391, figs. ; **114**, figs. ; **277**; **395**). On leaves and culms of *Saccharum officinarum* throughout India and Burma.

——**gloeosporioides** Penzig (Sacc. **III**:735 ; **344**). On *Citrus aurantium* and *C. medica*, Andaman Islands (Mitra) ; and throughout India.

——**gossypii** Southw. (Sacc. **X**:469). See entry under *Glomerella gossypii*.

——**graminicolum** (Ces.) Wilson (**29**:283 ; **89**:46 ; **93**:13 ; **111**:217, fig. ; **392**:2). On leaves of *Andropogon sorghum* throughout India and Burma.

——**heveae** Petch (Sacc. **XXII**:1203 ; **344**). On *Hevea brasiliensis*, Andaman Islands (Mitra).

——**jatrophae** (Speg.) comb. nov. (Sacc. **XXII**:940 as *Vermicularia jatrophae* Speg. ; **28**, **III**:31, figs.). On leaves of *Jatropha integerrima*, Calcutta (Bal).

——**lindemuthianum** (Sacc. & Mag.) Br. & Cav. (Sacc. **III**:717 as *Gloeosporium* ; **111**:256, 267, figs., as *Glomerella lindemuthiana*). On *Phaseolus vulgaris*, northern India and Nilgiris.

——**necator** Massee (**434** ; **319**, **XIII**:190). On *Piper nigrum*, Madras (McRae).

——**nigrum** Ell. & Hals. (Sacc. **XXII**:1203 ; **111**:356). On *Capsicum annuum*, Burma. This may be *Glomerella cingulata*, q.v.

——**punctiforme** (Niessl) comb. nov. (**388**:146 as *Vermicularia punctiformis* Niessl ; Sacc. **III**:231). On *Oxalis stricta* and O. *corniculata*, Royal Botanic Garden, Calcutta (Kurz).

——**zingiberis** (Sundarar.) comb. nov. (**428**, figs., as *Vermicularia zingiberis* Sundarar. ; **431**). On leaves, petioles, and rhizomes of *Zingiber officinale*, Amalapur, Godavari (Sundararaman).

Coniothyrina agaves (Dur. & Mont.) Petrak & Syd. (**384**:322 ; Sacc. **III**:318 as *Coniothyrium agaves* (Dur. & Mont.) Sacc. ; **456**:193). On leaves of *Agave* sp., Bilikere, Mysore (Butler).

Coniothyrium coffeae A. Zimm. (Sacc. **XVIII**:310 ; **456**:193). On leaves of *Coffea liberica*, Wynaad (McRae). (See **384**:356 for *C. coffeae* P. Henn., which was described later).

——**indicum** Cke & Massee (**167**:70 ; Sacc. **X**:265). On twigs of *Salix*, Hindu Kush, 11,500 ft. (Gilgit Expedition, possibly collected outside of India).

Petrak and Sydow (**384**:319) give a detailed account of this fungus, which they think will form the type of a distinct genus when more fully studied.

[Coniothyrium] palmicolum (Fr., pro parte) Starb. (**418**:78, figs.; Sacc. **XI**:515). On palm leaves, Nicobar Islands (Didrichsen). This fungus was included by Fries (**237, II**:466) in *Sphaeria palmicola*, without locality.

Cryptosporium calami Niessl (**387**:176; Sacc. **III**:744). On leaves of *Acorus calamus*, Royal Botanic Garden, Calcutta. Saccardo states that the specimen (Fungi Europ. 2454) he examined consisted of a *Pestalozzia*.

Cylindrosporium oxyacanthae (Kze & Schm.) Diedicke (Sacc. **III**:578 as *Phleospora*; **456**:220). On leaves of *Crataegus* sp., Harwan, Kashmir (Inayat).

Cytospora ? atra (Bon.) Sacc. (Sacc. **III**:257; **456**:192; **250**:7). On branches of *Morus alba*, Harwan, Kashmir (Butler).

——**bambusina** Diedicke (**456**:193). On dead stems of *Bambusa* sp., Pusa (Butler).

——**cedri** Syd. & Butler (**456**:193). On branches of *Cedrus libani* var. *deodara*, Kulu. Associated with *Stagonospora cedri*.

——**chrysosperma** (Pers.) Fr. (Sacc. **III**:260; **456**:192; **250**:9). On branches of *Populus ciliata*, Harwan, Kashmir (Butler).

——**citri** Diedicke (**456**:193). On living or fading leaves of *Citrus* sp., Pusa (Mitter).

——**sacchari** Butler (**93**:30, figs.; Sacc. **XXII**:962; **456**:193). On culms and sheaths of *Saccharum officinarum*, Madras and Central Provinces (Butler). Petrak (**383**:316) suggests that this may be the same as his *Leeina philippinensis*.

——**salicis** (Cda) Rabenh. (Sacc. **III**:261; **456**:192; **250**:24). On branches of *Salix alba*, Achibal, Kashmir (Inayat).

Darluca filum (Biv.) Cast. (Sacc. **III**:410; **403**:301). On uredo of *Puccinia ?polygoni-amphibii* on *Polygonum* sp., Mussoorie (Kar).

Dendrophoma jasmini Syd. (**456**:187). On dry branches of *Jasminum arborescens*, Pusa (Mitra), associated with *Microdiplodia jasmini*.

Dinemasporium graminum (Berk.) Lév. (Sacc. **III**:682; **185, III**:29, fig., as *Excipula graminum* Berk.). On grasses, India.

——**hispisdulum** (Schrad.) Sacc. (Sacc. **III**:685; **456**:217). On young, dead stem of *Thea sinensis*, Duars (Mann).

Diplodia andamensis Cke (**143**:95; Sacc. **III**:370). On stems of a creeping plant (vernacular "Pilita Dak"), Andaman Islands.

——**arachidis** Petch (Sacc. **XXII**:994; **456**:199). On stems of *Arachis hypogaea*, Pusa (Butler).

——**bambusina** Diedicke (**456**:201). On dead stems of *Bambusa* sp., Pusa (Sen).

——**butleri** Syd. (**456**:199). On dead branches of *Morus alba*, Harwan, Kashmir (Butler).

——**calami** Niessl (**387**:176; Sacc. **III**:372; **456**:200). On leaves of *Calamus* sp., Cawnpore (Butler).

[Diplodia] calecutiana Tassi (**460**:10, fig.; Sacc. **XIV**:937). On rotting fruits of *Ficus altissima*, " Calecuti " [(?) Calcutta], India.

~~——**catappae** Cke (**132** :114 ; Sacc. **III**:343). On nuts of *Terminalia catappa*,~~ Pondicherry.

——**catechu** Syd. & Butler (**456**:200). In the inflorescence of *Areca catechu*, Coimbatore (Butler).

——**citrina** Diedicke (**456**:197). On roots of *Citrus medica*, Sholapur, Bombay (Chibber).

——**corchori** Syd. (**456**:196 ; **113**(10):74 ; **113**(11):69 ; **412** (1), figs.). On stems of *Corchorus capsularis* and *C. olitorius*, Pusa, North Bihar, Kamrup, and various parts of Eastern Bengal and Assam (Shaw).

——**dalbergiae** Diedicke (**456**:198). On dead branches of *Dalbergia sissoo*, Pulliyanur, Travancore (Butler).

——**dracaenicola** Diedicke (**456** : 200). On languishing leaves of *Dracaena* sp., Pusa (Dastur). Petrak and Sydow (**384**:139) state that this is certainly a *Botryodiplodia*, and probably a small spored form of *B. dracaenicola* (P. Henn.) Pet. & Syd.

—— **embryopteridis** Cke (**140** :117 ; Sacc. **III**:361). On fruit of *Diospyros embryopteris* (*Embryopteris glutinifera*), Midnapore, Bengal.

——? **eriobotryae** Sacc. (Sacc. **III**:362 ; **456**:196). On living leaves of *Eriobotrya japonica*, Pusa, and Orai, United Provinces (Butler).

——**gossypina** Cke (**143**:95 ; Sacc. **III**:366). In old capsules of *Gossypium*, Bombay.

——**hibiscina** Cke & Ellis (Sacc. **III**:330 ; **456**:198). On stems of *Hibiscus cannabinus*, Dacca (Som) ; Cuttack (Butler).

——**indica** Diedicke (**456** :197). On bark of *Citrus aurantium*, Poona (Chibber).

——? **indigoferae** P. Brun. (Sacc. **XIV**:930 ; **456**:198). On dead stems and roots of *Indigofera arrecta*, Pusa (Butler) ; Bihar (Annett).

——**manihotis** Sacc. (**456**:197). On fruits of *Manihot utilissima*, Poona (Leather) ; on stems of same host, Poona (Chibber).

——**mori** West. (Sacc. **III**:351 ; **456**:199). On branches of *Morus* sp., Pusa (Butler) ; Wazirabad, Punjab (Mitter).

——**morina** Syd. (**456**:199). On dead branches of *Morus* sp. cult., Pusa (Butler).

——**musae** Diedicke (**456**:200). On dead fruits of *Musa sapientum*, Wahjain, Assam (Butler).

——**papayae** Thuem. (**488**:36 ; Sacc. **III**:350 ; **456**:198). On dead branches of *Carica papaya*, Malabar (Keck) ; perhaps this species on bark of the same host, Pusa (McRae) ; Dacca (Som). These later collections cannot be identified with certainty with Thuemen's species in the absence of material of the latter. They are perhaps to be referred to *Botryodiplodia*.

——**pedilanthi** Syd. (**456**:197). On dead stem of *Pedilanthus tithymaloides*, Pusa (Inayat).

[Diplodia] phyllostictae Cke (**138**:147 ; Sacc. **III**:364). On leaves of forest trees, associated with *Oedocephalum aurantiacum*, Mysore.

——**pithecolobii** Diedicke (**456**:199). On bark of *Pithecolobium* sp., Pusa (Subramaniam).

——**rheea** Cke (**144**:94 ; Sacc. **III**:369). On stems of *Boehmeria nivea*, Assam.

——**ricinicola** Sacc. (**456**:193 ; **403**:308). On stems of *Ricinus communis*, Pusa (Butler) ; Bellary (Burkill).

——**sansevieriae** Syd. (**456**:200). On leaves of *Sansevieria* sp., Pusa (Inayat).

——**sycina** Mont. (Sacc. **III**:350 ; **456**:198). On bark of *Ficus glomerata*, Pusa (Butler).

——**variispora** Diedicke (**456**:196). On leaves of *Eugenia jambolana*, Hunsur, Mysore (Butler).

Diplodina butleri Diedicke (**456**:195). On languishing leaves of *Stipa* sp., Harwan, Kashmir (Butler).

——**pedilanthi** Syd. (**456**:195). On stems of *Pedilanthus tithymaloides*, Pusa (Butler).

Diplozythiella bambusina Diedicke (**456**:215 ; **3**:29). On fading leaves of *Bambusa* sp., Dehra Dun (Butler) ; Central Provinces (Pearl). Pearl (**3**:29) reported that the perfect stage appeared to be a new genus of the Hypocreaceae.

Dothiorella himalayensis Diedicke (**456**:218). On living leaves of *Rhododendron campanulatum*, Ranikhet, Kumaon (Inayat).

——**mangiferae** Syd. (**456**:192). On dead branches, with bark, of *Mangifera indica*, Lucknow (Bahadur).

Entomosporium maculatum Lév. (Sacc. **III**:657 ; **456**:219). On living leaves of *Pyrus communis*, Kashmir (Inayat) ; of *P. pashia* and *Cydonia vulgaris*, Achibal, Kashmir (Butler).

Ephelis japonica P. Henn. (Sacc. **XVIII**:446 ; **456**:217). On inflorescences of *Paspalum kora*, Sylhet (Butler). Apparently the same species occurs on ? *Eragrostis* sp. in the same locality, and perhaps the same species on *Panicum sanguinale*, Dacca (Som). A specimen on *Cymbopogon martini* var. *sofia*, Dharmpur, Punjab (Burkill) is not well developed, but seems to be a distinct species.

——**oryzae** Syd. (**445**:489 ; **456**:218 ; **326**:109 ; **2**:21 ; **5**:6). On the inflorescence of *Oryza sativa*, Poona (Mitra) ; Madras (McRae ; Sydow, Fungi exot. exsicc. 439) ; Nagpur (Dastur).

Fusicoccum jatrophae Syd. (**456**:192). On bark of *Jatropha curcas*, Dehra Dun (Butler).

Gloeosporium ampelophagum (de Bary) Sacc. (Sacc. **III**:719 ; **89**:46 ; **3**:33). On *Vitis vinifera*, Poona.

——**kurzianum** Niessl (**387**:176 ; Sacc. **III**:705 ; **402**:541 ; Sacc. Fungi Italici fig. 1020). On some member of the Leguminosae, Royal Botanic Garden, Calcutta (Kurz).

[Gloeosporium] musarum Cke & Massee (Sacc. **X**:461 ; **201**, figs.). On fruits of *Musa paradisiaca* throughout India.

——**paradoxum** (de Not.) Fckl (Sacc. **III**:707 ; **456**:219). On leaves of *Hedera helix*, Harwan, Kashmir (Butler).

——**piperatum** Ell. & Ev. (Sacc. **XVI**:453 as *Gnomoniopsis* ; **393**:5). On *Capsicum* sp., Mandalay, Burma (Rhind). See *Glomerella cingulata*.

——**terminaliae** Syd. & Butler (**456**:219). On leaves of *Terminalia catappa*, Insein, Burma (Butler).

Haplosporella mangiferae (Died.) Petrak & Syd. (**384**:102 ; **456**:205 as *Cytosphaera mangiferae* Diedicke). On dead branches of *Mangifera indica*, Lucknow (Bahadur). See also *Pleosphaeropsis*.

Hendersonia creberrima Syd. & Butler (**456**:208, figs.). On ripe fruit of *Mangifera indica*, Pusa (Butler). The fungus appears to be the cause of a ripe rot.

——**heraclei** Sacc. (Sacc. **III**:432 ; **456**:207). On living leaves and stems of *Heracleum* sp., Harwan, Kashmir (Butler).

——**minutissima** Sacc. (Sacc. **III**:437 ; **387**: 176 as *H. microscopica* Niessl, not Fr.). On *Apluda mutica*, Royal Botanic Garden, Calcutta. Stated by Niessl to be the macrostylosporous form of a *Leptosphaeria*.

——**obtusa** Cke (**139**:20 ; Sacc. **III**:423 ; **132**:115 as *H. lonicerae* Cke, not Fr.). On twigs of *Lonicera diversifolia*, Saharanpur (Dr. Jameson).

——**tamarindi** Syd. (**456**:209). On living leaves of *Tamarindus indica*, Mozufferpore (Butler).

——**theicola** Cke (**129**:90 as *H.* "*theaecola*" ; Sacc. **III**:427). On leaves of *Thea sinensis*, Cachar.

Hendersonina sacchari Butler (**115**:191, figs.; **456**:209 ; **113** (17):51 ; **111**:388, figs.). On culms of *Saccharum officinarum*, Samalkota (Butler) ; Jorhat (Meggit) ; Bihar (McRae).

Lasmeniella globulifera (Rabenh.) Petrak & Syd. (**384**:303 ; **387**:60 as *Coniothyrium globuliferum* Rab.; Sacc. **III**:308 ; **264**, No. 744 as *Lasmenia globulifera* (Rab.) v. Hoehn.; **456**:194). On leaves of *Bauhinia vahlii*, Royal Botanic Garden, Calcutta (Kurz) ; Bhim Tal, Kumaon (Inayat). See also *Phyllachora bauhiniae*.

Leptothyrium leguminum (Cke) Sacc. (Sacc. **III**:632 ; **144**:93 as *Leptostroma leguminum* Cke). On legumes of *Clitoria* sp., Madras (? Hobson).

——**pomi** (Mont. & Fr.) Sacc. (Sacc. **III**:632 ; **113** (11): 74). On fruits of *Pyrus malus*, Kumaon (Shaw).

Macrophoma boussingaultiae Syd. & Butler (**456**:187). On living leaves of *Boussingaultia baselloides*, Kistna, Madras (Mitra). Petrak and Sydow (**384**:115) propose to restrict the genus *Macrophoma* to the single species *M. pinea* (Desm.) Pet. & Syd. Under their classification the species listed here would require to be transferred to other genera, which has not yet been done.

[Macrophoma] calophylli Syd. (**456**:188). On living or fading leaves of *Calophyllum inophyllum*, Kistna, Madras (Subramaniam).

——**caryotae** Syd. (**456**:189). On leaves of *Caryota urens*, Poona (Chibber).

——**cassiocarpa** (Cke) Berl. & Vogl. (Sacc. **X**:203 ; **144**:93 as *Sphaeropsis cassiaecarpum* Cke ; Sacc. **III**:147 as *Phoma cassiocarpa* (Cke) Sacc.). On legumes of *Cassia absus* and *Clitoria* sp., Madras.

——**celastrina** Diedicke (**456**:188). On living leaves of *Celastrus* sp., Pusa (Butler).

——**clitoricarpa** (Cke) Berl. & Vogl. (Sacc. **X**:197 ; **144**:93 as *Sphaeropsis clitoreaecarpum* Cke ; Sacc. **III**:147 as *Phoma clitoricarpa* (Cke) Sacc.). On legumes of *Clitoria* sp., Madras (Bidie).

——**musae** (Cke) Berl. & Vogl. (Sacc. **X**:199 ; **89**:48 ; **456**:189 ; **467**:153 ; **144**:93 as *Sphaeropsis ? musarum* Cke ; Sacc. **III**:163 as *Phoma musae* (Cke) Sacc., not *P. musarum* Cke ; **274**:489 as *Dothidea musae* Klotz.; **51**:398 ; Sacc. **II**:613 and 625 as *Phyllachora musae* (Klotz.) Sacc.). On leaves of *Musa* sp., India (Wight) ; of *M. paradisiaca*, Belgaum (Hobson) ; of *M. sapientum*, Anand, Bombay (Blatter) ; Dehra Dun, Pusa, Coimbatore, and Tellicherry in Malabar (Butler). Theissen and Sydow (**481**:570) examined the type of *Dothidea musae* Klotzsch and found it was this *Macrophoma* ; Cooke's statement (Grev., XIII, p. 64) " asci clavate " was presumably an error.

——**phaseolina** Tassi (Sacc. **XVIII**:268 ; **456**:189). On living leaves of *Vigna catjang*, Pusa (Subramaniam).

——**piperina** Syd. (**456**:188). On fading leaves of *Piper nigrum*, Sylhet (Butler).

——**sycophila** (Massee) Sacc. & D. Sacc. (Sacc. **XVIII**:273 ; **456**:198). On fading leaves of *Ficus religiosa*, Kangra (Mitter).

Macrophomina phaseoli (Maubl.) Ashby (**22**:145, fig.; **334** ; **113**(21) ; **434** ;**409**:119, figs., as " *Rhizoctonia solani* "; **413**:191 ; **5**:8 as *Sclerotium bataticola* Taub.; **456**:187 as *Macrophoma cajani* Syd. & Butler ; **385**:227 as *Dothiorella cajani* (Syd. & Butl.) Petrak & Syd.; **412**(2), figs., as *Macrophoma corchori* Saw.; **113**(12)). On *Solanum tuberosum*, *Gossypium* spp., *Corchorus capsularis*, *C. olitorius*, *Cajanus indicus*, *Arachis hypogaea*, *Alysicarpus* sp., *Carica papaya*, *Citrullus vulgaris*, *Crotalaria juncea*, *Cucurbita maxima*, *Dolichos biflorus*, *D. lablab*, *Hibiscus cannabinus*, *Lycopersicum esculentum*, *Medicago sativa*, *Morus alba*, *Nicotiana tabacum*, *Phaseolus lunatus*, *P. mungo* var. *radiatus*, *Sesamum indicum*, *Solanum melongena*, and *Vigna catjang*, as a root, stem, and tuber parasite throughout India. Haigh (**255**) has recently described within this species three groups, based on size of sclerotia, and reports that pycnidia have been connected only with the group with the smallest sclerotia. Petrak and Sydow (**385**:227 ; **384**:217) have abandoned the former's genus *Macrophomina* which was founded on this species on *Sesamum* in the Philippines, regarding it as synonymous with *Dothiorella*.

Microdiplodia agaves (Niessl) Tassi (Sacc. **XVIII**:323 ; **387**:176 as *Diplodia agaves* Niessl ; Sacc. **III**:371 ; **456**:200). On leaves of *Agave americana*, India (Kurz) ; of *A.* sp., Dharwar and *Yucca aloifolia*, Shillong and Pusa (Butler). Petrak (**383**:218) suggests that this is the pycnidial stage of his *Pleospora bataanensis.* He found also that Niessl's fungus was a *Microdiplodia*, but overlooked Tassi's earlier combination.

——**indica** Syd. (**456**:201). On living leaves of *Iris* sp., Srinagar, Kashmir (Butler).

——**jasmini** Syd. (**456**:201). On dry branches of *Jasminum arborescens*, Pusa (Mitra), associated with *Dendrophoma jasmini.*

Micropera dahliae Diedicke (**456**:214). On dead stem of *Dahlia variabilis*, Pusa (Dastur).

Microstroma juglandis (Bereng.) Sacc. (Sacc. **IV**:9 ; **455**:280). On leaves of *Juglans regia*, Kumaon Himalaya (Butler). The systematic position of this fungus is uncertain. Some have considered it to be a Basidiomycete near *Exobasidium*, but it is apparently one of the Fungi Imperfecti. Saccardo and others consider it to belong to the Mucedineae, while Maire, Briosi and Cavara, and Karakulin refer it to the Melanconiaceae, an opinion supported by the most recent study of the fungus by F. A. Wolf in Journ. Elisha Mitchell Sci. Soc., XLV, pp. 130-136, 1929.

Pestalozzia funera Desmaz. (Sacc. **III**:791 ; **456**:220 ; **326**:110). On leaves of *Cunninghamia sinensis*, Dehra Dun (Butler) ; of *Eucalyptus globulus*, near Coonoor, 5,000 ft. (McRae).

——**fuscescens** Sorauer (Sacc. **XXII**:1227). On young plants of *Livistona (Corypha) australis* exported from India.

——**mangalorica** Thuem. (**488**:37 ; Sacc. **III**:790). On living leaves of *Bridelia scandens*, Mangalore (Keck).

——**palmarum** Cke (**132**:115 ; Sacc. **III**:796 ; **96**:21 ; **2**:27 ; **344**). On *Cocos nucifera*, Bengal ; Travancore (Butler) ; Madras (McRae) ; Andaman Islands (Mitra).

——**phoenicis** Vize (**133**:14, 17, figs.; Sacc. **III**:796). On leaves of *Phoenix dactylifera*, ?India (? Hobson).

——**theae** Sawada (**111**:451, fig.; **507** ; **497**:40 ; **393**:4 ; **503** ; **318**:106 as "*P. guepini* Desm."). On leaves of *Thea sinensis*, Assam, Bengal, and Madras.

Phleospora mori (Lév.) Sacc. (Sacc. **III**:577 ; **456**:214 ; **98**:11 as *Septogloeum mori* (Lév.) Briosi & Cav.). On leaves of *Morus alba*, Dubgaon, Kashmir (Butler).

Phoma casuarinae Tassi (Sacc. **XIV**:885 ; **456**:186). On leaves of *Casuarina* sp., Karwar (Butler).

——**causcorae** Rabenh. (**387**:176 ; Sacc. **III**:131). On stem of "*Causcora decussata*", Royal Botanic Garden, Calcutta (Kurz).

——**desmonci** Rabenh. (**387**:60 ; Sacc. **III**:157). On leaves of *Desmoncus* "*melanochaetus*", Royal Botanic Garden, Calcutta.

[Phoma] foedata (Lév.) Sacc. (Sacc. **III**:104 ; **285**:63 as *Sphaeropsis foedata* Lév.). On leaves of *Hoya coronaria* (*H. wallichiana*), India.

——**fourcroyae** Thuem. (Sacc. **III**:160 ; **456**:186). On leaves of *Furcraea* (*Fourcroya*) *gigantea*, Shillong (Butler).

——**glumarum** Ellis & Tracy (Sacc. **X**:185 ; **89**:46 ; **105**:35 ; **456**:187). On glumes of *Oryza sativa* throughout India.

——**keckii** (Thuem.) Sacc. (Sacc. **III**:90 ; **488**:37). On stems of *Calotropis gigantea*, Kanara (Keck).

——**oryzae** Cke & Massee (**165**:15 ; Sacc. **X**:185). On stems of *Oryza sativa*, Calcutta.

——**pardanthi** Diedicke (**456**:186). On dry stems of *Pardanthus chinensis* (=*Belamcanda punctata*), Pusa (Butler).

——**polyanthis** Diedicke (**456**:186). On dead leaves and stems of *Polianthes tuberosa*, Pusa (Inayat).

——**saccharina** Syd. (**456**:187 ; **393**:5). On culms of *Saccharum officinarum*, Pusa (Butler) ; Mandalay (Rhind).

——**salicina** West. (Sacc. **III**:97 ; **456**:186). On branches of *Salix* sp., Achibal, Kashmir (Butler).

——**solani** Cke & Hark. (Sacc. **X**:175 ; **257**). Recorded as injurious to *Solanum melongena*, Comilla, Bengal (Hector).

Phomopsis artabotrydis Syd. (**456**:191). On living leaves of *Artabotrys odoratissimus*, Poona (Chibber).

——**cajani** Syd. (**456**:191). On dead branches of *Cajanus indicus*, Pusa (Butler).

——**heteronema** Sacc. (**403**:17). On epicarp of *Areca catechu*, Portuguese East Indies [? India] (comm. Alfr. Moller).

——**pandani** Diedicke (**456**:191). On leaves of *Pandanus* spp., Poona (Chibber).

Phyllosticta ambrosioides Thuem. (Sacc. **III**:55 ; **456**:184). On leaves of *Chenopodium album*, Mussoorie (Mitra).

——**buddleiae** Syd. (**456**:183). On living leaves of *Buddleia* sp., Dehra Dun (Butler).

——**buteae** Syd. (**456**:180). On living leaves of *Butea frondosa*, Pusa (Subramaniam).

——**cajani** Syd. (**456**:178). On leaves of *Cajanus indicus*, Mozufferpore (Butler).

——**capparidicola** Speg. (Sacc. **XVIII**:824 ; **456**:182). On leaves of *Capparis* sp., Burdwan (Butler).

——**catappae** Syd. (**456**:181). On leaves of *Terminalia catappa*, Maymyo (Butler).

——**chrysanthemi** Ell. & Dearn. (Sacc. **XI**:479 ; **456**:177). On leaves of *Chrysanthemum* sp., Pusa (Butler).

——**clerodendri** Syd. & Butler (**456**:183). On living leaves of *Clerodendron* sp., Nadiad, Bombay (Butler).

——**cocculi** Thuem. (**488**:36 ; Sacc. **III**:29). On living or languishing leaves of *Anamirta cocculus*, Kanara (Keck).

——**cocos** Cke (**144**:94 ; Sacc. **III**:59 ; **456**:185). On leaves of *Cocos nucifera*, Belgaum (Hobson). Apparently the same species on leaves of *Caryota* sp., Dacca (Som).

[**Phyllosticta**] **codiaeicola** Diedicke (**456**:184). On leaves of *Codiaeum* sp., Ganeshkhind, Poona (Mitra).

——?**coffeicola** Speg. (Sacc. **XIV**:857 ; **456**:182 ; not *P. coffeicola* Delacr., Sacc. **XVIII**:237). On leaves of *Coffea arabica*, Koppa, Mysore (Butler).

——**cycadina** Passerini (Sacc. **X**:124 ; **456**:185). On leaves of *Cycas revoluta*, Poona (Chibber).

——**desmodiicola** Diedicke (**456**:178). On leaves of *Desmodium* sp., Mussoorie (Mitra).

——**diospyri** Syd. & Butler (**456**:183). On living or fading leaves of *Diospyros embryopteris*, Pusa (Butler). An *Ascochyta* occurs in similar pycnidia on the same spots.

——**dolichi** Brun. (Sacc. **XI**:478 ; **456**:177). On leaves of *Dolichos biflorus*, Pusa (Butler).

——**eriodendri** Diedicke (**456**:179). On leaves of *Eriodendron anfractuosum*, Ganeshkhind, Poona (Mitra).

——**exigua** Syd. (**456**:183). On leaves of *Quercus* sp., Kumaon (Butler).

——**glycines** Thuem. (Sacc. **III**:11 ; **456**:178). On leaves of *Glycine hispida*, Verinag, Kashmir (Butler).

——**glycosmidis** Syd. & Butler (**456**:177 ; **28, III**:33, figs.). On living leaves of *Glycosmis pentaphylla*, Dehra Dun (Butler) ; Wahjain, Assam (Som) ; Calcutta (Chowdhuri).

——**grewiae** Diedicke (**456**:181). On leaves of *Grewia* sp., Dehra Dun (Butler).

——**hibisci** Peck (Sacc. **X**:103 ; **456**:182). On leaves of *Hibiscus cannabinus*, Cuttack (Butler).

——**hortorum** Speg. (Sacc. **III**:49 ; **456**:184). On leaves of *Solanum melongena*, Achibal, Kashmir (Inayat) ; Surat (Butler).

——**hoyae** Diedicke (**456**:180). On leaves of *Hoya* sp., Pusa (Inayat).

——**humuli** Sacc. & Speg. (Sacc. **III**:53 ; **456**:184). On leaves of *Humulus lupulus*, Dubgaon, Kashmir (Butler).

——**ingae-dulcis** Diedicke (**456**:178). On leaves of *Pithecolobium* (*Inga*) *dulce*, Cocanada (Butler).

——**ipomoeae** Ell. & Kellerm. (Sacc. **X**:127 ; **456**:182). On leaves of *Ipomoea* sp., Kirkee, Poona (Mitra).

——**marmorata** Cke (**145**:13 ; Sacc. **III**:36 ; **456**:182). On leaves of *Mallotus philippinensis*, Saharanpur (Duthie) ; Dehra Dun (Butler).

——**miurai** I.Miyake (Sacc. **XXII**:864 ; **456**:185). On leaves of *Oryza sativa*, Meharpur, Bengal (Butler).

——**persicae** Sacc. (Sacc. **III**:8 ; **456**:184). On leaves of *Prunus persica*, Darjeeling (McRae).

——**pirina** Sacc. (Sacc. **III**:7 ; **456**:184). On leaves of *Pyrus communis*, Lyallpur (Butler) ; Darjeeling (McRae).

[**Phyllosticta**] **pongamiae** Syd. (**456**:178). On leaves of *Pongamia glabra*, Dacca (Som).

——**prunicola** (Opiz?) Sacc. (Sacc. **III**:4 ; **456**:184). On leaves of *Prunus armeniaca*, Achibal, Kashmir (Butler) ; of *P. avium* and *P. communis*, Verinag, Kashmir (Butler).

——**religiosa** Syd. (**456**:183). On leaves of *Ficus religiosa*, Poona (Chibber).

——**sesbaniae** Syd. (**456**:179). On leaves of *Sesbania* sp., Pusa (Butler).

——**sissoo** Diedicke (**456**:179). On leaves of *Dalbergia sissoo*, Pusa (Butler).

——**sorghina** Sacc. (Sacc. **III**:61 ; **456**:185). On leaves of *Andropogon sorghum* (*Sorghum vulgare*), Manaparai, Trichinopoly (Butler).

——**symploci** Syd. (**456**:180). On leaves of *Symplocos* sp., Nilgiris (Butler).

——**tectonae** Syd. & Butler (**456**:181). On leaves of *Tectona grandis*, Puttimari, Assam (Taslim).

——**tricoloris** Sacc. (**456**:182 ; Sacc. **III**:38 as form). On leaves of *Viola odorata*, Ganeshkhind, Poona (Butler).

Phyllostictina arecae (Diedicke) Petrak & Syd. (**384**:188 ; **456**:185 as *Phyllosticta arecae* Died.). On leaves of *Areca catechu*, Bilin, Burma (Butler).

——**ar'ocarpina** (Syd. & Butler) Syd. (**384**:189 ; **456**:180 as *Phyllosticta artocarpina* Syd. & Butl.). On leaves of *Artocarpus integrifolia*, Bombay Presidency (Chibber). See *Septoria artocarpi*.

——**cruenta** (Fr.) Petrak & Syd. (**384**:209 ; Sacc. **III**:58 as *Phyllosticta cruenta* (Fr.) Kickx ; **456**:185). On leaves of *Polygonatum* sp., Harwan in Kashmir, and Kumaon (Butler).

——**murrayae** Syd. (**456**:186 ; **384**:188). On living leaves of *Murraya koenigii*, Dehra Dun (Butler). Von Hoehnel (Ann. Myc., XVIII:93, 1920) and Guba (Phytopath., **XIV** : 234, 1924) also discuss this fungus.

——**putranjivae** Syd. (**384**:201 ; **456**:182 as *Phyllosticta putranjivae* Syd.). On leaves of *Putranjiva roxburghii*, Pusa (Butler).

Plenozythia euphorbiae Syd. (**456**:215). On bark of *Euphorbia tirucalli*, Pusa (Butler).

Pleocyta sacchari (Massee) Petrak & Syd. (**384**:455 ; Sacc. **XIV**:1019 as *Melanconium* ; **93**:14, fig., as *Trichosph.ıria sacchari* Massee). On culms of *Saccharum officinarum*, Saharanpur, Bihar, and Madras. It has also been reported on *Zea mays* from the Philippines.

Pleosphaeropsis capparidis Diedicke (**456**:205 ; **384**:105 as *Haplosporella capparidis* (Died.) Petrak & Syd.). On dead branches of *Capparis* sp., Pusa (Butler). Petrak and Sydow (**384**:103) regard the genus *Pleosphaeropsis*, which was founded by Diedicke for these three Indian species, as a subgenus of *Haplosporella* Speg. The three species are fully described by Petrak and Sydow.

——**dalbergiae** Diedicke (**456**:203 ; **384**:103 as *Haplosporella dalbergiae* (Died.) Petrak & Syd.). On dead branches of *Dalbergia sissoo*, Pusa (Butler).

[Pleosphaeropsis] gossypii Diedicke (**456**:204 ; **384**:104 as *Haplosporella gossypii* (Died.) Petrak & Syd.) On dead branches of *Gossypium* sp. cult., Pusa (Butler).

Polystigmina rubra (Desm.) Sacc. (Sacc. **III**:622 ; **456**:217). On leaves of *Prunus domestica* var. *insititia* (given by error in **456** : 217 as *P. communis* var. *insititia*), Harwan, Kashmir (Butler) ; of *P. padus*, Murree (Butler).

Pyrenochaete oryzae Shirai (Sacc. **XXII**:934 ; **105**:35 ; **456**:189). On leaves and sheaths of *Oryza sativa*, Hmawbi, Burma (Butler).

Rhynchodiplodia citri Briosi & Farnetti (Sacc. **XVIII**:330 ; **456**:207). On *Citrus* fruits, Poona.

Robillarda scutata Syd. (**467**:153). On dead leaves of *Mimusops hexandra*, Anand, Bombay (Blatter).

——**sessilis** Sacc. (Sacc. **III**:408 ; **456**:195). On leaves of *Dalbergia* sp., Darjeeling (McRae).

Saccidium depazeoides Cke (**172**:7 ; Sacc. **X**:422). On leaves of *Aspidopterys* "*caudata*" [*A. ? cordata*], India.

Sclerophoma piceae (Fiedler) v. Hoehn. (**264**, No. 402 ; Sacc. **III**:101 as *Phoma* ; **456**:191). On branches of *Cedrus libani* var. *deodara*, Kulu (Parnell).

Septogloeum acaciae Syd. (**445**:489 ; **456**:220). On leaves of *Acacia arabica*, Coimbatore (McRae) ; Kalpi, Jhansi District (Carr).

——**poincianae** Syd. (**445**:490). On leaves of *Poinciana alata*, Coimbatore and Koilpatti (McRae).

Septoria aciculosa Ell. & Everh. (Sacc. **III**:511 ; **456**:213). On leaves of *Fragaria* sp., Harwan, Kashmir, and of *F. indica*, Shillong (Butler).

——**aitchisoni** Syd. (**443**, **I**:171 ; Sacc. **XVIII**:386). On leaves of *Jasminum humile*, Kurram Valley, Afghanistan frontier (Aitcheson).

——**alliacea** Cke (**144**:94 ; Sacc. **III**:572). On leaves of some Alliaceous plant, Belgaum (Hobson).

——**arcuata** Cke (**144**:94 ; Sacc. **III**:499 ; **456**:213). On leaves of *Ficus* sp., Belgaum (Hobson) ; Wynaad (McRae) ; of *F. indica*, Dehra Dun (Butler) ; of *F. benjamina*, Poona (Butler) ; of *F. bengalensis*, Kanara (Kulkarni).

——**artocarpi** Cke (**132**:114 ; **134**:5, fig.; Sacc. **III**:500). On leaves of *Artocarpus integrifolia*, Mysore. The description suggests *Phyllostictina artocarpina*.

——**bakeri** Syd. (**456**:214). On leaves of *Leucas* sp., Pusa (Kar).

——**brachyspora** Sacc. (Sacc. **III**:500 ; **456**:213). On leaves of *Ficus* sp., Khed, Bombay (Chibber).

——**brassicae** Ell. & Everh. (Sacc. **XIV**:968 ; **456**:211). On leaves of *Brassica campestris* var. *rapa*, Sopor and Achibal, Kashmir (Butler).

——**butleri** Diedicke (**456**:212). On leaves of *Viburnum* sp., Harwan and Verinag, Kashmir (Butler).

——**cannabis** (Lasch) Sacc. (Sacc. **III**:557 ; **456**:213). On leaves of *Cannabis sativa*, Pusa (Butler).

[**Septoria**] **cattanei** Thuem. (**488**:37 ; Sacc. **III**:477). On leaves of *Citrus medica*, Kanara (Keck).

——**chrysanthemella** Sacc. (Sacc. **XI** : 542 ; **456** : 211). On leaves of *Chrysanthemum indicum*, Pusa and Begumserai, Bihar, and Dehra Dun (Butler).

——**conyzae** Diedicke (**456**:210). On leaves of *Conyza* sp., Pusa (Butler).

——**cordiae** Syd. (**456**:211). On leaves of *Cordia rothii*, Ganeshkhind, Poona (Chibber).

——**diversimaculans** Diedicke (**456**:210). On leaves of *Cnicus argyracanthus*, Shadipore, Kashmir (Butler).

——**dolichi** Berk. & Curt. (Sacc. **III**:509 ; **456**:212). On leaves of *Dolichos lablab*, Wynaad (McRae).

——**hyalina** Ell. & Everh. (Sacc. **XI**:538 ; **456**:213). On leaves of *Viola patrinii*, Mussoorie (Butler).

——**lactucae** Peck (Sacc. **III**:552 ; **459**:210). On leaves of *Lactuca* sp., Harwan, Kashmir (Butler). Saccardo considers *S. lactucae* Pass. to be apparently the same.

——**lupulina** Ell. & Kellerm. (Sacc. **X**:380 ; **456**:213). On leaves of *Humulus lupulus*, Dubgaon, Kashmir (Butler).

——**macropora** Sacc. (Sacc. **III**:526 ; **456**:211). On leaves of *Paeonia emodii*, Kumaon (Butler).

——**myriactidis** Syd. (**456**:210). On leaves of *Myriactis nepalensis*, Achibal, Kashmir (Butler). Sometimes accompanied by a *Phyllosticta*.

——**pipula** Cke (**143**:95 ; Sacc. **III**:499). On leaves of *Ficus religiosa*, Belgaum (Hobson). Cooke spelled the specific name "*Pipula*" from the vernacular name of the host, but it is given in Saccardo as "*Pipulae*."

——**plantaginea** Passerini (Sacc. **III**:554 ; **456**:211). On leaves of *Plantago* sp. (near *P. lanceolata*), Harwan, Kashmir (Inayat). Often accompanied by a *Phyllosticta*.

——**polygonicola** (Lasch) Sacc. (Sacc. **X**:380 ; **456**:213). On leaves of *Polygonum* sp., Harwan, Kashmir (Butler).

——**rosae** Desmaz. (Sacc. **III**:485 ; **456**:213). On leaves of *Rosa* sp., Darjeeling (Hafiz Khan).

——**rosarum** West. (Sacc. **III**:486 ; **456**:213). On leaves of *Rosa* sp., Kangra, Punjab (Mitter) ; near Srinagar, Kashmir (Butler).

——**rubi** West. (Sacc. **III**:486 ; **456**:213). On leaves of *Rubus idaeus*, Shillong (Butler) ; of *R. lasiocarpus*, Harwan, Kashmir (Butler).

——**scrophulariae** Peck (Sacc. **III**:534 ; **456**:212). On leaves of *Scrophularia* sp., Harwan and near Srinagar, Kashmir (Butler).

——**sojae** Syd. & Butler (**456**:212). On living or fading leaves of *Glycine soja*, Verinag, Kashmir (Inayat) ; Achibal, Kashmir (Butler).

[Septoria] sordidula (Lév.) Sacc. (Sacc. **III**:548 ; **286**:277 as *Ascospora sordidula* Lév.). On leaves of *Saussurea* (*Aplotaxis*) sp., India (Jacquemont). The type at Paris consists of four leaves, each with several spots darkened by pycnidia.

——**stipina** Diedicke (**456**:214). On leaves of *Stipa* sp., Harwan, Kashmir (Butler).

——**tabacina** Diedicke (**456**:210). On leaves of *Artemisia* sp., Darjeeling (McRae).

——**tritici** Desmaz. (Sacc. **III**:561 ; **456**:214 ; **111**:175, figs.). On leaves of *Triticum vulgare*, Lyallpur (Mitra). *S. gramineum* Desmaz. (mentioned in **111**:175) is found by Weber (Phytopath., XII:337-385) to be different.

——**variegata** Vize (**133**:14 ; Sacc. **III**:507). On large coriaceous leaves, India (? Hobson).

——**verbenae** Rob. & Desmaz. (Sacc. **III**:537 ; **456**:212). On leaves of *Verbena officinalis*, Achibal, Kashmir (Butler).

——**viburni** West. (Sacc. **III**:493 ; **456**:212). On leaves of *Viburnum* sp., Harwan, Kashmir (Butler).

——**vicoae** Diedicke (**456**:211). On leaves of *Vicoa vestita*, Pusa (Butler).

——**villarsiae** Desmaz. (Sacc. **III**:541 ; **456**:214). On leaves of *Limnanthemum* sp., Kashmir (Butler) ; of *L. nymphoides*, Woolar Lake, Kashmir (Butler).

——**violae** West. (Sacc. **III**:518 ; **456**:213). On leaves of *Viola* sp., Kumaon (Butler).

Sirococcus butleri Syd. (**456**:191). On dead branches of *Jasminum* sp., Pusa (Butler).

——**calophylli** Syd. (**456**:190). On living or fading leaves of *Calophyllum inophyllum*, Kistna, Madras (Subramaniam).

Sirothyrium taxi Syd. (**456**:218). On living leaves of *Taxus baccata*, Murree (Butler).

Sphaeronema bengalensis Diedicke (**456**:190). On wood of *Ficus bengalensis*, Pusa (Chibber).

Sphaeropsis absus Cke (**144**:93 ; Sacc. **III**:292). On twigs of *Cassia absus*, Madras (? Hobson). Petrak and Sydow (**384**:16) change the generic name *Sphaeropsis* Sacc. to *Haplosporella* Speg., since Léveillé's earlier genus *Sphaeropsis* is an ascigerous fungus.

——? **malorum** Berk. (Sacc. **III**:294 ; **348**, figs.; **384**:148 as *Botryodiplodia*). Isolated from fruit of *Pyrus malus*, Allahabad (Mitter and Tandon).

——**palmarum** Cke (Sacc. **III**:157 as *Phoma* ; **456**:194 ; **384**:164 as *Botryodiplodia*.) On petioles of *Cocos nucifera*, Poona (Chibber).

Stagonospora arecae Diedicke (**456**:207). On leaves of *Areca catechu*, Chittagong (Sen).

——**cedri** Syd. & Butler (**456**:207). On dead branches of *Cedrus libani* var. *deodara*, Kulu (Butler). Associated with *Cytospora cedri*.

Stauronema sacchari Syd. & Butler (**456**:217). On rotting culms of *Saccharum officinarum*, Dehra Dun (Butler.)

Thyrostroma mori (Nomura) v. Hoehn. (**264**, No. 718 ; Sacc. **XXII**:1230 as *Steganosporium mori* (Nom.) Sacc. & Trott.; **98**:1, figs., as *Coryneum mori* Nom.). On branches of *Morus alba* and *Celtis australis* (*C. caucasica*), Kashmir (Butler).

Urohendersonia indica Syd. (**456**:209). On living leaves of *Erythrina* sp., Darjeeling (McRae).

LIST OF SYNONYMS AND OTHER NAMES FOUND TO HAVE BEEN APPLIED TO INDIAN FUNGI.

(Variant spellings not included here.)

List of synonyms and other names found to have been applied to Indian Fungi—*contd.*

List of synonyms and other names found to have been applied to Indian Fungi—*contd.*

List of synonyms and other names found to have been applied to Indian Fungi—*contd.*

List of synonyms and other names found to have been applied to Indian Fungi—*contd.*

List of synonyms and other names found to have been applied to Indian Fungi—*contd.*

List of synonyms and other names found to have been applied to Indian Fungi—*concld.*

BIBLIOGRAPHY.

1. ANON. . . . . Proceedings of the Mycological Conference held at Pusa on 5th Feb. 1917 and following days. Calcutta, Supdt. Govt. Printing, 17 pp., 1917.

2. —— . . . . Proceedings of the Second Meeting of Mycological Workers in India held at Pusa on 20th Feb. 1919 and following days. Calcutta, Supdt. Govt. Printing, 68 pp., 1919.

3. —— . . . . Proceedings of the Third Meeting of Mycological Workers in India held at Pusa on 7th of Feb. 1921 and following days. Calcutta, Sudpt. Govt. Printing, 45 pp., 1921.

4. —— . . . . Appendix M. Annual Report of the plant pathologist to Government of Bombay, Poona, for the year 1923-24. Ann. Rept. Dept. of Agric. Bombay Presidency for the year 1923-24, pp. 162-164, 1925.

5. —— . . . . Fungi received at the Imperial Bureau of Mycology, List I, pp. 1-8. Suppl. to the Rev. Appl. Myc., 1925.

6. ACTON, H. W. & McGUIRE, C. . Tinea cruris : its manifestations, diagnosis, and treatment. Indian Med. Gaz., LXII, pp. 419-428, 8 pls. (2 col.), 1927.

7. —— & PANJA, G. Seborrhoeic dermatitis or pityriasis capitis : a lesion caused by the Malassezia ovale. Indian Med. Gaz., LXII, pp. 603-614, 10 pls. (2 col.), 1927.

8. AJREKAR, S. L. . . A note on the life-history of Cystopsora oleae Butl. Ann. Myc., X, pp. 307-309, 3 figs., 1912.

9. —— . . The castor rust (?Melampsorella ricini de Toni). Jour. Bombay Nat. Hist. Soc., XXI, pp. 1092-1095, 2 pls., 1912.

10. —— . . On the mode of infection and prevention of the smut disease of sugarcane. Agric. Journ. India, XI, pp. 288-295, 1 pl., 1916.

11. —— . . On the identity of Blastospora butleri Syd. Journ. Bombay Nat. Hist. Soc., XXVI, pp. 696-697, 1 pl., 1919.

12. —— . . The problem of potato storage in Western India. Agric. Journ. India, XIX, pp. 35-44, 1924.

13. [AJREKAR, S. L.]. . The cause of cotton wilt in India. Journ. Indian Bot. Soc., V, pp. 1-8, 1926.

14. —— . . . Observations on a disease of Jowar (Sorghum vulgare) caused by Sphacelia (conidial stage of Claviceps). Journ. Ind. Bot. Soc., V, pp. 55-61, 1 pl., 1926.

15. —— & BAL, D. V. Observations on the wilt disease of cotton in the Central Provinces. Agric. Journ. India, XVI, pp. 598-617, 2 pls., 1921.

16. —— & KAMAT, M. N. The relationship of the species of Fusarium causing wilt and dry rot of potatoes in Western India. Agric. Journ. India, XVIII, 5, pp. 515-520, 1 pl., 1923.

17. —— & TONAPY, B. R. A note on the life history of Uromyces aloës (Cke) P. Magn. Journ. Indian Bot. Soc., III, pp. 267-269, 1 pl., 1 fig., 1923.

18. ARNAUD, G. . . Les Astérinées. Ann. Ecole Nat. Agric. Montpellier, XVI, pp. 1-288, 19 figs., 1 diag., 2 maps, 53 pls., 1918. (Printed also as a Thesis, Fac. des Sci., Paris, 1918.)

19. ARTHUR, J. C. . . New species of Uredineae II. Bull. Torrey Bot. Club, XXIX, pp. 227-231, 1902.

20. —— . . . Uredinales. North American Flora, VII, pp. 83-848, 1907-1927. [With collaborators.]

21. —— . . . Relationship of the genus Kuehneola. Bull. Torrey Bot. Club, XLIV, pp. 501-511, 1917.

22. ASHBY, S. F. . . Macrophomina phaseoli (Maubl.) comb. nov. the pycnidial stage of Rhizoctonia bataticola (Taub.) Butl. Trans. Brit. Myc. Soc., XII, pp. 141-147, 1 fig., 1927.

23. ASHPLANT, H. T. . . Annual report for 1923-25. Planters' Chron., XX, pp. 440-444, 465-469, 1925; reprinted in Bull. Rubber Growers' Assoc., VII, pp. 545-554, 1925.

24. ASHWORTH, J. H. . On Rhinosporidium seeberi (Wernicke 1903), with special reference to its sporulation and affinities. Trans. Roy. Soc. Edinburgh, LIII, II, 16, pp. 301-338, 5 pls., 1 fig., 1923.

25. AYYANGAR, C. R. . A leaf spot and blight disease of onions caused by Alternaria palandui nov. sp. Agric. Res. Inst., Pusa, Bull. 179, 14 pp., 2 pls., 1 graph, 1928.

26. BAGCHEE, K. D. . . Investigations on the infestations of 'Peridermium complanatum', Barclay, on the needles, and of Peridermium himalayense n. sp., on the stem of Pinus longifolia Roxb. Part I. Distribution, pathological study of the infections, and morphology of the parasites. Indian Forest Records (Bot. Ser.), XIV, 3, 24 pp., 14 pls., 1929.

27. —— . . A new species of Cronartium from the Himalayas. Nature, CXXIV, pp. 691-692, 2 figs., 1929.

28. BAL, S. N., CHOWDHURY, H. P., BANERJEE, K. G., & DUTTA, A. C. Commentationes Mycologicae. Journ. Dept. Sci., Univ. Calcutta, I, No. 1, Vol. I, pp. 1-4, 1 pl., 1919; II, Nos. 2-4, Vol. II. pp. 1-9, 3 pls., 1919-20; III, Nos. 5-7, Vol. II. pp. 31-36, 2 pls., 1921; IV, Nos. 8-10, Vol. III, pp. 1-8, 1 pl., 1921; V, (A few species of Meliola growing on different host plants). Ibid., Vol. IV, pp. 1-7, 1922.

29. BARBER, C. A. . . Diseases of Andropogon sorghum in the Madras Presidency. Bull. Dept. Land Records and Agric., Madras, II, No. 49, pp. 273-288, 1904.

30. BARCLAY, A. . . On a Uredine affecting the Himalayan spruce-fir (Abies smithiana, Forbes). Journ. Asiatic Soc. Bengal, LV, pp. 1-11, 3 pls., 1886.

31. —— . . On a second species of Uredine affecting Abies smithiana Forbes. Journ. Asiatic Soc. Bengal, LV, pp. 140-143, 2 pls., 1886.

32. —— . . On a new species of Uredine parasitic on Cedrus deodara. Journ. Asiatic Soc. Bengal, LV, pp. 223-226, 2 pls., 1886.

33. —— . . On the life-history of a new Aecidium on Strobilanthes dalhousianus Clarke. Scient. Mem. Med. Officers Army of India, II, pp. 15-27, 2 pls., 1887.

34. —— . . Aecidium urticae Schum. var. himalayense. Scient. Mem. Med. Officers Army of India, II, pp. 29-38, 3 pls., 1887.

35. —— . . Descriptive list of the Uredineae occurring in the neighbourhood of Simla (Western Himalaya). I, Journ. Asiatic Soc. Bengal, LVI, pp. 350-375, 4 pls., 1887; II, ibid., LVIII, pp. 232-251, 3 pls., 1889; III, ibid., LIX, pp. 75-112, 4 pls., 1890.

36. —— . . On the life-history of a new Caeoma on Smilax aspera Linn. Scient. Mem. Med. Officers Army of India, IV, pp. 37-46, 2 pls., 1889.

37. [BARCLAY A.] . . On the life-history of Puccinia geranii-sylvatica Karst. var. himalensis. Ann. Bot., V, pp. 27-36, 1 pl., 1890.

38. —— . . On the life history of a Himalayan Gymnosporangium (G. cunninghamianum n. sp.). Scient. Mem. Med. Officers Army of India, V, pp. 71-78, 3 pls., 1890.

39. —— . . On a Chrysomyxa on Rhododendron arboreum Sm. (Chr. himalense n. sp.). Scient. Mem. Med. Officers Army of India, V, p. 79-86, 2 pls., 1890.

40. —— . . On the life-history of a Uredine on Rubia cordifolia L. (Puccinia collettiana nov. sp.). Scient. Mem. Med. Officers Army of India, V, pp. 87-91, 1 pl., 1890.

41. —— . . On some rusts and mildews in India. Journ. Bot., XXVIII, pp. 257-261, 1 pl., 1890.

42. —— . . Description of a new fungus, Aecidium esculentum n. sp., on Acacia eburnea. Journ. Bombay Nat. Hist. Soc., V, pp. 161-165, 1 pl. ; note added by D. Prain, pp. 165-167, 1 pl., 1890.

43. —— . . Additional Uredineae from the neighbourhood of Simla. Journ. Asiatic Soc. Bengal, LX, p. 211-230, 2 pls., 1891.

44. —— . . On two autoecious Caeomata in Simla. Scient. Mem. Med. Officers Army of India, VI, pp. 65-69, 1 pl., 1891.

45. —— . . Rhododendron Uredineae. Scient. Mem. Med. Officers Army of India, VI, pp. 71-74, 2 pls., 1891.

46. —— . . On the life-history of a remarkable Uredine on Jasminum grandiflorum L. (Uromyces cunninghamianus nov. sp.). Trans. Linn. Soc. London, II Ser., Bot., III, pp. 141-151, 2 pls., 1891.

47. —— . . On the life-history of Puccinia coronata var. himalayensis. Trans. Linn. Soc. London, II Ser., Bot., III, pp. 227-236, 13 figs., 1891.

48. —— . . On the life-history of Puccinia jasmini-chrysopogonis, nov. sp. Trans. Linn. Soc. London, II Ser., Bot., III, pp. 237-242, 10 figs., 1891.

49. —— . . Rust and mildew in India. Journ. Bot., XXX, pp. 1-8, 40-49, 1892.

50. BECK VON MANNAGETTA, G. Fungi, *in* Itinera principum Sax Coburgi, II, pp. 144-148, 1 pl. 1888.

51. BERKELEY, M. J. . Descriptions of exotic fungi in the collection of Sir W. J. Hooker, from memoirs and notes of J. F. Klotzsch, with additions and corrections. Ann. Nat. Hist., III, pp. 375-401, 1 pl., 1839.

52. —— . . . Enumeration of fungi, collected by Dr. Hostmann, in Surinam. London Journ. Bot., I, pp. 138-157, 2 pls., 1842.

53. —— . . . Notices of fungi in the Herbarium of the British Museum. Ann. & Mag. Nat. Hist., X, Suppl., pp. 370-385, 4 pls., 1843.

54. —— . . . [Fir disease]. Gard. Chron., No. 40, p. 627, 1 fig., 1852.

55. —— . . . [On Ravenelia]. Gard. Chron., 1853, p. 132, 1853.

56. —— . . . *in* Hooker, J. D., Flora Novae-zeylandiae, II, pp. 172-210. 2 pls., 1855.

57. —— . . . Decades of fungi, Decas 1-62, Nos. 1-620. In Hooker's London Jour. Bot., III-VIII, 1844-1856.

58. —— . . . On two tuberiform vegetable productions from Travancore. Trans. Linn. Soc. London, XXIII, pp. 91-92, 1862.

59. —— . . . On some new fungi from Mexico. Journ. Linn. Soc. London, Bot., IX, pp. 423-425, 1 pl., 1865.

60. —— . . . [Polyporus anthelminticus]. Gard. Chron., 1866, p. 753, 1866.

61. —— . . . Fungi of the Plains of India. Intellectual Observer, XII, pp. 18-21, 5 figs., 1867.

62. —— . . . Australian fungi, received principally from Baron von Mueller and Dr. R. Schomburgh. Jour. Linn. Soc. London, Bot., XIII, pp. 155-177, 1873.

63. —— . . . Thread blight in tea plantations. Gard. Chron., 1873, pp. 810-811, June 14, 1873.

64. —— . . . On the thread blight of tea. Quart. Jour. Micro. Sci., N. S. XV, pp. 130-133, 4 figs., 1875.

65. —— . Three fungi from Kashmir. Grev., IV, pp. 137-138, 1876.

66. —— . . . Three new Indian fungi. Grev., XI, pp. 39-40, 1882.

67. —— & BROOME, C. E. The fungi of Ceylon. I. Jour. Linn. Soc. London, Bot., XI, pp. 494-572, 1871 ; II, ibid., XIV, pp. 29-140, 9 pls., 1875 ; supplement, ibid., XV, pp. 82-86, 1877.

68. Blatter, E. . . A list of Indian fungi, chiefly of the Bombay Presidency, with the description of two new species. Journ. Bombay Nat. Hist. Soc., XXI, pp. 146-152, 1 pl., 1911.

69. Blumer, S. . . Variationsstatistische Untersuchungen an Erysiphaceen. Ann. Myc., XXIV, pp. 179-193, 1926.

70. Bose, S. R. . . I. Description of fungi in Bengal. Proc. Indian Assoc. Cult. Sci., IV, pp. 109-114, pls., 1919. II. Ibid. Proc. Sci. Convention Indian Assoc. Cult. Sci. for the year 1918, pp. 136-143, 12 pls., 1920. III. Fungi of Bengal. Polyporaceae of Bengal, Part III. Bull. Carmichael Med. Coll., Belgachia, I, pp. 1-5, 7 pls., 1920. IV. Polyporaceae of Bengal, Part IV. Ibid., II, pp. 1-5, 13 pls., 1921. V, ibid., Part V. Ibid., III, pp. 20-25, 9 pls., 1922. VI, ibid., Part VI, Proc. Sci. Convention, Indian Assoc. Cult. Sci. for the year 1919. VII, ibid., Part VII. Ibid., for the year 1920-21, pp. 27-36, 1923. VIII, ibid., Part VIII. Journ. Dept. Sci., Calcutta Univ., IX, pp. 27-34, 5 pls., 1928. IX, ibid., Part IX. Ibid., IX, pp. 35-44, 5 pls., 1928.

71. —— . . Records of Agaricaceae from Bengal. Journ. Asiatic Soc. Bengal, N. S., XVI, pp. 347-354, 1 pl., 1920.

72. —— . . Spore-culture of Panaeolus cyanescens B. & Br. Calcutta Univ. Sci. Ser., Sir A. Mukerji Silver Jubilee, III, pp. 81-82, 1 col. pl., ? 1920.

73. —— . . One new species of Polyporaceae and some Polypores new to Bengal. Ann. Myc., XIX, pp. 129-131, 3 pls., 1921.

74. —— . . Possibilities of mushroom industry in India by cultivation. Agric. Journ. India, XVI, pp. 643-647, 1921.

75. —— . . Two new species of Polyporaceae. Journ. Indian Bot., II, pp. 300-301, 2 pls., 1921.

76. —— . . Une Polyporacée nouvelle du Bengale. Bull. Soc. Myc. de France, XXXVIII, p. 173, 1922. (Figure, Pl. VIII, vol. XXXIX, 1923).

77. —— . . Une Polyporacée nouvelle de l'Inde. Bull. Soc. Myc. de France, XXXIX, p. 226, 1 pl., 1923.

78. [Bose, S. R.] . . The fungi cultivated by the termites of Barkuda. Records Indian Museum, XXV, 2, pp. 253-258, 1 fig., 1 pl., 1923.

79. —— . . Les Polyporacées du Bengale. Revue Path. Veg. et Ent. Agric., XI, pp. 134-149, 1924.

80. —— . . A new species of Polyporaceae from Bengal. Ann. Myc., XXIII, pp. 179-181, 2 figs., 1925.

81. —— . . Revival of an old fruit body of Hexagonia discopoda Pat. & Hariot, and successful spore culture from its fresh spore-discharge. Ann. Myc., XXVII, pp. 321-323, 2 figs., 1929.

82. Brefeld, O. . . Untersuchungen aus dem Gesammtgebiete der Mykologie, Heft XII, pp. 99-236, 7 pls., 1895.

83. Bresadola, G. . . Selecta mycologica. I. Diagnoses specierum novarum. Ann. Myc., XVIII, pp. 26-58, 1920.

84. —— . . Synonymia et adnotanda mycologica. Ann. Myc., XIV, pp. 221-242, 1916 ; ibid., XVIII, pp. 58-70, 1920.

85. —— . . Selecta mycologica. II. Studi Trentini, Ser. II, Sci. Nat. ed Econ., VII, 1, pp. 1-31 [of reprint], 1 col. pl., 1926.

86. Burns, W. . . . First experiments in the treatment of grapevine mildew in the Bombay Presidency. Bull. Dept. Agric. Bombay, 1910, 14 pp., 1 pl., 1910.

87. Butler, E. J. . . Potato diseases of India. Agric. Ledger, Crop Disease and Pest Series, No. 7, pp. 87-124, 8 figs., 1903.

88. —— . . A deodar disease in Jaunsar. Indian Forester, XXIX, Appendix Ser., 8 pp., 2 pls., 1903.

89. —— . . Pilzkrankheiten in Indien im Jahre 1903. Zeitsch. für Pflanzenkr., XV, pp. 44-48, 1905.

90. —— . . Some Indian forest fungi. Indian Forester, XXXI, pp. 487-496, 548-556, 611-617, 670-679, 14 figs., 1905. [Reprint, pp. 1-32, 1905.]

91. —— . . The wilt disease of pigeon pea and pepper. Agric. Journ. India, I, 1, pp. 25-36, 5 pls., 1906.

92. —— . . Some diseases of palms. Agric. Journ. India, I, 4, pp. 299-310, 2 pls., 1906.

93. —— . . Fungus diseases of sugarcane in Bengal. Mem. Dept. Agric. India, Bot. Ser., I, 3, pp. 1-53, 11 pls. (2 col.), 1906.

94. [BUTLER, E. J.] . An account of the genus Pythium and some Chytridiaceae. Mem. Dept. Agric. India, Bot. Ser., I, 5, 160 pp., 10 pls., 1907.

95. —— . . Some diseases of cereals caused by Sclerospora graminicola. Mem. Dept. Agric. India, Bot. Ser., II, 1, 24 pp., 5 pls., 1907.

96. —— . . Report on coconut palm disease in Travancore. Agric. Res. Inst., Pusa, Bull. 9, 23 pp., 1908.

97. —— . . Fomes lucidus (Leys.) Fr., a suspected parasite. Indian Forester, XXXV, pp. 514-518, 1 pl., 1909.

98. —— . . The mulberry disease caused by Coryneum mori Nom. in Kashmir, with notes on other mulberry diseases. Mem. Dept. Agric. India, Bot. Ser., II, 8, 18 pp., 2 figs., 3 pls., 1909.

99. —— . . The bud-rot of palms in India. Mem. Dept. Agric. India, Bot. Ser., III, 5, 64 pp., 5 pls., 1 map, 1910.

100. —— . . A new genus of the Uredinaceae. Ann. Myc., VIII, pp. 444-448, 1 pl., 1910.

101. —— . . The wilt disease of pigeon-pea and the parasitism of Neocosmospora vasinfecta Smith. Mem. Dept. Agric. India, Bot. Ser., II, 9, 64 pp., 6 pls. (2 col.), 1910.

102. —— . . The leaf spot of tumeric (Taphrina maculans sp. nov.) Ann. Myc., IX, pp. 36-39, 1 pl., 1 fig., 1911.

103. —— . . On Allomyces, a new aquatic fungus. Ann. Bot., XXV, pp. 1023-1034, 1 pl., 1911.

104. —— . . The rusts of wild vines in India. Ann. Myc., X, pp. 153-158, 1 fig., 1912.

105. —— . . Diseases of rice. Agric. Research Inst., Pusa, Bull. 34, pp. 1-37, 3 pls., 3 figs., 1913.

106. —— . . Pythium de Baryanum Hesse. Mem. Dept. Agric. India, Bot. Ser., V, pp. 262-267, 1 pl., 1913.

107. —— . . The downy mildew of maize (Sclerospora maydis (Rac.) Butler). Mem. Dept. Agric. India, Bot. Ser., V, 5, pp. 275-280, 2 pls. (1 col.), 1913.

108. —— . . Notes on some rusts in India. Ann. Myc., XII, pp. 76-82, 1914.

109. —— . . Rotting of pomegranates. Agric. Journ. India, IX, pp. 205-206, 1914.

110. [BUTLER, E. J.] . Tikka disease and the introduction of exotic groundnuts into the Bombay Presidency. Agric. Journ. India, IX, pp. 59-70, 1 pl., 1914.

111. —— . . Fungi and disease in plants. Thacker, Spink & Co., Calcutta, VI+547 pp., 206 figs., 1918.

112. —— . . The wilt diseases of cotton and sesamum in India. Agric. Journ. India, XXI, 4, pp. 268-273, 1 pl., 1926.

113. —— ET AL. . Reports of the Imperial Mycologist, in Reports of the Agricultural Research Institute and College, Pusa, (1) 1907-09, by E. J. Butler and W. McRae, pp. 63-68, 1909 ; (2) 1909-10, by E. J. Butler, pp. 40-47, 1910 ; (3) 1910-11, by E. J. Butler, pp. 50-57, 1912 ; (4) 1911-12, by E. J. Butler, pp. 54-64, 1913 ; (5) 1912-13, by E. J. Butler, pp. 55-68, 1914 ; (6) 1913-14, by F. J. F. Shaw, pp. 48-60, 1914 ; (7) 1914-15, by F. J. F. Shaw, pp. 51-58, 1916 ; (8) 1915-16, by E. J. Butler, pp. 44-57, 1916 ; (9) 1916-17, by E. J. Butler, pp. 52-70, 1917 ; (10) 1917-18, by F. J. F. Shaw, pp. 71-83, 1918 ; (11) 1918-19, by E. J. Butler, pp. 68-85, 1919 ; (12) 1919-20, by E. J. Butler, pp. 58-67, 1920 ; (13) 1920-21, by F. J. F. Shaw, pp. 34-40, 1921 ; (14) 1921-22, by W. McRae, pp. 44-50, 1922 ; (15) 1922-23, by W. McRae, pp. 53-60, 1923 ; (16) 1923-24, by W. McRae, pp. 41-51, 1924 ; (17) 1924-25, by M. Mitra, pp. 45-57, 1925 ; (18) 1925-26, by W. McRae, pp. 54-69, 1926 ; (19) 1926-27, by W. McRae, pp. 45-55, 1 graph, 1928 ; (20) 1927-28, by W. McRae, pp. 56-70, 1928 ; (21) 1928-29, by W. McRae, pp. 51-66, 1930.

114. —— & HAFIZ KHAN, A. Red rot of sugarcane. Mem. Dept. Agric. India, Bot. Ser., VI, 5, pp. 151-178, 1 pl., 1913.

115. —— —— . . Some new sugarcane diseases. Mem. Dept. Agric. India, Bot. Ser., VI, 6, pp. 181-208, 6 pls. (1 col.), 1 fig., 1913.

116. —— & HAYMAN, J. M. Indian wheat rusts. Mem. Dept. Agric. India, Bot. Ser., I, 2, 58 pp., 1 graph, 5 pls. (4 col.), 1906.

117. —— & KULKARNI, G. S. Colocasia blight caused by Phytophthora colocasiae Rac. Mem. Dept. Agric. India, Bot. Ser., V, 5, pp. 233-261, 4 pls. (1 col.), 1913.

118. CHAMPION, H. G. . Notes on the death of the Chir (Pinus longifolia) poles in the Almora plantations of Kumaon. Indian Forester, XLVIII, pp. 168-174, 232-246, 1922.

119. CHAUDHURI, H. . . Annual partial wilting in Hibiscus tiliaceus. Journ. Indian Bot. Soc., VI, pp. 109-112, 1927.

120. CHIBBER, H. M. . . A working list of diseases and vegetable pests of some of the economic plants occurring in the Bombay Presidency. Poona Agric. College Mag., II, 3, 18 pp., 1911.

121. CIFERRI, R. . . Quarta contribuzione allo studio degli Ustilaginales. Ann. Myc., XXVI, pp. 1-68, 1 pl., 1928.

122. COLEMAN, L. C. . . Diseases of the areca palm (Areca catechu L.). I. Koleroga or rot-disease. Ann. Myc., VIII, pp. 591-626, 3 pls. (1 col.), 1910.

123. —— . . Diseases of the areca palm. I. Koleroga. Mysore State Dept. Agric., Myc. Ser., Bull., II, 92 pp., 5 figs., 18 pls. (2 col.), 1910.

124. —— . . The control of koleroga of the areca palm, a disease caused by Phytophthora omnivora var. arecae. Agric. Journ. India, X, pp. 129-136, 1915.

125. —— & KANNAN, K. K. Some scale insect pests of coffee in south India. Mysore State Dept. Agric., Ent. Ser., Bull. 4, 1918.

126. ——, VENKATA RAO, M. K., & NARASIMHAN, M. J. Black rot or koleroga of coffee in Mysore. Mysore State Dept. Agric., Myc. Ser., Bull. 5, 12 pp., 4 pls., 1 fig., 1923.

127. COLLEY, R. H. & TAYLOR, MINNIE W. Peridermium kurilense Dict., on Pinus pumila Pall., and Peridermium indicum n. sp. on Pinus excelsa Wall. Journ. Agric. Res., XXXIV, pp. 327-330, 1 fig., 1927.

128. COOKE, M. C. . . Kashmir morels. Trans. Bot. Soc. Edinburgh, X, pp. 439-443, 2 figs., 1870 ; also in Pharmac. Journ. and Trans., London, III Ser., I, pp. 345-346, 2 figs., 1870.

129. —— . . Blights on tea and cotton. Grev., I, p. 90, 1872.

130. —— . . Himalayan leaf fungi. Grev., III, pp. 75-76, 1874.

131. —— . . Mycographia, seu Icones Fungorum. London, 1875-78.

132. —— . . Some Indian fungi. Grev., IV, pp. 114-118, 1 pl., 1876.

133. [COOKE, M. C.] . . Some Indian fungi. Grev., V, pp. 14-17, 1 pl., 1876.
134. —— . . Reporton diseased leaves of coffee and other plants. Indian Museum Report. pp. 1-7, 1 col. pl., 1876.
135. —— . . Two coffee diseases. Popular Sci. Review, XV, pp. 161-168, 1 pl., 1876.
136. —— . . Affinities of Pellicularia. Grev., IV, pp. 134-135, 1876.
137. —— . . Some parasites of coniferae. Indian Forester, III, pp. 88-96, 1877.
138. —— . . Orange mould on forest trees. Grev., V, pp. 145-147, 1877.
139. —— . . Praecursor ad monographiam Hendersoniae. Nuovo Giorn. Bot. Ital., X, pp. 17-27, 1878.
140. —— . . Some Indian fungi. Grev., VI, pp. 117-118, 1878.
141. —— . . Some extra-European fungi. Grev., VII, pp. 13-15, 1878.
142. —— . . Some Himalayan fungi. Grev., VII, p. 61, 1878.
143. —— . . Some exotic fungi. Grev., VII, pp. 94-96, 1879.
144. —— . . Fungi of India. Grev., VIII, pp. 93-96, 1880.
145. —— . . Exotic fungi. Grev., IX, pp. 10-15, 1880.
146. —— . . The genus Ravenelia. Journ. Royal Microsc. Soc., III, pp. 384-389, 1 pl., 1880.
147. —— . . On Hymenochaete and its allies. Grev., VIII, pp. 145-150, 1880.
148. —— . . Some exotic fungi. Grev., IX, pp. 97-101, 1881.
149. —— . . Exotic fungi. Grev., X, pp. 121-130, 1882.
150. —— . . On Xylaria and its allies. Grev., XI, pp. 81-94, 1883.
151. —— . . Hypoxylon and its allies. Grev., XI, pp. 121-140, 1883.
152. —— . . Nummularia and its allies. Grev., XII, pp. 1-8, 1883.
153. —— . . The genus Anthostoma. Grev., XII, pp. 49-53, 1883.
154. —— . . On Sphaerella and its allies. Journ. Bot., XXI, pp. 67-71, 106-110, 136-139, 1883.
155. —— . . Notes on Hypocreaceae. Grev., XII, pp. 77-83, 100-101, 1884.
156. —— . . Some exotic fungi. Grev., XII, p. 85, 1884.
157. —— . . Fungi of Perak. Grev., XIII, pp. 1-4, 1884.
158. —— . . Some exotic fungi. Grev., XIII, p. 6, 1884.

159. [COOKE, M. C.] . . Synopsis Pyrenomycetum. Grev., XIII, pp. 8-16, 41-45, 61-72, 100-109, 1884-1885.

160. —— . . Some remarkable moulds. Journ. Quekett Microsc. Club, Ser. 2, II, pp. 138-143, 2 pls., 1885.

161. —— . . Some exotic fungi. Grev., XIV, pp. 11-14, 1885.

162. —— . . Some exotic fungi. Grev., XIV, pp. 89-90, 1886.

163. —— . . Praecursores ad monographia Polypororum. Grev., XV, pp. 19-27, 50-60, 1886.

164. —— . . Synopsis Pyrenomycetum. Grev., XV, pp. 80-86, 122-125, 1887.

165. —— . . Some exotic fungi. Grev., XVI, pp. 15-16, 1887.

166. —— . . Some exotic fungi. Grev., XVI, pp. 25-26, 1887.

167. —— . . Some exotic fungi. Grev., XVI, pp. 69-72, 1888.

168. —— . . Exotic Agarics. Grev., XVI, pp. 105-106, 1888.

169. —— . . Some exotic fungi. Grev., XVI, p. 121, 1888.

170. —— . . New exotic fungi. Grev., XVII, pp. 42-43, 1888.

171. —— . . Some exotic fungi. Grev., XVIII, pp. 34-35, 1889.

172. —— . . Some Asiatic fungi. Grev., XIX, p. 7, 1890.

173. —— . . Omitted diagnoses. Grev., XIX, pp. 71-75, 1891.

174. —— . . Additions to Daedalea. Grev., XIX, pp. 92-93, 1891.

175. —— . . Trametes and its allies. Grev., XIX, pp. 98-103, 1891.

176. —— . . Some omitted diagnoses. Grev., XIX, p. 104, 1891.

177. —— . . Favolus and Laschia. Grev., XIX, p. 105, 1891.

178. —— . . Species of Hydnei. Grev., XX, pp. 1-4, 1891.

179. —— . . Notes on Tremellini. Grev., XX, p. 15, 1891.

180. —— . . Himalayan truffles. Grev., XX, p. 67, 1892. [With G. MASSEE.]

181. —— . . Neglected diagnoses. Grev., XX, pp. 81-85, 1892.

182. —— . . New exotic fungi. Grev., XX, pp. 90-92, 1892.

183. —— . . Exotic fungi. Grev., XXI, pp. 73-75, 1893.

184. —— . . Omitted diagnoses. Grev., XXI, p. 76, 1893.

185. CORDA, A. C. J. . . Icones fungorum hucusque cognitorum. Prag., I-VI, many plates, 1837-1854.

186. CUNNINGHAM, D. D. . *In* Sanitary Commissioner with the Government of India, Seventh Annual Report, 1870, Appendix B, p. 230, pl. XI, fig. 13, 1871.

187. —— . On the occurrence of conidial fructification in the Mucorini, illustrated by Choanephora. Trans. Linn. Soc., II Ser., Bot., I, pp. 409-422, 1 pl., 1879.

188. [CUNNINGHAM, D. D.] . On a new genus of the family Ustilagineae. Sci. Mem. Med. Officers Army of India, III, pp. 27-32, 2 col. pls., 1887.

189. —— . Notes on the life-history of Ravenelia sessilis B. and Ravenelia stictica B. and Br. Scient. Mem. Med. Officers Army of India, IV, pp. 20-36, 2 col. pls., 1889.

190. —— . A new and parasitic species of Choanephora. Ann. Royal Bot. Gard. Calcutta, VI, pp. 163-172, 2 pls., 1895.

191. —— On certain diseases of fungal and algal origin affecting economic plants in India. Sci. Mem. Med. Officers Army of India, X, pp. 95-130, 6 col. pl., 1897.

192. —— & PRAIN, D. A note on Indian wheat rusts. Records Bot. Survey India, I, pp. 99-124, 1896.

193. CUNNINGHAM, G. H. . A revision of the New Zealand Nidulariales, or "Bird's-nest fungi." Trans. New Zealand Inst., LV, pp. 59-66, 2 pls., 1924.

194. CURREY, F. . . Mycological notes. Quart. Journ. Microscop. Sci., VII, pp. 225-235, 1 pl., 1859.

195. —— . . On a new genus in the order Mucedines. Journ. Linn. Soc. London, Bot., XIII, pp. 333-334 and 578, 1 pl., 1873.

196. —— . . On a collection of fungi made by Mr. Sulpiz Kurz, Curator of the Botanic Garden, Calcutta. Trans. Linn. Soc. London, II Ser., Bot., I, pp. 119-131, 3 col. pls., 1874.

197. —— & HANBURY, D. Remarks on Sclerotium stipitatum Berk. & Curr., Pachyma cocos Fr. and some similar productions. Trans. Linn. Soc. London, XXIII, pp. 93-97, 2 pls., 1862.

198. DASTUR, J. F. . . On Phytophthora parasitica nov. spec., a new disease of the castor oil plant. Mem. Dept. Agric. India, Bot. Ser., V, 4, pp. 177-231, 10 pls. (1 col.), 1913.

199. —— . . The potato blight in India. Mem. Dept. Agric. India, Bot. Ser., VII, 3, pp. 163-176, 1 pl., 1915.

200. —— . . Black thread disease of Hevea in Burma. Burma Dept. Agric., Bull. 14, 4 pp., 1 pl., 1916.

201. —— . . Spraying for ripe rot of the plantain fruit. Agric. Journ. India, XI, pp. 142-149, 2 figs., 1916.

202. [DASTUR, J. F.] . . Phytophthora sp. on Hevea brasiliensis. Mem. Dept. Agric. India, Bot. Ser., VIII, 5, pp. 217-232, 2 figs., 1916.

203. —— . . Phytophthora on Vinca rosea. Mem. Dept. Agric. India, Bot. Ser., VIII, 6, pp. 233-242, 14 figs., 1916.

204. —— . Conditions influencing the distribution of potato blight in India. Agric. Journ. India, Special Indian Science Congress Number, 1917, pp. 90-96, 1917.

205. —— . . Glomerella cingulata (Stoneman) Spauld. and v. Sch. and its conidial forms, Gloeosporium piperatum E. & E. and Colletotrichum nigrum E. and Hals., on chillies and Carica papaya. Ann. Appl. Biol., VI, pp. 245-268, 1 pl., 1920.

206. —— . . The mode of infection by smut in sugar-cane. Ann. Bot., XXXIV, pp. 391-397, 1920.

207. —— . . Choanephora cucurbitarum (B. & Rav.) Thaxter, on Chillies (Capsicum spp.). Ann. Bot., XXXIV, pp. 399-403, 1 pl., 1920.

208. —— . . Die-back of chillies (Capsicum spp.) in Bihar. Mem. Dept. Agric. India, Bot. Ser., XI, 5, pp. 129-144, 2 pls., 1921.

209. —— . . A short note on the foot-rot disease of Pan in the Central Provinces. Agric. Journ. India, XXII, pp. 105-108, 1927.

210. DIETEL, P. . . . Ueber die Gattung Pileolaria Cast. Jena Geog. Ges. Mitth. VIII, (Bot. Ver. Gesamtthuringen), pp. 20-25, 1890.

211. —— . . . Uredineen aus dem Himalaya. Hedw., XXIX, pp. 259-270, 1 pl., 1890.

212. —— . . . Die Gattung Ravenelia. Hedw., XXXIII, pp. 22-69, 5 pls., 1894.

213. —— . . . Drei neue Uredineengattungen, Masseeella, Phakopsora, und Schizospora. Ber. deut. Bot. Ges., XIII, pp. 332-335, 2 figs., 1895.

214. —— . . . Uredineae japonicae I.—Engler's Bot. Jahrb., XXVII, pp. 564-576, 1 pl., 1899.

215. —— . . . Uredinales, *in* Engler u. Prantl, Die naturlichen Pflanzenfamilien, I**, pp. 24-81, many figs., 1900.

216. —— . . . Über die Uredineengattung Pucciniostele Tranzschel et Komarov. Ann. Myc., II, pp. 20-26, 1904.

217. [DIETEL, P.] . . . Über die Arten der Gattung Phragmidium. Hedw., XLIV, pp. 112-132, 1 pl., 1905.

218. —— . . . . Monographie der Gattung Ravenelia Berk. Beih. Bot. Centralbl., XX, pp. 343-413, 2 pls., 1906.

219. —— . . . . Beschreibungen einiger neuer Uredineen. Ann. Myc., IV, pp. 303-308, 1906.

220. —— . . . . Über Chnoopsora, eine neue Uredineen-Gattung. Ann. Myc., IV, pp. 421-423, 1 fig., 1906.

221. —— . . . . Uredineen aus Japan. I, Ann. Myc., V, pp. 70-77, 1907 ; II, ibid., VI, pp. 222-229, 1 fig., 1908 ; III, ibid., VIII, pp. 304-314, 1910.

222. —— . . . . Zwei neue Arter der Gattung Phakopsora. Ann. Myc., VIII, p. 469, 1910.

223. —— . . . . Eine Bemerkung über Uredo cronartiiformis Barcl. Ann. Myc., X, pp. 385-386, 1912.

224. —— . . . . Hemibasidii, *in* Engler u. Prantl., Die Natürlichen Pflanzenfamilien, II Aufl., Band 6, pp. 1-98, 80 figs., 1928.

225. DIETZ, S. M. . . . The alternate hosts of crown rust, Puccinia coronata Corda. Journ. Agric. Res., XXIII, pp. 953-970, 4 figs., 1926.

226. DODGE, C. W. . . . The higher Plectascales. Ann. Myc., XXVII, pp. 145-184, 2 pls., 2 figs., 1929.

227. ELLIS, J. B. & EVERHART, B. M. New species of North American fungi from various localities. Bull. Torrey Bot. Club, XXIV, pp. 277-292, 1897.

228. FAIRLEY, N. H., MACKIE, F. P., CHITRE, G. D., GOKHALE, S. K., GORE, S. N., MALANDKAR, M. A., & SACASA, F. J. A progress report on researches in sprue (1924-1925). Indian Journ. Med. Res., XIV, pp. 105-123, 1 diag., 1926.

229. FINLOW, R. S. . . . " Heart damage " in baled jute. Mem. Dept. Agric. India, Chem. Ser., V, 2, pp. 33-68, 1918.

230. FERRARIS, T. . . . Agricoltura e fitopatologia nel Kashmir. Curiamo le Piante !, VI, 5, pp. 81-86, 1928.

231. FISCHER, A. (DE WALDHEIM). Les Ustilaginées et leurs plantes nourricières. Ann. Soc. Nat., VI Sér., IV, pp. 190-276, 1890.

232. FISCHER, ED. . . . Beiträge zur Kenntnis exotischer Pilze. Hedw., XXIX, pp. 161-171, 1 col. pl., 1890.

233. —— . . . Zur Kenntnis von Graphiola und Farysia. Ann. Myc , XVIII, pp. 188-197, 7 figs., 1920.

234. [FISCHER, ED.] . . Weitere Beiträge zur Kenntnis der Gattung Graphiola. Ann. Myc., XX, pp. 228-237, 4 figs., 1922.

235. FITZPATRICK, H. M. . Monograph of the Coryneliaceae. Mycologia, XII, pp. 206-237, 239-267, 7 pls., 1 table, 1920.

236. FLEMING, J. . . On some microscopic leaf fungi from the Himalayas. Monthly Microsc. Journ., XII, pp. 270-274, 1874.

237. FRIES, E. . . . Systema mycologicum. Vol. I, LVII+520 pp., 1821, Vol. II, 624 pp., 1823 ; Vol. III, 524+202 pp., 1829.

238. —— . . . Eclogae fungorum praecipue ex herbariis Germanorum descriptorum. Linnaea, V, pp. 497-553, 1830.

239. —— . . . Epicrisis systematis mycologici, seu synopsis Hymenomycetum. Upsala, XII+612 pp., 1836-1838.

240. —— . . . Novae symbolae mycologicae. Nova Acta Regiae Societatis Sci., Upsala, III Ser., I, pp. 17-136, 1855.

241. GAILLARD, A. . . Le genre Meliola, pp. 1-163, supplement pp. 1-15, 16 pls., Paris, 1892.

242. GAMBLE, J. S. . . On the determination of the fungi which attack forest trees in India. Indian Forester, XXV, pp. 431-438, 1899.

243. GANDHI, S. R. . . A grape vine disease new to India, Dematophora necatrix Hartig. Poona Agric. Coll. Mag., XIX, 4, pp. 205-207, 1928.

244. GÄUMANN, E. . . Über die Gattung Kordyana Rac. Ann. Myc., XX, pp. 257-271, 7 figs., 1922.

245. —— . . Mykologische Mitteilungen. III. Ann. Myc., XXV, pp. 167-177, 1927.

246. GIESENHAGEN, K. . Ueber Hexenbesen an tropischen Farnen. Flora, LXXVI, pp. 130-156, 2 pls., 1892.

247. GLASSON, A. K. . . Mortality of Sal in Buxa Division, Bengal. Indian Forester, XLVIII, pp. 22-31, 1922.

248. GRAHAM, R. J. D. . Report of the Economic Botanist, in Report of Agricultural College at Nagpur, of Botanical and Chemical Research, etc., Dept. of Agric., Central Provinces and Berar, for the year 1914-15. Govt. Press, Nagpur, pp. 11-17, 1915.

249. GROVE, W. B. . . Species placed by Saccardo in the genus Phoma. Kew Bull., 1919, pp. 425-445, 6 figs., 1919.

250. —— . . The British species of Cytospora. Kew Bull., 1923, pp. 1-30, 1923.

251. HAFIZ KHAN, A. . . Root infection of Trametes pini (Brot.) Fr. Indian Forester, XXXVI, pp. 559-562, 2 pls., 1910.

252. —— . . The artificial development of sporophores of Polyporus gilvus (Schw.) Fr. & Pat. Indian Forester, LI, pp. 205-207, 1925.

253. —— . . Inoculation of Chir (Pinus longifolia) with Coleosporium campanulae (Pers.) Lév., on Campanula canescens Wall., and Coleosporium inulae (Kunze) Ed. Fisch. on Inula cappa DC. Indian Forester, LIV, pp. 176-178, 1928.

254. —— . . A preliminary report on the Peridermiums of India and the occurrence of Cronartium ribicola Fisch. on Ribes rubrum Linn. Indian Forester, LIV, pp. 431-443, 8 pls., 1928.

255. HAIGH, J. C. . . Macrophomina phaseoli (Maubl.) Ashby and Rhizoctonia bataticola (Taub.) Butler. Ann. Royal Bot. Gard., Peradeniya, XI, pp. 213-249, 7 pls., 4 figs., 1930.

256. HARIOT, P. & KARSTEN, P. A. Fungi novi. Revue Myc., XII, pp. 128-129, 1890.

257. HECTOR, G. P. . . Annual Report of the Economic Botanist to the Government of Bengal for the year 1923-24. Ann. Rept. Dept. Agric. Bengal for the year 1923-24, Appendix 1, pp. I-V, 1925.

258. —— . . Annual Report of the Economic Botanist to the Government of Bengal for the year 1925-26. Ann. Rept. Dept. of Agric. Bengal for the year 1925-26, Appendix I, pp. I-VI, 1927.

259. —— . . Annual Report of the 1st Economic Botanist, Bengal, for the year 1926-27. Ann. Rept. Dept. of Agric. Bengal for the year 1926-27, Appendix I, pp. 35-38, 1927.

260. HENNINGS, P. . . Fungi Indiae orientalis. Hedw., XXXIX, pp. (150)-(153), 1900.

261. —— . . Einige neue Uredineen aus verschiedenen Gebieten. Hedw., XXXIX, pp. (153)-(155), 1900.

262. —— . . Fungi. Monsunia, I, pp. 1-38, 1 pl., 1900.

263. —— . Fungi Indiae orientalis II, cl. W. Gollan a. 1900 collecti. Hedw., XL, pp. 323-342, 1901.

264. HÖHNEL, F. v. . . Fragmente zur Mykologie: Mitteil.I-XXV, Nos. 1-1225. Sitzungsber. Akad. Wissensch. Wien, Mathem.-Naturw. Klasse, B. 111-132, 1902-1923.

265. HOLE, R. S. . . A manual of botany for Indian forest students. 282 pp., 20 pls., 3 maps, Calcutta, 1909.

266. —— . . Mortality of spruce in the Jaunsar forests, United Provinces. Indian Forester, LIII, pp. 434-443, 1927.

267. HUTCHINSON, C. M. & RAM AYYAR, C. S. The Indian rice beer ferment. Mem. Dept. Agric. India, Bact. Ser., I, 6, pp. 137-168, 2 pls. (1 col.), 1915.

268. ITO, SEIYA . . . On the Uredineae parasitic on the Japanese Gramineae. Journ. Sapporo Agric. College, III, pp. 179-262, 3 pls., 1909.

269. JACZEWSKI, A. A.. . A rare rust fungus, Chrysomyxa (Barclayella) deformans Jacz. Bull. Leningrad Forestry Inst. CXXXIII, pp. 131-149, 5 figs., 1926. [In Russian]

270. JOSHI, S. D. . . The wilt disease of safflower. Mem. Dept. Agric. India, Bot. Ser., XIII, 2, pp. 39-46, 3 pls., 1924.

271. KAR, P. C. . . . Bud-rot of Palmyra palm. Bengal Agric. Journ., I, pp. 110-111, 1921 [1922].

272. KLÖCKER, A. . . Recherches sur les organismes de fermentation. Comptes Rendus des Trav. Lab. de Carlsberg, X, 4, pp. 285-346, 7 pls., 1913.

273. KLOTZSCH, J. F. . . Mycologische Berichtigungen. Linnaea, VII, pp. 193-204, 3 pls., 1832.

274. —— . . Fungi exotici e collectionibus britannorum. Linnaea, VIII, pp. 478-490, 2 pls., 1833.

275. KNOWLES, R. AND DAS GUPTA, B. M. On the nature of Blastocladia hominis. Indian Journ. Med. Res., XII, pp. 31-37, 4 pls., 1924.

276. KOMAROV, W. L. . . Ueber Pucciniostele clarkiana (Barcl.) Tranz. et Kom. Hedw., XXXIX, pp. (121)-(123), 1900.

277. KULKARNI, G. S. . Preliminary study of the red rot of sugarcane in the Bombay Presidency. Bombay Dept. Agric., Bull. 44, 8 pp., 3 pls. (1 col.), 1911.

278. —— . Observations on the downy mildew [Sclerospora graminicola (Sacc.) Schroet.] of Bajri and Jowar. Mem. Dept. Agric. India, Bot. Ser., V, 5, pp. 268-273, 2 pls. (1 col.), 1913.

279. [KULKARNI, G. S.] . Smuts of Jowar (Sorghum) in the Bombay Presidency. Agric. Res. Inst., Pusa, Bull. 78, 26 pp., 6 pls., 1918.

280. —— . The smut of Nachani or Ragi (Eleusine coracana Gaertn.). Ann. Appl. Biol., IX, pp. 184-186, 2 figs., 1922.

281. —— . Conditions influencing the distribution of grain smut (Sphacelotheca sorghi) of Jowar (Sorghum) in India. Agric. Journ. India, XVII, pp. 159-162, 1922.

282. —— . Smut (Ustilago paradoxa Syd. and Butl.) on Sawn (Panicum frumentaceum Roxb.). Journ. Indian Bot. Soc., III, pp. 10-11, 1922.

283. —— & MUNDKUR, B. B. Studies on the wilt disease of Cotton in the Bombay Karnatak. Series I. Mem. Dept. Agric. India, Bot. Ser., XVII, 2, 27 pp., 4 pls., 1928.

284. LÉVEILLÉ, J. H. . . Fungi, *in* Voyage dans l' Inde par Victor Jacquemont pendant les années 1829-1832, Tome IV, p. 179, 2 figs., 1844.

285. —— . . . Champignons exotiques. Ann. Sci. Nat., III Sér., II, pp. 167-221, 1844, and ibid., III, pp. 38-71, 1845.

286. —— . . . Description des champignons de l'herbier du Muséum de Paris. Ann. Sci. Nat., III Sér., V, pp. 111-167, 249-304, 1846.

287. LLOYD, C. G. . . . Mycological Notes, Nos. 1-75, pp. 1-1364, many pls. and figs., Cincinnati, Ohio, 1898-1925.

288. —— . . . Mycological Letters, 1-69, (each separately paged), Cincinnati, Ohio, 1904-1919.

289. —— . . . The Lycoperdaceae of Australia, New Zealand, and neighbouring islands. pp. 1-42, 15 pls., 49 figs., 1905.

290. —— . . . The Tylostomeae. pp. 1-28, 12 pls., 6 figs., 1906.

291. —— . . . The Nidulariaceae, or "Bird's-nest fungi". pp. 1-32, 10 pls., 20 figs., 1906.

292. —— . . . Synopsis of the known Phalloids. pp. 1-96, 109 figs., 1909.

293. —— . . . Synopsis of the genus Hexagonia. pp. 1-46, 55 figs., 1910.

294. —— . . . Synopsis of the sections Microporus, Tabacinus, and Funales of the genus Polystictus. pp. 49-70, 21 figs., 1910.

295. [LLOYD, C. G.] . . Synopsis of the section Ovinus of Polyporus. pp. 71-94, 14 figs., 1911.

296. —— . . Synopsis of the stipitate Polyporoids. pp. 95-208, 106 figs., 1912.

297. —— . . Synopsis of the stipitate Stereums. pp. 15-44, 34 figs., 1913.

298. —— . . Synopsis of the genus Cladoderris. 12 pp., 11 pls., 1913.

299. —— . . Synopsis of the genus Fomes. pp. 211-288, 41 figs., 1915.

300. —— . . Synopsis of the Section Apus of the genus Polyporus. pp. 291-392, 76 figs., 1915.

301. —— . . The Geoglossaceae. pp. 1-24, 27 figs., 1916.

302. —— . . The myths of mycology. pp. 1-16, 5 figs., 1917.

303. —— . . The genus Radulum. pp. 1-12, 24 figs., 1917.

304. —— . . Synopsis of some genera of the large Pyrenomycetes. 1st paper 1917, 2nd paper 1919, paged continuously, pp. 1-32, many figs.

305. —— . . Xylaria notes, Nos. 1 and 2, pp. 1-32, 70 figs., 1918.

306. MACKIE, F. P. & CHITRE, G. D. Animal experiments and sprue. Indian Journ. Med. Res., XVI, pp. 49-76, 1 pl., 1928.

307. MAGNUS, P. . . Einige Beobachtungen zur näheren Kenntniss der Arten von Diorchidium und Triphragmium. Ber. deut. Bot. Gesell., IX, pp. 118-124, 1 pl., 1891.

308. —— . . Eine neue Uredineengatung Schroeteriaster, gegründet auf Uromyces alpinus. Ber. deut. Bot. Gesell., XIV, 129-133, 1 pl., 1896.

309. —— . . *in* J. Bornmüller, Iter Syriacum, 1897. Fungi. Verhandl. der K. K. Zool. Bot. Gesell. Wien, L, pp. 432-499, 2 pls., 1900.

310. MANN, H. H. . . Blister blight of tea. Indian Tea Assoc. Bull. 3, 13 pp., 5 pls., Calcutta, 1906.

311. MASON, E. W. . . Annotated account of fungi received at the Imperial Bureau of Mycology. List II, 43 pp., 1928.

312. MASSEE, G. . . A monograph of the genus Podaxis Desv. (=Podaxon Fr.). Journ. Bot., XXVIII, pp. 33-39, 69-77, 2 pls., 1890.

313. —— . . A monograph of the Thelephoraceae. Journ. Linn. Soc. London, Bot., XXV, pp. 107-155, 3 pls., 1890, and ibid., XXVII, pp. 95-205, 3 pls., 1891.

314. —— . . Notes on exotic fungi in the Royal Herbarium, Kew. Grev., XXI, pp. 1-6, 1892.

315. [MASSEE, G.] . . Notes on fungi in the Royal Herbarium, Kew. Grev., XXI, pp. 33-35, 1892.

316. —— . . Revised descriptions of type specimens in Kew Herbarium. Grev., XXII, pp. 33-35, 1893.

317. —— . . Disease of pepper plants in Mysore. Kew Bull., 1895, pp. 178-180, 1895.

318. —— . . Tea blights. Kew Bull., 1898, pp. 105-112, 1898.

319. —— . . Fungi Exotici—

I. Kew Bull., 1898, pp. 113-136, 1898.
II. ibid., 1899, pp. 164-185, 1899.
III. ibid., 1901, pp. 150-169, 1901.
IV. ibid., 1906, pp. 91-94, 1906.
VI. ibid., 1907, pp. 121-124, 1907.
VIII. ibid., 1908, pp. 216-219, 1908.
X. ibid., 1910, pp. 1-6, 1 pl., 1910.
XI. ibid., 1910, pp. 249-253, 2 pls., 1910.
XIII. ibid., 1912, pp. 189-191, 1912.
XIV. ibid., 1912, pp. 253-255, 1912.

320. —— . . Revision of the genus Hemileia Berk. Kew Bull., 1906, pp. 35-42, 1 pl., 1906.

321. MCALPINE, D. . . The smuts of Australia. 288 pp., 10 figs., 56 pls., Melbourne, 1910.

322. MCRAE, W. . . Report on the outbreak of blister blight of tea in the Darjeeling District. Agric. Res. Inst. Pusa, Bull. 18, 20 pp., 6 pls. (1 col.), 1910.

323. —— . . The outbreak of blister blight on tea in the Darjeeling District in 1908-9. Agric. Journ. India, V, pp. 126-137, 1 fig., 4 pls. (1 col.), 1910. (Reprinted as Bull. No. 32, Indian Tea Assoc., 1910).

324. —— . . The edible mushroom. Agric. Journ. India, V, pp. 197-204, 1 col. pl., 1910.

325. —— . . Soft-rot of ginger in the Rangpur district, Eastern Bengal. Agric. Journ. India, VI, pp. 139-146, 1 col. pl., 1911.

326. —— . . Notes on south Indian fungi. Madras Agric. Dept. Year Book, 1917, pp. 108-111, 1917.

327. —— . . Phytophthora meadii n. sp. on Hevea brasiliensis. Mem. Dept. Agric. India, Bot. Ser., IX, 5, pp. 219-273, 3 figs., 3 pls., 1918.

328. [McRae, W.] . . A new species of Phytophthora parasitic on the Para rubber tree. Journ. Bombay Nat. Hist. Soc., XXV, p. 760, 1918.

329. —— . . A disease of the Para rubber tree caused by Phytophthora meadii McR. Agric. Journ. India, XIV, pp. 566-577, 1 map, 1 fig., 1 pl., 1919.

330. —— . . Blast of paddy. Agric. Journ. India, XIV, pp. 65-70, 1919.

331. —— . . Fomes lucidus (Leys.) Fr. on Acacia melanoxylon R. Br., and Pongamia glabra Vent. Madras Agric. Dept. Year Book, 1919, pp. 91-95, 1920.

332. —— . . I. History of the operations against bud-rot of palms in south India. II. Inoculation experiments with Phytophthora palmivora Butl. on Borassus flabellifer Linn. and Cocos nucifera Linn. Mem. Dept. Agric. India, Bot. Ser., XII, 2, pp. 21-70, 1 map, 1 graph, 1923.

333. —— . . Economic Botany. Part III. Mycology. Ann. Rept. Board Scientific Advice, India, 1922-23, pp. 31-35, 1924.

334. —— . . India :—new plant diseases reported during the year 1928. Intern. Bull. of Plant Protect., III, 2, pp. 21-22, 1929.

335. —— & Shaw, F. J. F. Report on experiments with Cajanus indicus (Rahar) for resistance to Fusarium vasinfectum (wilt disease). Sci. Repts. Agric. Res. Inst. Pusa, 1925-26, pp. 208-212, 1 graph, 1926.

336. Mehta, K. C. . . On the mode of infection and perennation of the smut of "Doob", Cynodon dactylon Pers. Journ. Indian Bot. Soc., III, pp. 243-251, 1 pl., 1923.

337. Mello, F. de . . Trichophyton viannai (n. sp.), the infecting agent in a case of Dermatomycosis. Indian Journ. Med. Res., V, pp. 220-233, 7 figs., 1 pl., 1917.

338. —— Contribution to the study of the Indian Aspergilli. Journ. Indian Bot., I, pp. 158-161, 1920.

339. —— & Pais, Ana. Un caso de nocardiose pulmonar simulando a tisica. Arq. Higiene e Pat. Exot., VI, pp. 132-206, 1918.

340. —— & Carmo Vás, J. A. Contribução para o estudo das Aspergilaceas indianas. Arq. Indo-Portugueses de Med. e Hist. Nat., I, pp. 43-99, pls., 1921.

341. MITRA, M. . . . Morphology and parasitism of Acrothecium penniseti n. sp. (a new disease of Pennisetum typhoideum). Mem. Dept. Agric. India, Bot. Ser., XI, 3, pp. 57-74, 4 pls. (1 col.), 1921.

342. —— . . . Helminthosporium spp. on cereals and sugarcane in India. Part I. Diseases of Zea mays and Sorghum vulgare caused by species of Helminthosporium. Mem. Dept. Agric. India, Bot. Ser., XI, pp. 219-242, 3 pls., 1923.

343. —— . . . Gall formation on the roots of mustard due to a smut (Urocystis coralloides Rostrup). Agric. Journ. India, XXIII, 2, pp. 104-106, 2 pls., 1928.

344. —— . . . Some diseases of crops in the Andaman Islands. Agric. Res. Inst., Pusa, Bull. 195, 14 pp., 1929.

345. —— . . . Phytophthora parasitica Dast. causing "damping off" disease of cotton seedlings and "fruit-rot" of guava in India. Trans. Brit. Mycol. Soc., XIV, pp. 249-254, 2 figs., 1929.

346. —— & SUBRAMANIAM, L. S. Fruit rot disease of cultivated Cucurbitaceae caused by Pythium aphanidermatum (Eds.) Fitz. Mem. Dept. Agric. India, Bot. Ser., XV, pp. 79-84, 3 pls., 1928.

347. MITRA, S. K. . . Mycology. Ann. Rept. Dept. Agric., Assam, for the year 1927-28, pp. 36-37, 1928.

348. MITTER, J. H. & TANDON, R. N. A cultural study of two fungi found in an Indian hill apple. Journ. Indian Bot. Soc., VIII, 3, pp. 212-218, 1 pl., 1 diag., 2 graphs, 1929.

349. MONTAGNE, J. F. C. . Cryptogamae Nilgherenses. Ann. Sci. Nat., II Sér., XVIII, pp. 12-23, 1 pl., 1842.

350. —— . Champignons, *in* Bélanger, Voyage aux Indes orientales pendant les années 1825-1829, II part., pp. 145-159, 1846.

351. —— . Sylloge generum specierumque Cryptogamarum. XXIV+498 pp., Paris, 1856.

352. MURRILL, W. A. . . Kashmir fungi. Mycologia, XVI, p. 133, 1924.

353. NARASIMHAN, M. J. . The areca koleroga work during 1921. Mysore Agric. Calendar, pp. 4-8, 1922.

354. —— . Wild plants affected by koleroga. Mysore Agric. Calendar, pp. 36-37, 1927.

355. NECHITSCH, A. . Sur les ferments de deux levains de l'Inde, le Mucor Praini et le Dematium Chodati. Inst. de Bot., Univ. de Genève, VI Sér., V fasc., pp. 1-38, 6 figs., 1 tab., 1904.

356. NIESSL, G. . . . Einige neue Pyrenomyceten. Hedw., XX, pp. 97-100, 1881.

357. NISBET, J. . . . The diseases of trees. Indian Forester, XXI, pp. 126-133, 1 pl., 1895.

358. NORRIE, F. H. B. . Rhinosporidium infection of the nose. Journ. of Laryngology and Otology, XLIV, pp. 505-513, 1929.

359. PAOLI, G. . . . Nuovi Laboulbeniomiceti parassiti di Acari. Redia, VII, pp. 283-295, 1 pl., 1911.

360. PATOUILLARD, N. . Champignons parasites des Phanérogams exotiques. Rev. Myc., VIII, pp. 80-85, 1 pl., 1886.

361. —— . Contributions à l'étude des champignons extra-européens. Bull. Soc. Myc. France, III, pp. 119-131, 1 fig., 1887.

362. —— . Quelques champignons du Tonkin (Suite). Bull. Soc. Myc. France, XLI, pp. 337-342, 2 figs., 1925.

363. PEARL, R. T. . . Report of the Mycologist to the Government of the Central Provinces and Berar. Rept. Dept. Agric., Central Provinces and Berar, for the year ending 30th June, 1922, pp. 19-20, 1923.

364. PETCH, T. . . . The fungi of certain termite nests. Ann. Royal Bot. Gardens, Peradeniya, III, pp. 185-270, 17 pls., 1906.

365. —— . . . Sclerotium stipitatum Berk. and Curr. Ann. Myc., V, pp. 401-403, 1 fig., 1907.

366. —— . . . The genus Endocalyx, Berkeley and Broome. Ann. Bot., XXII, pp. 391-400, 1 pl., 1908.

367. —— . . . Revisions of Ceylon fungi—

I. Ann. Royal Bot. Gard. Peradeniya, IV, pp. 21-68, 1907.
II. ibid., IV, pp. 373-444, 1910.
III. ibid., V, pp. 265-301, 1912.
IV. ibid., VI, pp. 153-183, 1916.
V. ibid., VI, pp. 307-355, 1917.
VI. ibid., VII, pp. 1-44, 1919.
VII. ibid., IX, pp. 119-184, 1924.
VIII. ibid., X, pp. 161-180, 1927.

368. [PETCH, T.] . . The diseases of the tea bush. Macmillan, London, XII+220 pp., 69 figs., 3 col. pls., 1923.

369. —— . . Ustilagineae and Uredineae of Ceylon. Ann. Royal Bot. Gard., Peradeniya, V, pp. 223-256, 1912.

370. —— . . Horse-hair blights. Ann. Royal Bot. Gard., Peradeniya, VI, pp. 43-68, 6 pls., 1915.

371. —— . . Ceylon Lentini. Ann. Royal Bot. Gard., Peradeniya, VI, pp. 145-152, 1916.

372. —— . . A preliminary list of Ceylon Polypori. Ann. Royal Bot. Gard., Peradeniya, VI, pp. 1-58, 1916.

373. —— . . Additions to Ceylon fungi—
I. Ann. Royal Bot. Gard., Peradeniya, VI, pp. 195-256, 1917.
II. ibid., VII, pp. 279-322, 1922.
III. ibid., IX, pp. 313-328, 1925.
IV. ibid., X, pp. 131-138, 1926.

374. —— . . Gasteromycetae Zeylanicae. Ann. Royal Bot. Gard., Peradeniya, VII, pp. 57-78, 1919.

375. —— . . Studies in entomogenous fungi—
I. Trans. Brit. Myc. Soc., VII, pp. 89-167, 3 pls. (2 col.), 1921.
II. Ann. Royal Bot. Gard., Peradeniya, VII, pp. 167-278, 4 pls. (2 col.), 1921.
III. Trans. Brit. Myc. Soc., IX, pp. 108-128, 5 figs., 1 col. pl., 1923.
IV. ibid., X, pp. 28-45, 3 figs., 1 pl., 1924.
V. ibid., X, pp. 46-80, 1 fig., 2 pls., 1924.
VI. ibid., X, pp. 152-182, 1 fig., 1 pl., 1925.
VII. ibid., X, pp. 183-189, 1 fig., 1925.
VIII. ibid., X, pp. 244-271, 1 fig., 1926.
IX. ibid., XI, pp. 50-66, 1 fig., 1 pl., 1926.
X. ibid., XI, pp. 251-254, 1 fig., 1926.
XI. ibid., XI, pp. 254-258, 1 fig., 1926.
XII. ibid., XII, pp. 44-52, 1 pl., 1927.
XIII. ibid., XII, pp. 105-113, 1927.

376. —— . . Xylariaceae Zeylanicae. Ann. Royal Bot. Gard., Peradeniya, VIII, pp. 119-166, 1924.

377. —— . . Thread blights. Ann. Royal Bot. Gard., Peradeniya, IX, pp. 1-45, 9 pls., 1924.

378. —— . . Entomogenous fungi: additions and corrections.
I. Trans. Brit. Myc. Soc., X, pp. 190-201, 1925.
II. ibid., XI, pp. 258-266, 1 fig., 1926.

379. [PETCH, T.] . . Notes on Ceylon Thelephoraceae, etc. Ann. Royal Bot. Gard., Peradeniya, IX, pp. 259-298, 1925.

380. —— . . Matula. Trans. Brit. Myc. Soc., XI, pp. 67-81, 2 figs., 2 pls., 1926.

381. —— . . Septobasidium rameale. Trans. Brit. Myc. Soc., XII, pp. 276-282, 2 pls., 1927.

382. PETHYBRIDGE, G. H. . On the rotting of potato tubers by a new species of Phytophthora having a method of sexual reproduction hitherto undescribed. Sci. Proc. Royal Dublin Soc., XIII, pp. 529-565, 3 pls., 1913.

383. PETRAK, F. . . Mykologische Notizen—
VI. Ann. Myc., XXI, pp. 182-335, 1923.
X. ibid., XXVII, pp. 324-410, 1929.

384. —— & SYDOW, H. Die Gattungen der Pyrenomyzeten, Sphaeropsideen, und Melanconieen. Feddes Repertorium, Beihefte, XLII, 551 pp., Dahlem, 1926-1927.

385. —— —— Kritisch-systematische Originaluntersuchungen über Pyrenomyzeten, Sphaeropsideen, und Melanconieen. Ann. Myc., XXIII, pp. 209-294, 1925, and XXVII, pp. 87-115, 1929.

386. PINCHING, H. C. . . Rubber in Burma. Bull. Rubber Growers' Assoc., X, pp. 647-655, 1 diag., 1923.

387. RABENHORST, L. . . Fungi europaei exsiccati. Hedw., XVII, pp. 31, 44-47, 59-63, 88-90, 171-176, 1878.

388. —— . . Fungi europaei Cent. XXVI. Hedw., XX, pp. 145-151, 1881.

389. RAMAKRISHNA AYYAR, T. S. Pythium aphanidermatum (Eds.) Fitz. on Opuntia dillenii Haw. Mem. Dept. Agric. India, Bot. Ser., XVI, 7, pp. 191-201, 3 pls., 2 graphs, 1929.

390. REHM, H. . . . Ascomycetes novi. II. Ann. Myc., VI, pp. 313-325, 1908.

391. REICHARDT, H. W. . Fungi, Hepaticae, et Musci frondosi, *in* Bot. Teil, Reise der Oesterreichischen Fregatte Novara um die Erde in den Jahren 1857, 1858, 1859, pp. 133-196, 17 pls., 1870.

392. RHIND, D. . . . Report of the Mycologist, Burma, for the period ending 30th June 1924. Rangoon, Supdt. Govt. Printing, Burma, 6 pp., 1924.

393. [RHIND, D.] . . Annual Report of the Mycologist, Burma, for the year ending the 30th June, 1925. Rangoon, Supdt. Govt. Printing, Burma, 5 pp., 1926.

394. —— . . . The diseases of Para rubber in Burma. Dept. of Agric. Burma, Bull. 14, 16 pp., 9 figs., 1926.

395. —— . . . Annual Report of the Mycologist, Burma, for the year ending the 30th June, 1926. Rangoon, Supdt. Govt. Printing, Burma, 7 pp., 1927.

396. —— . . . Preliminary note on an internal boll disease of cotton in Burma. Agric. Journ. India, XXII, pp. 34-38, 1927.

397. —— . . . India : mycological notes on Burma. Internat. Rev. of Agric., N. S., XIX, pp. 744-745, 1928.

398. RICKER, P. L. . . Notes on fungi II. With new species from various localities. Journ. Myc., XI, pp. 111-115, 1905.

399. ROBERTSON, H. F. . Annual Report of the Mycologist, Burma, for the year ending 30th June, 1928. Rangoon, Supdt. Govt. Printing, Burma, 10 pp., 1928.

400. ROUMEGUÈRE, C. . Fungi in reg. div. Australiae et Asiae a Jul. Remy collecti, 1863-1866. Rev. Myc., 1880, pp. 152-154, 1 pl., 1880.

401. RYAN, RUTH . . Asterina spp. from India. Mem. Dept. Agric. India, Bot. Ser., XV, 5, pp. 103-105, 1928.

402. SACCARDO, P. A. . Fungi veneti novi v. critici v. mycologiae venetae. Michelia, II, pp. 528-563, 1882.

403. —— . . Notae mycologicae. XV. Ann. Myc., XI, pp. 14-21; 1913 ; XVIII, ibid., XII, pp. 282-314, 1914.

404. SALMON, E. S. . . A monograph of the Erysiphaceae. Mem. Torr. Bot. Club, IX, 292 pp., 9 pls., 1900.

405. —— . . On the variation shown by the conidial stage of Phyllactinia corylea (Pers.) Karst. Ann. Myc., III, pp. 493-505, 3 pls., 1905.

406. —— . . Notes on some species of Erysiphaceae from India. Ann. Myc., V, pp. 476-479, 1907.

407. SETCHELL, W. A. . Notes on Ustilagineae. Bot. Gaz., XIX, pp. 185-190, 1 pl., 1894.

408. SHAW, F. J. F. . . The copper blight of tea. Agric. Journ. India, VI, pp. 78-79, 1911.

409. —— . . The morphology and parasitism of Rhizoctonia. Mem. Dept. Agric. India, Bot. Ser., VI, 6, pp. 115-153, 5 figs., 11 pls. (5 col.), 1912.

410. [SHAW, F. J. F.] . A sclerotial disease of rice. Mem. Dept. Agric. India, Bot. Ser., VI, 2, pp. 11-23, 3 pls. (1 col.), 1913.

411. —— . . Anthracnose of sisal hemp. Agric. Journ. India, VIII, 1, 4 pp., 3 pls., 1913.

412. —— . . Studies in diseases of the jute plant. (1) Diplodia corchori Syd. Mem. Dept. Agric. India, Bot. Ser., XI, 2, pp. 37-58, 11 pls. (1 col.), 1921. (2) Macrophoma corchori Saw. Ibid., XIII, 6, pp. 193-199, 2 pls., 1924.

413. —— & AJREKAR, S. L. The genus Rhizoctonia in India. Mem. Dept. Agric. India, Bot. Ser., VII, 4, pp. 177-194, 4 pls., 1915.

414. —— & SUNDARARAMAN, S. The bud rot of coconut palms in Malabar. Ann. Myc., XII, pp. 251-262, 1 fig., 1 pl., 1914.

415. —— —— The bud-rot of coconut palms in Malabar. Agric. Journ. India, IX, pp. 111-117, 3 pls. (2 col.), 1914.

416. SINGH, S. . . . The liability of deodar to the attack of Trametes pini (Brot.) Fr. in Lolab, Kashmir. Indian Forester, L, pp. 361-365, 1924.

417. SPEGAZZINI, C. . . Laboulbeniales ritrovate nelle collezioni di alcuni Musei Italiani. Anal. del Mus. Nac. Hist. Nat. de Buenos Aires, XXVI, pp. 451-511, 49 figs., 1915.

418. STARBÄCK, K. . . Studier i Elias Fries' svampherbarium. I. Sphaeriaceae imperfecte cognitae. Bihang Kongl. Svenska Vetenskaps-Akad. Handl., XIX, Afd. III, 2, 114 pp., 4 pls., 1894.

419. STEVENS, F. L. . . The Meliolineae—
I. Ann. Myc., XXV, pp. 405-469, 2 pls., 1927.
II. ibid., XXVI, pp. 165-383, 5 pls., 1928.

420. —— . . Meliola spp. from India and one from Malay. Mem. Dept. Agric. India, Bot. Ser., XV, 5, pp. 107-111, 3 pls., 1928.

421. STEVENS, N. E. . . Two species of Physalospora on Citrus and other hosts. Mycologia, XVIII, pp. 206-217, 2 figs., 1926.

422. SUBRAMANIAM, L. S. . A Pythium disease of ginger, tobacco, and papaya. Mem. Dept. Agric. India, Bot. Ser., X, 4, pp. 181-194, 6 pls. (3 col.), 1919.

423. [SUBRAMANIAM, L. S.] The genus Cerebella in India. Journ. and Proc., Asiatic Soc. Bengal, N. S., XVII, pp. 205-208, 1 pl., 1921.

424. —— . Root rot and sclerotial diseases of wheat. Agric. Res. Inst., Pusa, Bull. 177, 7 pp., 1 fig., 1 pl., 1928.

425. SUNDARARAMAN, S. . Ustilago crameri Koern. on Setaria italica Beauv. Agric. Res. Inst., Pusa, Bull. 97, 11 pp., 2 pls., 1 map, 1921.

426. —— . The coconut-bleeding disease. Agric. Res. Inst. Pusa, Bull. 127, 8 pp., 5 pls., 1922. (See also Sundararaman, S., Krishnan Nayar, C., and Ramakrishnan, T. S., ibid., Bull. 169, 1928.)

427. —— . Helminthosporium disease of rice. Agric. Res. Inst., Pusa, Bull. 128, 7 pp., 4 pls. (2 col.), 1922.

428. —— . A new ginger disease in Godavari District. Mem. Dept. Agric. India, Bot. Ser., XI, 9, pp. 209-217, 4 pls. (2 col.), 1923.

429. —— . Bud-rot of coconuts caused by Phytophthora palmivora. Agric. Journ. India, XIX, 1, pp. 84-85, 1924.

430. —— . Vermicularia curcumae Syd. on Curcuma longa. Madras Agric. Dept. Year Book, 1925, pp. 18-19, 1926.

431. —— . Some Vermicularias of economic importance in South India. Madras Agric. Dept. Year Book, 1926, pp. 10-12, 1927.

432. —— . Administration Report of the Government Mycologist, Coimbatore, for 1926-27. Rept. Dept. of Agric., Madras Presidency, for the official year 1926-27, pp. 326-344, 1927.

433. —— . Administration Report of the Government Mycologist, Coimbatore, for 1927-28. Rept. Dept. of Agric., Madras, for the official year 1927-28, pp. 355-372, 1928.

434. —— . Administration Report of the Government Mycologist, Coimbatore, for 1928-29. Reprinted from Rept. Dept. Agric. Madras Presidency, for the official year 1928-29, 27 pp., 1929.

435. [SUNDARARAMAN, S.] & MARUDARAJAN, D. — Some Polyporaceae of the Madras Presidency. Madras Agric. Dept. Year Book, 1924, pp. 69-75, 1925.

436. —— & RAMAKRISHNAN, T. S. — The 'Mahali' disease of coconuts in Malabar. Mem. Dept. Agric. India, Bot. Ser., XIII, 4, pp. 87-97, 2 col. pls., 1 map, 1924.

437. —— —— — Foot-rot and wilt of Antirrhinums. Mem. Dept. Agric. India, Bot. Ser., XVI, 3, pp. 83-100, 7 pls. (2 col.), 2 graphs, 1928.

438. —— —— — A leaf-spot disease of safflower (Carthamus tinctorius) caused by Cercospora carthami, nov. sp. Agric. Journ. India, XXIII, 5, pp. 383-389, 5 pls. (2 col.), 1928.

439. SURI, P. N. . . Trametes pini on deodar in the Baspa Valley, Bushahr State, Punjab. Indian Forester, LII, pp. 327-330, 1 pl., 1926.

440. SYDOW, H. & P. . . Diagnosen neuer Uredineen und Ustilagineen nebst Bemerkungen zu einigen bereits bekannter Arter. Ann. Myc., I, pp. 15-23, 1903.

441. —— — Neue und Kritische Uredineen. Ann. Myc., I, pp. 324-334, 1903.

442. —— — Monographia Uredinearum, Leipzig—
I. XXV+972 pp., 45 pls., 1902-04.
II. XIX+396 pp., 14 pls., 1909-10.
III. 726 pp., 32 pls., 1912-15.
IV. IV+671 pp., 1923-24.

443. —— — Novae fungorum species—
I. Ann. Myc., II, pp. 162-174, 1904.
VI. ibid., IX, pp. 142-146, 1 pl., 1911.
VIII. ibid., X, pp. 405-410, 1 fig., 1912.
XII. ibid., XII, pp. 195-204, 1914.
XIII. ibid., XIII, pp. 35-43, 2 figs., 1915.
XIV. ibid., XIV, pp. 256-262, 1 fig., 1916.
XV. ibid., XV, pp. 143-148, 1917.
XVI. ibid., XVIII, pp. 154-160, 1 fig., 1920.
XVII. ibid., XIX, pp. 304-309, 1921 (by H. Sydow).

444. —— — Über eine Anzahl aus der Gattung Uromyces auszuschliessender resp. unrichtig beschreibender Arten. Ann. Myc., VI, pp. 133-143, 1908.

445. SYDOW, H. . . Beiträge zur Kenntnis der Pilzflora des südlichen Ostindiens. I, Ann. Myc., XI, pp. 326-330, 1913. II, ibid., XII, pp. 484-490, 1914.

446. SYDOW, H. & P. . Mykologische Mitteilungen. Ann. Myc., XVI, pp. 240-248, 1918.

447. —— Mykologische Mitteilungen. Ann. Myc., XVII, pp. 33-47, 2 figs., 1919.

448. SYDOW, H. . . Die Verwertung der Verwandtschaftsverhältnisse und des gegenwärtigen Entwicklungsganges zur Umgrenzung der Gattungen bei den Uredineen. Ann. Myc., XIX, pp. 161-175, 1921.

449. —— . . Über einige wenig bekannte Uredineen aus dem Kew Herbar. Ann. Myc., XX, pp. 54-60, 1922.

450. —— . . Notizen über Ustilagineen. Ann. Myc., XXII, pp. 277-291, 1924.

451. —— . . Fungi chinenses. Ann. Myc., XXVII, pp. 418-434, 1929.

452. SYDOW, H. & P. & BUTLER, E. J. Fungi Indiae orientalis. Pars I. Ann. Myc., IV, pp. 424-445, 1906.

453. —— —— Fungi Indiae orientalis. Pars II. Ann. Myc., V, pp. 485-515, 5 figs., 1907.

454. —— —— Fungi Indiae orientalis. Pars III. Ann. Myc., IX, pp. 372-421, 1 pl., 9 figs., 1911.

455. —— —— Fungi Indiae orientalis. Pars IV. Ann. Myc., X, pp. 243-280, 11 figs., 1912.

456. —— —— Fungi Indiae orientalis. Pars V. Ann. Myc., XIV, pp. 177-220, 4 figs., 1916.

457. SYDOW, H. & MCRAE, W. Hyphomycetes Indiae orientalis. Annales Crypt. Exot., II, pp. 262-271, 1929 (1930).

458. —— & PETRAK, F. Micromycetes philippinenses. Ann. Myc., XXVI, pp. 414-446, 1928.

459. TASLIM, M. . . Stem-rot of berseem caused by Rhizoctonia solani Kühn. Agric. Res. Inst., Pusa, Bull. 180, 8 pp., 2 pl. (1 col.), 1928.

460. TASSI, F. . . . Novae Micromycetum species descriptae et iconibus illustratae. Bull. Laborat. Bot. Univ. Siena, I, 1, pp. 6-15, 3 pls., 1898 ; ibid., III, 1, pp. 14-21, 2 pls., 1900.

461. THAKUR, A. K. & NORRIS, R. V. A biochemical study of some soil fungi with special reference to ammonia production. Journ. Indian Inst. Sci., XI A, 12, pp. 141-160, 4 graphs, 1928.

462. THAXTER, R. . . . Preliminary diagnoses of new species of Laboulbeniaceae—
I. Proc. Amer. Acad. Arts and Sci., XXXV, pp. 153-209, 1900.
II. ibid., XXXV, pp. 409-450, 1900.
III. ibid., XXXVI, pp. 397-414, 1901.
IV. ibid., XXXVII, pp. 19-45, 1902.
V. ibid., XXXVIII, pp. 7-57, 1903.

463. —— . . . Contribution towards a monograph of the Laboulbeniaceae. Memoirs Amer. Acad. Arts and Sci., XII, pp. 187-429, 26 pls., 1902.

464. —— . . . Laboulbeniales parasitic on Chrysomelidae. Proc. Amer. Acad. Arts and Sci., L, 2, pp. 17-50, 1914.

465. —— . . . New Indo-Malayan Laboulbeniales. Proc. Amer. Acad. Arts and Sci., LI, 1, pp. 3-51, 1915.

466. THEISSEN, F. . . . Xylariaceae austro brasiliensis. I. Xylaria. Denkschr. Kais. Akad. Wissensch. Wien, Mathem. Naturwiss. Kl., LXXXIII, pp. 47-86, 7 figs., 11 pls., 1909.

467. —— . . . Fungi aliquot Bombayenses a Rev. Ed. Blatter collecti. Ann. Myc., IX, pp. 153-159, 1 fig., 1911.

468. —— . . . Fragmenta brasilica IV. Ann. Myc., X, pp. 1-32, 5 figs., 1912 ; ibid., V. Ann. Myc., X, pp. 159-204, 1912.

469. —— . . . Zur Revision der Gattung Dimerosporium. Beih. Bot. Centralb., XXIX, 2 Abt., pp. 45-73, 1912.

470. —— . . . Zur Revision der Gattungen Microthyrium und Seynesia. Oesterr. Bot. Zeitschr., LXII, pp. 216-221, 275-280, 327-329, 395-396, 430-434, 1912.

471. —— . . . Le genre Asterinella. Broteria, Ser. Bot., X, pp. 101-123, 20 figs., 1912.

472. —— . . . The fungi of India. Part I. Journ. Bombay Nat. Hist. Soc., XXI, 4, pp. 1273-1303, 4 pls., 10 fig., 1912. Part II. Ibid., XXII, 1, pp. 144-159, 5 pls., 1913.

473. —— . . . Die Gattung Asterina. Abhandl. K. K. Zool. Bot. Gesellsch. Wien, VII, 3, pp. 1-130, 8 pls., 1913.

474. —— . . . Lembosia-Studien. Ann. Myc., XI, pp. 425-467, 1 pl., 1913.

475. —— . . . Die Trichothyriazeen. Beih. Bot. Centralbl., XXXII, 2 Abt., pp. 1-16, 3 figs., 1 pl., 1914.

476. [THEISSEN, F.] . . Annotaçoes a mycoflora brasileira. Broteria, Ser. Bot., XII, pp. 13-31, 7 figs., 1914.

477. —— . . De Hemisphaerialibus notae supplendae. Broteria, Ser. Bot., XII, pp. 73-96, 1914.

478. —— . . Trichopeltaceae n. fam. Hemisphaerialium. Centr. für Bakt., 2 Abt., XXXIX, pp. 625-640, 7 figs., 1 pl., 1914.

479. —— . . Studie über Botryosphaeria. Ann. Myc., XIV, pp. 297-340, 1916.

480. —— . . Mycologische Mitteilungen. Ann. Myc., XVI, pp. 174-188, 4 figs., 1918.

481. —— & SYDOW, H. Die Dothideales. Ann. Myc., XIII, pp. 149-746, 6 pls., 1915.

482. —— —— Einige nachträgliche Mitteilungen über Dothideen sowie über Erikssonia und verwandte Formen. Ann. Myc., XIV, pp. 444-453, 1916.

483. —— —— Die Gattung Parodiella. Ann. Myc., XV, pp. 125-142, 1917.

484. —— —— Synoptische Tafeln. Ann. Myc., XV, pp. 389-491, 38 figs., 1917.

485. THOM, C. & CHURCH, M. B. The Aspergilli. Williams & Wilkins, Baltimore, 272 pp., 14 figs., 4 pls., 1926.

486. —— The Penicillia. Ballière, Tindall & Cox, London, 643 pp., 99 figs., 1930.

487. THOMAS, K. M. . . Some coffee diseases of South India and their control. Planters' Chron., XIX, 41, pp. 697-704, 1924.

488. THUEMEN, F. v. . . Fungorum novorum exoticorum. Decas altera. Rev. Myc., II, pp. 36-38, 1880.

489. —— . . Die Blasenrost-Pilze der Coniferen. Monographie der Gattung Peridermium Lév. Mitteil. forstl. Versuchsw. Oesterr., II, pp. 297-323, 1880.

490. TROUP, R. S. . . A destructive fungus on Xylia dolabriformis. Indian Forester, XXVI, pp. 19-21, 1900.

491. —— . . Peridermium cedri as a destructive fungus. Indian Forester, XXXVIII, pp. 222-223, 1 pl., 1912; ibid., XL, pp. 469-472, 1 pl., 1914.

492. TUNSTALL, A. C. . A stem disease of tea caused by Nectria cinnabarina (Tode) Fr. Pamphlet of the Indian Tea Assoc., Calcutta, 1918, 6 pp., 4 pls., 1918.

493. —— . . Tea roots. Part II. Pamphlet No. 44, Indian Tea Assoc., Calcutta, 17 pp., 7 pls., 1918.

494. [TUNSTALL, A. C.] . Notes on some fungus diseases prevalent during season of 1922. Quart. Journn. Sci. Dept. Indian Tea Assoc., 1922, 3, pp. 115-123, 1922.

495. —— . . Some observations on stem and root diseases. Quart. Journ. Indian Tea Assoc., 1923, 3, pp. 86-91, 1923.

496. —— . . Some observations on micro-organisms associated with tea fermentation. Quart. Journ. Indian Tea Assoc., 1923, 4, pp. 126-131, 1924.

497. —— . . Some observations on stem diseases in Cachar. Quart. Journ. Indian Tea Assoc., 1925, 1, pp. 37-44, 1 pl., 1925.

498. —— . . Some notes on tea diseases caused by Corticium spp. Quart. Journ. Indian Tea Assoc., 1925, 2, pp. 53-59, 4 pls., 1925.

499. —— . . The distribution of blister blight (Exobasidium vexans). Quart. Journ. Indian Tea Assoc., 1924, 4, pp. 255-256, 1925.

500. —— . . Stem disease caused by Corticium sp. Quart. Journ. India Tea Assoc., 1924, 4, pp. 256-258, 1925.

501. —— . . An outbreak of blister blight in the Surma Valley. Quart. Journ. Indian Tea Assoc., 1927, 1, pp. 20-24, 1927.

502. —— . . Some observations on violet root rot (Sphaerostilbe repens, B. & Br.). Quart. Journ. Indian Tea Assoc., 1927, 2, pp. 69-71, 1927.

503. —— . . Vegetable parasites of the tea plant. The blights. Quart. Journ. Indian Tea Assoc., 1927, 3, pp. 73-86 ; 4, pp. 173-182, 1927.

504. —— . . Vegetable parasites of the tea plant (continued). Blights on the stem. Quart. Journ. Indian Tea Assoc., 1928, 1, pp. 45-57 ; 4, pp. 220-231, 1928.

505. —— . . Fungi on tea roots. Quart. Journ. Indian Tea Assoc., 1929, 1, pp. 1-10, 1929.

506. —— . . Vegetable parasites of the tea plant (continued). Blights on the root. Quart. Journ. Indian Tea Assoc., 1929, 2, pp. 68-75, 1929, and 1930, 1, pp. 28-36, 1930.

507. —— & BOSE, S. C. Fungus diseases of the tea leaf. Brown blight. Quart. Journ. Indian Tea Assoc., 1920, 1, pp. 37-43, 5 pls. (3 col.), 1920.

508. [TUNSTALL, A. C. & BOSE, S. C.] Fungus diseases of the tea leaf. Grey blight. Quart. Journ. Indian Tea Assoc., 1920, 3, pp. 150-154, 4 pls. (2 col.), 1920.

509. —— —— Fungus diseases of the tea leaf. Copper blight. Quart. Journ. Indian Tea Assoc., 1921, 3, pp. 168-171, 1921.

510. —— —— The fungus diseases of the tea leaf. Quart. Journ. Indian Tea Assoc., 1921, 4, pp. 209-213, 4 pls. (3 col.), 1921.

511. —— —— The fungus diseases of the tea leaf. Quart. Journ. Indian Tea Assoc., 1922, 1, pp. 35-43, 2 pls. (1 col.), 1922.

512. TURNER, W. B. . . The fresh water algae of East India. Kongl. Svenska Vetenskaps-Akad. Handl., XXV, 5, pp. 1-187, 23 pls., 1892.

513. UPPAL, B. N. . . India: Sclerotium rot of Betel vine (Piper betle) in Bombay Presidency. Internat. Rev. of Agric., N. S., XIX, p. 1076, 1928.

514. —— . . Appendix M. Summary of the work done under the Plant Pathologist to Government, Bombay Presidency, Poona, for the year 1927-28. Ann. Rept. Dept. of Agric. Bombay Presidency, for the year 1927-28, pp. 203-206, 1929.

515. —— & KAMAT, M. N. Artificial infection of Pennisetum typhoideum by Sclerospora graminicola. Agric. Journ. India, XXIII, 4, pp. 309-310, 1928.

516. VAHID, S. A. . . Damage to Acacia arabica by Fomes pappianus Bres. Indian Forester, LIV, pp. 662-664, 2 pls., 1928.

517. VENKATARAYAN, S. V. . New hosts for Corticium koleroga (Cooke) von Höhnel, with some observations on the basidiospore formation. Journ. Mysore Agric. Exper. Union, VII, pp. 23-28, 4 figs., 1925.

518. WAGLE, P. V. . . Studies in the shedding of mango flowers and fruits, Part I. Mem. Dept. Agric. India, Bot. Ser., XV, 8, pp. 219-249, 5 pls., 1928.

519. WAKEFIELD, E. M. . Fungi exotici—

XX. Kew Bull., 1916, pp. 71-77, 4 figs. 1 pl., 1916.

XXIV. ibid., 1918, pp. 207-210, 7 figs., 1918.

XXVI. ibid., 1922, pp. 161-165, 8 figs., 1922.

520. WATT, G. . . Fungi, *in* Dictionary of the economic products of India, I, pp. 130-133, 1889; III, pp. 259, 455-458, 1890.

521. —— . . . Indian fungi. Agric. Ledger, II, No. 20, pp. 1-131 (251-381), 11 figs., 1895.

522. —— & MANN, H. H. The pests and blights of the tea plant (second edition). XV+429 pp., 44 figs., 24 pls., Calcutta, Supdt. Govt. Printing, India, 1903.

523. WINTER, G. . . Exotische Pilze. II. Hedw., XXIV, pp. 21-35, 1885.

524. WOODROW, G. M. . Four interesting Bombay plants. Journ. Bombay Nat. Hist. Soc., XV, pp. 363-364, 2 pls., 1903.

525. WOLLENWEBER, H. W. Fusaria autographice delineata. Ann. Myc., XV, pp. 1-56 [illustrations issued separately], 1917.

526. ZUNDEL, G. L. I. . Monographic studies on the Ustilaginales attacking Andropogon. Mycologia, XXII, pp. 125-158, 1930.

HOST INDEX.

HOST INDEX—*contd.*

Amaranthus viridis—
Cystopus bliti

Amaryllis sp.—
Aecidium amaryllidis

Amorphophallus campanulatus—
Sclerotium rolfsii

Amphicarpaea edgeworthii—
Woroninella aecidioides

Anamirta cocculus—
Phyllosticta cocculi

Ananas sativus—
Asterinella stuhlmanni

Anaphalis sp.—
Puccinia investita

Andrachne cordifolia—
Monosporidium andrachnis

Andropogon aciculatus—
Balansia andropogonis

Andropogon annulatus—
Entyloma obesum
Uromyces andropogonis-annulati
Ustilago andropogonis-annulati

Andropogon assimilis—
Phyllachora assimilis
Puccinia pusilla

Andropogon bladhii—
Ustilago duthiei

Andropogon caricosus var. *molicomous*—
Sphacelia sp.

Andropogon contortus—
Balansia sclerotica
Cerebella andropogonis-contorti
Endodothella dispar
Puccinia versicolor
Sorosporium contortum
Ustilago warneckeana

Andropogon echinulatus—
Puccinia hookeri

Andropogon foveolatus—
Ustilago superflua

Andropogon halepensis—
Puccinia purpurea
Sclerospora graminicola var. andropogonis-sorghi
Sorosporium reilianum
Sphacelotheca cruenta
S. sorghi

Andropogon intermedius—
Puccinia duthiae

Andropogon laniger—
Puccinia nakanishikii

Andropogon martini—
Sorosporium wildemanianum

Andropogon micranthus—
Phyllachora assimilis
Puccinia kozukensis

Andropogon muricatus—
Phyllachora cyperi var. donacis
Ustilago effusa

Andropogon nardus—
Balansia sclerotica
Cerebella nardi
Puccinia nakanishikii
Ustilago nardi

Andropogon pertusus—
Puccinia duthiae
Ustilago tenuis

Andropogon schoenanthus—
Balansia sclerotica
Uromyces schoenanthi
Ustilago schoenanthi

Andropogon serratus—
Phyllachora graminis
Puccinia prunicolor

Andropogon sorghum—
Aerothecium lunatum
Cerebella sorghi-vulgaris
Colletotrichum graminicolum
Helminthosporium turcicum
Phyllachora sorghi
Phyllosticta sorghina
Puccinia purpurea
Sclerospora graminicola, var. andropogonis-sorghi
Sorosporium reilianum
Sphacelia sorghi
Sphacelotheca cruenta
S. sorghi
Tolyposporium ehrenbergii

Andropogon tristis—
Puccinia ? andropogonis

Andropogon triticeus—
Phyllachora graminis

Andropogon tuberculatus—
Ustilago andropogonis-tuberculati

Andropogon sp.—
Puccinia propinqua
Sorosporium geminellum
Ustilago amadelpha

Aneilema nudiflorum—
Ustilago burkilli

Anemone rivularis—
Aecidium ? leucospermum
Puccinia fusca

Angelica glauca—
Puccinia ? ellisii

Anisomeles ovata—
Synchytrium rytzii

Anthistiria anathera—
Puccinia anthistiriae

Anthistiria arundinacea—
Ustilago bursa

Anthistiria imberbis—
Puccinia burmanica

HOST INDEX—*contd.*

HOST INDEX—*contd.*

HOST INDEX—*contd.*

HOST INDEX—*contd.*

HOST INDEX—*contd.*

HOST INDEX—*contd.*

HOST INDEX—*contd.*

HOST INDEX—*contd.*

HOST INDEX—*contd.*

HOST INDEX—*contd.*

HOST INDEX—*contd.*

HOST INDEX—*contd.*

HOST INDEX—*contd.*

Ixora parviflora—
Phyllachora ixorae

Ixora polyantha—
Phyllachora ixorae

Ixora undulata—
Daldinia concentrica (wood)

Jambosa : see *Eugenia*

Jasminum arborescens—
Dendrophoma jasmini (branches)
Fusicladium butleri
Microdiplodia jasmini
Uromyces hobsoni

Jasminum auriculatum—
Meliola ? jasminicola

Jasminum grandiflorum—
Uromyces hobsoni

Jasminum humile—
Puccinia chrysopogi
Septoria aitchisoni

Jasminum malabaricum—
Asterina spissa
Blastospora butleri
Echinosphaeria profusa
Meliola ? jasminicola
Uromyces hobsoni

Jasminum officinale—
Uromyces hobsoni

Jasminum pubescens—
Uromyces comedens

Jasminum sambac—
Meliola jasminicola

Jasminum sp.—
Alternaria dianthi
Corticium koleroga
Sirococcus butleri

Jatropha curcas—
Fusicoccum jatrophae

Jatropha glandulifera—
Phytophthora arecae

Jatropha integerrima—
Colletotrichum jatrophae

Jatropha nana—
Cercospora ajrekari

Juglans regia—
Microstroma juglandis
Phyllactinia corylea

Juglans sp.—
Fomes fomentarius (wood)

Justicia diffusa—
Puccinia peraffinis

Justicia gendarussa—
Puccinia thwaitesii

Justicia procumbens—
Synchytrium rytzii

Justicia sp.—
Plasmopara wildemaniana
Synchytrium rytzii

Kyllinga triceps—
Puccinia mysorensis

Lactarius sp.—
Hypomyces floccosus

Lactuca dissecta—
Bremia lactucae

Lactuca macrorhiza—
Puccinia prenanthis-purpureae var. himalensis

Lactuca scariola—
Bremia lactucae

Lactuca sp.—
Septoria lactucae

Lagenaria vulgaris—
Pythium aphanidermatum
Sphaerella citrullina
Sphaerotheca humuli var. fuliginea
Synchytrium rytzii

Lagerstroemia lanceolata—
Rhytisma lagerstroemiae

Lannea : see *Odina*

? *Lasianthus* sp.—
Aecidium flavidum

Lathyrus sativus—
Peronospora viciae
Sclerotinia sclerotiorum

Lathyrus sphaericus—
Uromyces fabae

Lathyrus sp.—
Erysiphe polygoni

Launaea asplenifolia—
Puccinia butleri
see P. macrorrhynchi

Launaea nudicaulis—
Bremia lactucae
Puccinia inayati
P. sonchi

Launaea pinnatifida—
Aecidium microrhynchi

Laurus sp.—
Stomiopeltis aspersa

Lawsonia alba—
Asterina lawsoniae
Corticium koleroga

Lecanium colemanii (insect)—
Empusa lecanii

Lecanium hemisphaericum (insect)—
Cephalosporium lecanii

Lecanium hemisphaericum var. *coffeae* (insect)—
Hypocrella javanica

Lecanium marsupiale (insect)—
Hypocrella reineckeana

HOST INDEX—*contd.*

HOST INDEX—*contd.*

Manihot utilissima—
Botryodiplodia manihoticola (bark)
Cercospora henningsii
Diplodia manihotis
Ophiobolus manihotis

Manisuris granularis—
Ustilago erythraeensis

Marsdenia sp.—
Asterina travancorensis

Medicago denticulata—
Peronospora trifoliorum
Uromyces striatus

Medicago lupulina—
Parodiella perisporioides
Peronospora trifoliorum
Pseudopeziza medicaginis
Sclerotinia sclerotiorum

Medicago sativa—
Helicobasidium purpureum
Macrophomina phaseoli
Oidiopsis taurica
Peronospora trifoliorum
Pseudopeziza medicaginis
Sclerotium rolfsii
Uromyces striatus
Urophlyctis alfalfae

Melastoma sp.—
Munkiodothis melastomata

Melia azedarach—
Cercospora subsessilis

Melilotus alba—
Erysiphe polygoni
Peronospora trifoliorum

Melilotus indica—
Erysiphe polygoni
Peronospora trifoliorum

Melilotus parviflora—
Peronospora trifoliorum

Meliola sp.—
Spegazzinia meliolae

Meliosma simplicifolia—
Aecidium meliosmatis-myrianthi

Memecylon edule—
Asterina memecylonicae
Glenospora uromycoides
Meliola memecyli
Phyllachora aliena

Menispermum cordifolium : see *Tinospora*

Mentha sylvestris—
Puccinia menthae—

Merendera aitchisoni—
Aecidium merenderae

Mesua ferrea—
Diaporthe curvatispora (bark)

Michelia nilagirica—
Phyllachorella micheliae

Microchloa setacea—
Ustilago microchloae

Microrhynchus : see *Launaea*

Mimosa sp.—
Phyllachora rhytismoides

Mimusops elengi—
Uromyces mimusops

Mimusops hexandra—
Robillarda scutata

Momordica balsamina—
Erysiphe cichoracearum

Momordica charantia—
Cercospora momordicae

Morinda tinctoria—
Cercospora morindae

Moringa sp.—
Botryodiplodia hypoxyloidea (bark)

Morus alba—
Aecidium mori
Cytospora ? atra
Diplodia butleri
Macrophomina phaseoli
Phleospora mori
Phyllactinia corylea
Thyrostroma mori

Morus indica—
·Aecidium mori
Trichosporium aterrimum (bark)

Morus serrata—
Aecidium mori

Morus sp.—
Cryptovalsa rabenhorstii
Diplodia mori
D. morina
Ganoderma lucidum (wood)
Polyporus hispidus

Mucuna deeringiana—
Uromyces mucunae

Mucuna ? pruriens—
Uromyces mucunae

Murraya koenigii—
Phyllostictina murrayae

Musa paradisiaca—
Gloeosporium musarum
Macrophoma musae

Musa sapientum—
Diplodia musae
Macrophoma musae

Musa sp.—
Fusarium cubense

Musca sp. (insect)—
Empusa muscae

Myriactis nepalensis—
Aecidium myriactidis
Septoria myriactidis

HOST INDEX—*contd.*

HOST INDEX—*contd.*

HOST INDEX—*contd.*

HOST INDEX—*contd.*

HOST INDEX—*contd.*

HOST INDEX—*contd.*

HOST INDEX—*contd.*

HOST INDEX—*contd.*

HOST INDEX—*contd.*

HOST INDEX—*contd.*

HOST INDEX—*concld.*

Errata to the Scientific Monograph No. 1 of the Imperial Council of Agricultural Research entitled "The Fungi of India."

Page 102, line 18. A rectangular bracket is to be inserted in the beginning.

Page 120, line 32. *For* " Benga " *read* " Bengal ".

Page 124, line 19. *For* " Da dalea " *read* " Daedalea ".

Page 129, line 19. *For* " massee " *read* " Massee ".

Page 131, line 1. The rectangular bracket and the letter " T " should be on the line.

Page 139, line 16. *For* " Sporodesmium brassicaeae " *read* " Sporodesmium brassicae ".

Page 171, line 45. *For* " galli Pers ", *read* " galii Pers ".

MGIPC—M—38 CPB—19-4-1932—500.